Post-transcriptional Regulation through Long Non-coding RNAs (lncRNAs)

Post-transcriptional Regulation through Long Non-coding RNAs (lncRNAs)

Editors

Michael Ladomery
Giuseppina Pisignano

MDPI • Basel • Beijing • Wuhan • Barcelona • Belgrade • Manchester • Tokyo • Cluj • Tianjin

Editors
Michael Ladomery
Department of Applied Sciences
University of the West of
England
Bristol
United Kingdom

Giuseppina Pisignano
Department of Biology and
Biochemistry
University of Bath
Bath
United Kingdom

Editorial Office
MDPI
St. Alban-Anlage 66
4052 Basel, Switzerland

This is a reprint of articles from the Special Issue published online in the open access journal *Non-coding RNA* (ISSN 2311-553X) (available at: www.mdpi.com/journal/ncrna/special_issues/Post_trans_regul).

For citation purposes, cite each article independently as indicated on the article page online and as indicated below:

LastName, A.A.; LastName, B.B.; LastName, C.C. Article Title. *Journal Name* **Year**, *Volume Number*, Page Range.

ISBN 978-3-0365-1217-4 (Hbk)
ISBN 978-3-0365-1216-7 (PDF)

Contents

Preface to "Post-transcriptional Regulation through Long Non-coding RNAs (lncRNAs)"

Long non-coding RNAs (lncRNA) are increasingly prominent in the field of molecular biology. Growing evidence suggests that they are involved in a bewildering array of molecular processes. In this Special Issue we present a collection of articles that, together, highlight the involvement of lncRNAs in post-transcriptional regulation of gene expression.

Michael Ladomery, Giuseppina Pisignano

Editors

Editorial

Post-Transcriptional Regulation through Long Non-Coding RNAs (lncRNAs)

Giuseppina Pisignano [1,*] and Michael Ladomery [2,*]

[1] Department of Biology and Biochemistry, University of Bath, Claverton Down, Bath BA2 7AY, UK
[2] Faculty of Health and Applied Sciences, University of the West of England, Coldharbour Lane, Frenchay, Bristol BS16 1QY, UK
* Correspondence: gp529@bath.ac.uk (G.P.); Michael.Ladomery@uwe.ac.uk (M.L.)

Citation: Pisignano, G.; Ladomery, M. Post-Transcriptional Regulation through Long Non-Coding RNAs (lncRNAs). *Non-coding RNA* **2021**, *7*, 29. https://doi.org/10.3390/ncrna7020029

Received: 26 April 2021
Accepted: 27 April 2021
Published: 29 April 2021

Publisher's Note: MDPI stays neutral with regard to jurisdictional claims in published maps and institutional affiliations.

The discovery of thousands of non-coding RNAs (ncRNAs) pervasively transcribed from the eukaryotic genome has revolutionized the "central dogma" of biology and shifted the attention on the role of RNAs as regulatory molecules, more than simply traditional mediators of genomic information. Non-coding RNAs are transcripts that do not encode proteins and are generally classified as short or long depending on their average size (< or >200 nt). Non-coding RNAs are found in nearly all organisms. Among them, the long non-coding RNAs (lncRNAs) play key roles in many biological processes in development and disease. Since their discovery, the lncRNA field has exploded, and new roles for lncRNAs are constantly emerging, making their investigation a priority in studying gene expression regulation at any level.

This Special Issue encases seven review papers and one original research article from experts in the ncRNA field and illustrates the main mechanisms through which lncRNAs modulate gene expression at the post-transcriptional level. This collection of articles provides a complete overview of their multifunctional roles and presents an additional layer of complexity in the regulation of gene expression and associated cellular processes.

LncRNA length, low expression, and lack of sequence conservation have frequently represented a major technical limitation in their identification and characterization. In their review, Carter et al. provide an exhaustive guide of both in silico and low-to-high throughput experimental approaches to assist researchers to face this challenge. They also offer critical insights to advance our understanding of how lncRNAs are involved in tumorigenesis [1].

A wide range of RNA-binding proteins (RBPs) have been shown to cooperate with lncRNAs to regulate gene expression. In their review, Briata and Gherzi draw attention to the complexity of lncRNA–RBP associations [2]. They illustrate the variety of mechanisms through which lncRNA–RBP complexes can control essentially all post-transcriptional processes in the cell. Sadeq et al. discuss how endogenous lncRNA-associated dsRNA structures are tolerated, whereas viral-derived dsRNA triggers a complex defence network; and further examine the potential implications in the context of autoimmune disease and cancer treatments [3].

In their review, Pisignano and Ladomery describe multiple mechanisms through which lncRNAs contribute to the regulation of alternative splicing and how their action further enhances the expression of mRNA-splicing variants, thereby increasing proteomic diversity in complex organisms [4].

In a more cytoplasmic context, Karakas and Ozpolat discuss how lncRNAs can affect mRNA translation by controlling translation factors and signalling pathways in normal and tumour conditions [5], while Sebastian-delaCruz et al. highlight the importance of lncRNAs in the regulation of mRNA stability and turnover as the basis for the correct functionality of cellular processes and homeostasis [6]. In this regard, in another work presented in this Special Issue, Munz et al. found in a diffuse, large B cell lymphoma cell

1

line a lncRNA (*lncTNK2-2:1*) associated with the increased stability of transcripts that are affected by mTOR inhibition and responsible for the DNA damage response [7].

Fonouni-Farde et al. conclude this Special Issue by describing how plant lncRNAs use sophisticated mechanisms to regulate RNA degradation, alternative splicing, translation, post-translational modifications and even protein localisation [8].

Taken together, this Special Issue highlights the relevance of lncRNAs as crucial regulatory molecules in most post-transcriptional regulation mechanisms, both in animals and plants, and aims to encourage research groups and young researchers to further develop new studies in the field. A more comprehensive understanding of the molecular mechanisms of post-transcriptional regulation by lncRNAs will certainly advance our understanding of the many intricate cellular processes that are still far from being fully elucidated.

Conflicts of Interest: The authors declare no conflict of interest.

References

1. Carter, J.; Ang, D.; Sim, N.; Budiman, A.; Li, Y. Approaches to Identify and Characterise the Post-Transcriptional Roles of lncRNAs in Cancer. *Non-Coding RNA* **2021**, *7*, 19. [CrossRef]
2. Briata, P.; Gherzi, R. Long Non-Coding RNA-Ribonucleoprotein Networks in the Post-Transcriptional Control of Gene Expression. *Non-Coding RNA* **2020**, *6*, 40. [CrossRef] [PubMed]
3. Sadeq, S.; Al-Hashimi, S.; Cusack, C.; Werner, A. Endogenous Double-Stranded RNA. *Non-Coding RNA* **2021**, *7*, 15. [CrossRef] [PubMed]
4. Pisignano, G.; Ladomery, M. Epigenetic Regulation of Alternative Splicing: How LncRNAs Tailor the Message. *Non-Coding RNA* **2021**, *7*, 21. [CrossRef] [PubMed]
5. Karakas, D.; Ozpolat, B. The Role of LncRNAs in Translation. *Non-Coding RNA* **2021**, *7*, 16. [CrossRef] [PubMed]
6. Sebastian-delaCruz, M.; Gonzalez-Moro, I.; Olazagoitia-Garmendia, A.; Castellanos-Rubio, A.; Santin, I. The Role of lncRNAs in Gene Expression Regulation through mRNA Stabilization. *Non-Coding RNA* **2021**, *7*, 3. [CrossRef] [PubMed]
7. Munz, N.; Cascione, L.; Parmigiani, L.; Tarantelli, C.; Rinaldi, A.; Cmiljanovic, N.; Cmiljanovic, V.; Giugno, R.; Bertoni, F.; Napoli, S. Exon–Intron Differential Analysis Reveals the Role of Competing Endogenous RNAs in Post-Transcriptional Regulation of Translation. *Non-Coding RNA* **2021**, *7*, 26. [CrossRef]
8. Fonouni-Farde, C.; Ariel, F.; Crespi, M. Plant Long Noncoding RNAs: New Players in the Field of Post-Transcriptional Regulations. *Non-Coding RNA* **2021**, *7*, 12. [CrossRef] [PubMed]

non-coding **RNA**

MDPI

Review

Approaches to Identify and Characterise the Post-Transcriptional Roles of lncRNAs in Cancer

Jean-Michel Carter [1,*], Daniel Aron Ang [1], Nicholas Sim [1], Andrea Budiman [1] and Yinghui Li [1,2,*]

[1] School of Biological Sciences (SBS), Nanyang Technological University (NTU), 60 Nanyang Drive, Singapore 637551, Singapore; DANI0045@e.ntu.edu.sg (D.A.A.); YANGUNIC001@e.ntu.edu.sg (N.S.); andrea.budiman@ntu.edu.sg (A.B.)

[2] Institute of Molecular and Cell Biology (IMCB), A*STAR, Singapore 138673, Singapore

* Correspondence: jm.carter@ntu.edu.sg (J.-M.C.); liyh@ntu.edu.sg (Y.L.)

Abstract: It is becoming increasingly evident that the non-coding genome and transcriptome exert great influence over their coding counterparts through complex molecular interactions. Among non-coding RNAs (ncRNA), long non-coding RNAs (lncRNAs) in particular present increased potential to participate in dysregulation of post-transcriptional processes through both RNA and protein interactions. Since such processes can play key roles in contributing to cancer progression, it is desirable to continue expanding the search for lncRNAs impacting cancer through post-transcriptional mechanisms. The sheer diversity of mechanisms requires diverse resources and methods that have been developed and refined over the past decade. We provide an overview of computational resources as well as proven low-to-high throughput techniques to enable identification and characterisation of lncRNAs in their complex interactive contexts. As more cancer research strategies evolve to explore the non-coding genome and transcriptome, we anticipate this will provide a valuable primer and perspective of how these technologies have matured and will continue to evolve to assist researchers in elucidating post-transcriptional roles of lncRNAs in cancer.

Keywords: lncRNA; cancer; post-transcription; RNA-binding; ribonucleoprotein; RNAi; interactome; prediction; database; CLIP

check for **updates**

Citation: Carter, J.-M.; Ang, D.A.; Sim, N.; Budiman, A.; Li, Y. Approaches to Identify and Characterise the Post-Transcriptional Roles of lncRNAs in Cancer. *Non-coding RNA* **2021**, *7*, 19. https://doi.org/10.3390/ncrna7010019

Received: 29 January 2021
Accepted: 5 March 2021
Published: 9 March 2021

Publisher's Note: MDPI stays neutral with regard to jurisdictional claims in published maps and institutional affiliations.

1. Introduction

Transcription is at the forefront of the conversion of stable genomic information into reactive biochemical agents that form and modulate dynamic biological systems. This fundamental process relentlessly transcribes at least 62% of the human genome, resulting in a variety of non-coding RNA (ncRNAs) species that outnumbers the selection of more stable RNAs concerned with translation that accumulate in the cell such as ribosomal RNA (rRNA), transfer RNA (tRNA) and messenger RNA (mRNA) [1,2]. Far from being redundant transcriptional byproducts, ncRNAs can also act as pleiotropic reactive biochemical agents interacting with both RNAs and proteins and are first to propagate any genome level information changes to the biological network state that shapes cellular behaviour [3,4].

Diseased states such as cancer arise from cumulative corruption of the genomic source code, resulting in the opportunistic dysregulation of the conserved transcriptional [5], post-transcriptional [6], translational [7] and post-translational processes [8] that ultimately allow them to escape systemic control. Among the hundreds of thousands of genetic abnormalities that may occur, tremendous progress has been made in understanding how specific key "driver" mutations affect important protein coding genes (oncogenes and tumour suppressors) and influence the aforementioned processes to provide a selective advantage to cancerous cells to develop in a given tissue or microenvironment [9]. However, many mutations found in cancer also accumulate in the non-coding genome [10,11].

The large non-coding transcriptional contributions such as small non-coding RNA (sncRNA), enhancer RNA (eRNA) and long non-coding RNAs (lncRNA) have come under increased interest for cancer research. They feature prominently among the rapidly increasing list of non-coding regulatory elements vulnerable to mutations (promoters, enhancers) and contributing to dysregulating the critical processes aforementioned [12,13]. Although some of these ncRNAs, such as microRNAs (miRNAs, a subset of sncRNAs) have garnered plenty of research momentum [14], others such as lncRNAs and circular RNAs (circRNAs) are still burgeoning especially in the context of cancer biology [15].

Long non-coding RNAs ($\geq$200 bp; lncRNAs) encompass the largest and perhaps most intriguing category of ncRNAs in cancer currently known, as they exhibit highly dynamic and tissue specific expression patterns [16], a trait shared with most oncogenes/tumour suppressors [17]. The majority of these transcripts are localised in the nucleus and transcribed by RNA Polymerase II (RNA Pol II). Long ncRNAs share similar characteristics to messenger RNAs (mRNAs), such as having a $5'$-cap and $3'$ poly-A tail. Alternatively, circular forms may assemble through non-coding splicing of exons and introns (circRNA), bolstering their resistance to degradation further [18,19]. Classification of the diversity of ncRNAs, especially lncRNAs, is a work in progress. GENCODE has classified lncRNA annotation on the basis of their genomic context, however this offers little indication of their diverse functional potential aside from identifying possible antisense lncRNAs [20–22]. Interestingly, the length of an RNA has been found to correlate with their propensity to interact with other biochemical molecules, such as proteins [23]. Hence, the "longer" spectrum of ncRNAs may well be more prone to assume varied roles in regulatory processes through cross-molecular interactions. Indeed, many have already been found to play a pivotal role in those exploited by cancers, such as development and differentiation of cells [24,25]. In fact, since the discovery of MALAT1 (Metastasis-associated lung adenocarcinoma transcript 1) in the early 2000s [26], dysregulation of more than a dozen other lncRNAs, such as *H19* [27], *XIST* [28,29], and *HOTAIR* [30,31] have steadily been found to be associated with cancer progression and drug resistance [32–34]. There is also evidence of circRNAs playing a role in cancer progression [35] and chemotherapy resistance along with biomarker potential [36]. With approximately 16,000 lncRNA genes (28,000 transcripts) identified in Gencode 27, this may well represent a potential "goldmine" of hidden tumour suppressor/oncogenic targets [37].

A major challenge remains in uncovering these lncRNAs and revealing their functional roles in cancer-exploited regulatory processes. This is principally due to the novelty of the field, which is further complicated by their multifunctional interactive potential. At least four major mechanisms have been suggested to mediate their effects: (1) Act as signals to regulate transcription, (2) as decoys recruiting binding partners away from their other targets, (3) as guides directing the targeting of a ribonucleoprotein complex, for example; (4) as scaffolds bringing together multiple biomolecules together [38,39]. Such mechanisms are directly susceptible to propagating abnormalities in lncRNA expression or sequence typical of cancer to the post-transcriptional regulatory networks [40]. Mechanisms 2 and 3 are known to influence one of the most prominent post-transcriptional regulatory pathways: miRNA mediated RNA interference (RNAi) [41]. Effectively, lncRNAs may act as either target or ordnance for the RNA-induced silencing complex (RISC) that perturb stability of various RNAs including lncRNAs themselves [42]. Additionally, lncRNAs may interact with RNAs directly as antisense lncRNAs, such as KRT7-AS, which promotes gastric cancer progression [43]. In summary, aberrant lncRNAs may exert substantial influence over post-transcriptional dysregulation in cancer through RISC dependent and independent mechanisms mediated via their interactions with RNA-binding proteins (RBPs) or RNAs [44]. Identifying and characterising these interactive mechanisms utilising appropriate approaches is therefore critical to overcoming the aforementioned challenge in elucidating their roles in major regulatory processes such as post-transcription.

In this review, we will provide a survey of the most useful tools and techniques developed to help place lncRNAs on the post-transcriptional interactome map and reveal their roots to cancer. In the first part, we will provide an overview of such resources valued for primary identification and characterisation of lncRNAs, especially those capable of highlighting cancer relevant contextualisation. In the second part, we will cover how more advanced resources have been developed to help characterise how lncRNAs may interact with RNAs and proteins (Figure 1). Ultimately, we hope this will serve as a useful primer for new cancer research strategies interested in identifying and validating further lncRNAs as oncogenic/tumour suppressor-like players by mechanistically uncovering their post-transcriptional roles.

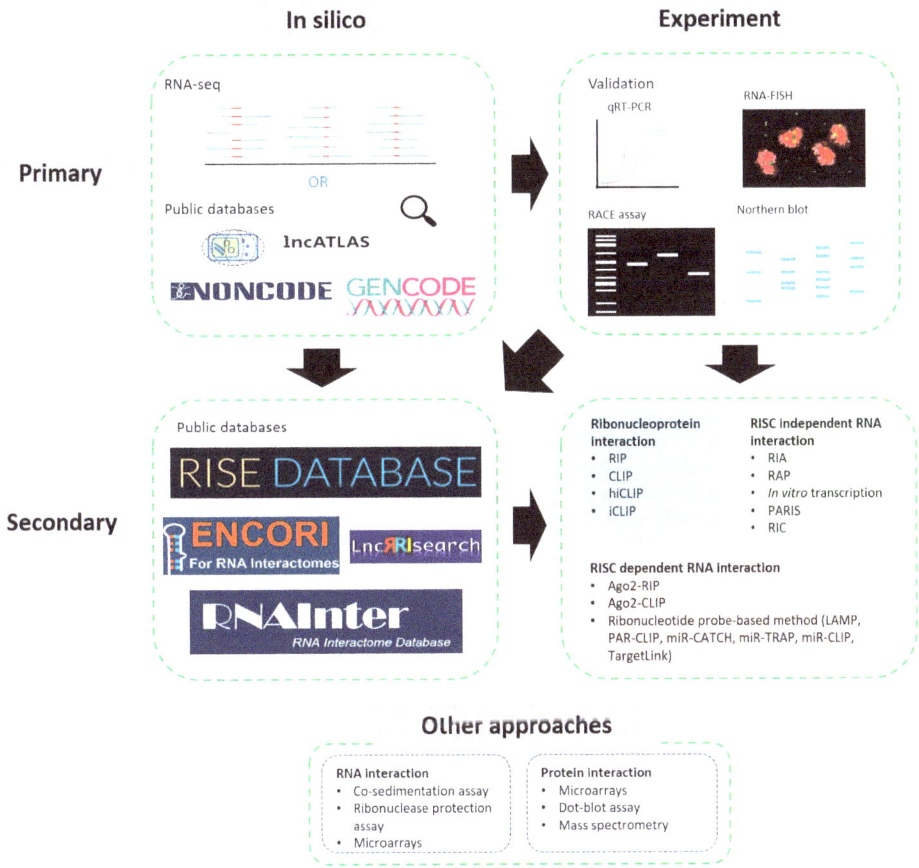

Figure 1. Workflow(s) for the detection and functional characterisation of a lncRNA of interest and its interacting partners. Primary approaches focus on identifying and assigning basic properties based on existing knowledge, predictions or biochemical experiments to validate expression or localisation for example. Secondary approaches focus on the identification of interactions with RNAs or RBPs utilising sequencing-based techniques, see Section 3.2. Further approaches may also be useful to validate high throughput or predictive results.

2. Identification and Primary Characterisation

2.1. Predictions, Identification from High-Throughput Data and Databases

Most cancer research strategies begin by identifying potential candidate genes/loci involved in the dysregulatory state under study. Screening for such candidates usually involves the intersection of high-throughput screening experiments, especially next-generation (NGS) or third-generation sequencing [45]. A number of RNA-sequencing techniques can directly provide valuable information on identity, expression and/or stability of RNAs including lncRNAs. These include capturing a sample of the total transcriptome via standard or single cell RNA-sequencing (RNA-seq/scRNA-seq); capturing nascent transcription using tagged nucleosides or analogs through Global or Precision Run-On sequencing (GRO-seq/PRO-seq; [46,47]) as well as Bru-seq/Bru-Chase-seq [48]; capturing full-length transcripts through Cap-analysis gene expression sequencing (CAGE-seq) and nano-cap analysis of gene expression (nanoCAGE+CAGEscan; [49]) or Oxford Nanopore native RNA sequencing [50]. Provided such sequencing datasets do not undergo selective library preparations these can allow the identification of lncRNAs directly by sequence. Other scenarios, such as Chromatin immunoprecipitation (ChIP) followed by high-throughput DNA sequencing or Assay for Transposase-Accessible Chromatin using sequencing (ChIP-seq or ATAC-seq) may lead towards selecting a putative non-coding region of the cancer genome [51,52]. Depending on the model system, experimental designs and resources, it will be worth considering whether performing additional sequencing experiments is really necessary to fulfil research objectives considering the large amounts of publicly available second-generation data already available.

Regardless of which experimental strategies have been implemented, orthogonal lines of evidence will always fall back on a specific locus or multiple loci of the non-coding genome. From this point, it is possible to infer the identity of potential linear lncRNAs by cross-referencing the loci coordinates with several large databases dedicated to cataloguing lncRNAs or predicting the coding potential of the region. If the coding potential has not already been evaluated, numerous machine-learning tools are available to perform this computation de novo. CPAT, FEELnc and PLEK may be particularly suitable for working with human cancer datasets as extensively evaluated alongside numerous other solutions in [53]. Further comparative reviews of the features of such tools can also be found in [54].

In the case of circular RNAs, most achieve their circular conformation via "backsplicing". This refers to the covalent linkages between a downstream 3' and an upstream 5' splice sites, which results in a reversal of exon sequences relative to the annotated transcript [55]. This unique mechanism can be exploited for their identification and therefore a number of tools have been developed to perform this on RNA-seq datasets, which are extensively evaluated and reviewed in [56,57]. CIRI and KNIFE are among some of the tools that showed robust performance even among background noise [58,59]. CIRI2 has also recently been released offering significant performance improvements over CIRI [60].

In the vast majority of cases—especially for cancer studies based on the human genome—a wealth of sequencing data, pre-generated predictions and annotations relevant for targeted loci are available from a number of public and restricted access databases. Aside from familiar initiatives such as RefSeq, Ensembl and FANTOM [61], many more specialised resources dedicated to allocating lncRNA identity and valuable annotations have emerged—some of which are tailored to cancer research, such as Lnc2Cancer or CSCD. Beyond simply determining whether the loci of interest is transcribed as a lncRNA, many of the resources presented offer insights into transcript localisation, expression as well as gene conservation, mutation and links to diseases, such as cancer. Much of the information is integrated from other public databases and projects such as ClinVar [62], COSMIC (Catalogue of Somatic Mutations in Cancer) [63], TCGA (The Cancer Genome Atlas) [64], 1000 Genomes Project (IGSR) [65], (G)ENCODE (Encyclopedia Of DNA Elements) [66,67], GEO (Gene Expression Omnibus) [68], dbSNP [69], UniProt [70], HPA (Human Protein Atlas) [71], GTEx (Genotype-Tissue Expression) [72], HBM2 (Human Body Map 2.0 GEO Dataset GSE30611), FANTOM (Functional Annotation of the Mammalian Genome) [73], CCLE (Cancer Cell Line Encyclopedia) [74], Disease Ontology [75], GO (Gene Ontology) [76], MeSH (Medical Subject Headings) [77] and TARGET (Therapeutically Applicable Research to Generate Effective Treatments) [78]. Additionally, many circRNA specialised databases integrate information and predictions from miRNA centric databases, since circRNAs often act miRNA sponges. These include Starbase [79], TargetScan [80], doRiNA [81], miRcode [82], miRTarBase [83], HMDD [84], OncomiRDB [85], dbDEMC [86] and miRecords [87]. Although most resources will require usage of the web interface, some offer more advanced programmatic access such as NONCODE [88]. All lncRNA and circRNA resources are summarized in Tables 1 and 2 respectively.

Table 1. Databases for identifying lncRNAs and their basic properties or associations. Overview of active databases cataloguing various properties (sequence conservation, mutation, expression, localisation) or associations attributed to lncRNA genes or transcripts. A link to the hosting website is provided followed by the latest known version as well as the most recent publication describing the database.

Database/Version/Ref.	Link	Conservation	Mutations	Expression	Localisation	Associations
LNCiPedia v5 (2019) [89]	https://lncipedia.org/ (accessed on 8 March 2021)	*H. sapiens, D. melanogaster, D. rerio, M. musculus, P. troglodytes*	NA	NA	NA	Relevant references
lncATLAS (2017) [90]	https://lncatlas.crg.eu/ (accessed on 8 March 2021)	NA	NA	**GENCODE**	**GENCODE**	NA
NONCODE v6 (2020) [88]	http://www.noncode.org/ (accessed on 8 March 2021)	*H. sapiens, M. musculus* and 15 more	**dbSNP**	**Human Body Map; NCBI GEO**	NA	**Gene Ontology**
lncWiki/Book (2019) [91,92]	https://bigd.big.ac.cn/lncrnawiki/index.php/Main_Page (accessed on 8 March 2021) https://bigd.big.ac.cn/lncbook/index (accessed on 8 March 2021)	NA	**ClinVar; COSMIC**	**HPA; GTEx;** Methylation	NA	**Gene Ontology; MeSH Ontology;** miRNA Interaction Prediction;
Lnc2Cancer v3 (2020) [93]	http://bio-bigdata.hrbmu.edu.cn/lnc2cancer/ (accessed on 8 March 2021)	NA	NA	Literature Mining	lncATLAS	Expression Correlation; Survival; TF Motif; *lncBook*
LncRNADisease v2 (2019) [94]	http://www.rnanut.net/lncrnadisease/ (accessed on 8 March 2021)	NA	NA	NA	NA	**Disease Ontology; MeSH Ontology;** Predictive Associations
LncMAP v2 (2018) [95]	http://bio-bigdata.hrbmu.edu.cn/LncMAP/ (accessed on 8 March 2021)	NA	NA	NA	NA	Associations with: TF, Genes, Drugs, Survival

Table 1. *Cont.*

Database/Version/Ref.	Link	Conservation	Mutations	Expression	Localisation	Associations
TANRIC v2 (2019) [96]	https://www.tanric.org (accessed on 8 March 2021)	NA	**TCGA** Somatic Mutations	**TCGA**	NA	**TCGA/CCLE** Correlations: Expression, Stage; Survival
MNDR v3.1 (2020) [97]	https://www.rna-society.org/mndr/ (accessed on 8 March 2021)	NA	NA	Mammalian	NA	Evidenced disease associations and Predictor
lncRNASNP v2 (2018) [98]	http://bioinfo.life.hust.edu.cn/lncRNASNP/#!/ (accessed on 8 March 2021)	NA	**TCGA** and **COSMIC** SNVs	NA	NA	miRNA binding & SNP effects; GWAS LD; Mutation effects
lncRNAMAP (2014) [99]	https://lncrnamap.mbc.nctu.edu.tw (accessed on 8 March 2021)	NA	NA	**NCBI GEO**	NA	miRNA and endo-siRNA predictors
LncTarD (2020) [100]	http://bio-bigdata.hrbmu.edu.cn/LncTarD/ (accessed on 8 March 2021)	NA	NA	NA	NA	Disease-related Target Prediction
EVLncRNAs (2017) [101]	http://biophy.dzu.edu.cn/EVLncRNAs/ (accessed on 8 March 2021)	NA	NA	NA	NA	Manually curated disease association
LncSPA (2020) [102]	http://bio-bigdata.hrbmu.edu.cn/LncSpA/ (accessed on 8 March 2021)	NA	NA	**GTEx, HPA, HBM2, FANTOM, TCGA, TARGET**	NA	Expression in diseased tissues

8

Table 2. Databases for identifying circRNAs and their basic properties or associations. The table summarises for each database what species the data are based on, as well as data sources, integrations or predictions of circRNA genes or transcripts. A link to the hosting website is provided followed by the latest known version as well as the most recent publication describing the database. IRES and MRE correspond to Internal Ribosome Entry Sites and miRNA Response Elements. See Section 3.2 for more information on CLIP and PAR-CLIP techniques.

Database/Version/Ref.	Link	Species	Data Sources	Integrations	Predictions
CircAtlas (2020) [103]	http://159.226.67.237:8080/new/index.php (accessed on 8 March 2021)	*H. sapiens, M. mulatta, M. musculus, R norvegicus, S. scrofa and G gallus*	1070 RNA-seq samples across 6 species	Integrates **circR2Disease** and **circRNADIsease** for disease associations	Co-expression network; Functional inference from **GO/KEGG**; RBP and miRNA binding
circRNAdb (2016) [104]	http://reprod.njmu.edu.cn/cgi-bin/circrnadb/circRNADb.php (accessed on 8 March 2021)	*H. sapiens*	Literature and RNA-seq dataset	**UniProt**	Protein domains, post-translational modifications, half-lifes
CircFunBase (2019) [105]	http://bis.zju.edu.cn/CircFunBase/ (accessed on 8 March 2021)	*H. sapiens, M. musculus + 13 more.*	Literature search	**CircInteractome** (CLIP data), **miRBase**	miRNA-circRNA interactions
circBase (2017) [106]	http://www.circbase.org/ (accessed on 8 March 2021)	*H. sapiens, C. elegans, D. melanogaster, M. musculus, L. chalumnae, L. menadoensis*	Various publications [18,107–111]	**doRiNA**	NA
Circbank (2019) [112]	http://www.circbank.cn/ (accessed on 8 March 2021)	*M. musculus, R. norvegicus, D. melanogaster*	**circBase, miRBase**	m6A literature, COSMIC somatic mutations	IRES, circRNA-miRNA prediction
CIRCpedia v2 (2018) [113]	https://www.picb.ac.cn/rnomics/circpedia/ (accessed on 8 March 2021)	*H. sapiens, M. musculus, R. norvegicus, D. rerio, D. melanogaster, C*	180 RNA-seq samples across 6 species	NA	Putative circRNAs
CircRNADisease (2018) [114]	http://cgga.org.cn:9091/circRNADisease/ (accessed on 8 March 2021)	*H. sapiens*	Manual curation of 800 publications	NA	Association to diseases
CircR2Disease (2018) [115]	http://bioinfo.snnu.edu.cn/CircR2Disease/ (accessed on 8 March 2021)	*H. sapiens*	Manual curation of literature	NA	Association to diseases

Table 2. *Cont.*

Database/Version/Ref.	Link	Species	Data Sources	Integrations	Predictions
TSCD (2017) [116]	http://gb.whu.edu.cn/tscd/ (accessed on 8 March 2021)	*H. sapiens, M. musculus*	**ENCODE** + NCBI **GEO** RNA-seq	**Starbase, Gene Ontology**	MRE, Protein binding sites
circad (2020) [117]	http://clingen.igib.res.in/circad/ (accessed on 8 March 2021)	*H. sapiens, M. musculus, R. rattus*	Manual curation of literature	NA	Asssociation to diseases
circVAR (2020) [118]	http://soft.bioinfo-minzhao.org/circvar/ (accessed on 8 March 2021)	*H. sapiens*	**circBase, circNet, circRNAdb**	**1000 Genomes, ClinVAR, GWASCatalog, ClinVAR, COSMIC**	Association to diseases/cancer
CSCD (2018) [119]	http://gb.whu.edu.cn/cscd/ (accessed on 8 March 2021)	*H. sapiens*	228 RNA-seq samples from **ENCODE**	**Starbase**	Cancer Association, MRE, RBP, ORFs
Circ2Traits (2013) [120]	http://gyanxet-beta.com/circdb/ (accessed on 8 March 2021)	*H. sapiens*	RNA-seq [107]	**Starbase, TargetScan, miRCode, dbSNP, GWAS catalog,** PAR-CLIP Data [121]	miRNA interactions
Circ2Disease (2018) [122]	http://bioinformatics.zju.edu.cn/Circ2Disease/index.html (accessed on 8 March 2021)	*H. sapiens*	Manual curation of literature	**HMDD, OncomiRDB, miRTarBase, dbDEMC, miRecords**	miRNA interactions
CircInteractome (2016) [123]	https://circinteractome.nia.nih.gov/ (accessed on 8 March 2021)	*H. sapiens*	**circBase**	**Starbase, miRBase**	IRES, RBP and miRNA binding sites

While these databases offer an increasingly large and reliable set of annotations, they remain putative until validated in the specific cancer model system under study using the experimental approaches covered below.

2.2. Experimental Approaches: Validation of Expression, Localisation & Structure

Before any advanced experiments take place, it is usually preferable to validate basic characteristics of the target lncRNAs of interest. We will therefore briefly highlight some key primary techniques valued for lncRNA characterisation as well as potential limitations in their application.

Northern blotting has long been applied for analysing expression of specific RNAs, enabling relative quantification, determination of sizes and providing an assessment of the RNA quality [124,125]. Modern protocols allow reduced chemical usage and good specificity and this core method still remains a vital tool for primary characterisation of lncRNAs [126]. In addition, this technique is still one of the most direct methods for demonstrating the circular configuration of circRNAs. Furthermore, the method is often integrated with more advanced procedures to investigate ribonucleoprotein complexes [127].

Providing the target RNA may be reverse transcribed, **RT-qPCR** (reverse transcriptase polymerase chain reaction) may offer a more convenient high sensitivity assay. The exponential nature of qPCR, however, requires careful consideration of confounding factors such as genomic DNA contamination and appropriate selection of reference genes [128]. The latter can still be quite problematic when the system of study features aberrant expression of multiple genes including house-keeping genes commonly chosen as references. The recent availability of large pan-cancer datasets may be able to help overcome this problem [129].

Another valuable technique, especially for characterising unknown isoforms of lncRNAs suspected of undergoing splicing is rapid amplification of cDNA ends (RACE), which produces full length sequences of RNA transcripts. RACE utilises reverse transcription with a 5′ or 3′ primer of a known sequence of the RNA of interest to produce a cDNA copy, this is then followed up with PCR amplification. The product can then be coupled with high-throughput screening in a technique called RACE-Seq to characterize the RACE fragments [130,131].

RNA fluorescence in situ hybridization (RNA-FISH) is the reference technique for visually detecting and determining the distribution of any type of RNA in cellular compartments as well as cells that express the RNA of interest. This technique uses fluorescently labelled probes specific for the target RNA [132,133]. While this technique has been crucial in uncovering the mechanism of several lncRNAs [134,135], the high amounts of repetitive elements in lncRNAs increases the challenge of detecting a legitimate lncRNA signal. This may result in the probe binding to high-abundance, off-target RNAs instead of the intended lncRNA. Furthermore, lncRNA signals in the nucleus appear as "bright blobs", which can be difficult to differentiate from non-specific background signals [136].

SHAPE (Selective 2′-hydroxyl acylation analysed by primer extension) involves the use of reagents, such as N-methylisatoic anhydride (NMIA) and 1-methyl-7-nitroisatoic anhydride (1M7) that react with the 2′-hydroxyl group of the RNA backbone, forming ribose 2′-O-adducts [137]. Adduct formation is dependent on nucleotide flexibility and is quantified at nucleotide resolution by performing RT and comparing the product against a control [138]. This can be further coupled with mutational profiling (MaP), which accounts for the occasional incorporation of noncomplementary nucleotides or deletions caused by reverse transcriptase enzymes, to generate SHAPE profiles where mutations are counted and facilitate the identification of RNA secondary structure formation at nucleotide resolution [139,140]. With valuable evidence supporting the expression, localisation and possible structure of the target lncRNAs, the next step is to estimate and conclusively identify what biomolecules may be interacting with it.

3. Secondary Characterisation: Predicting and Detecting Interactions

3.1. Predictions and Databases

As mentioned in the introduction, there are four types of molecular mechanisms suggested to mediate lncRNA effects through versatile interactions with DNA, RNA and protein molecules [141]. All three types of interactions have been studied and modelled. However, the interactions with the most direct effects on post-transcription are expected to involve only RNA and protein. For more information on predicting lncRNA:DNA interaction potential, we invite the reader to consult recent developments in this relatively new field [142–144].

In-silico prediction of RNA-RNA and RNA-protein interactions are active areas of research recently boosted by machine learning techniques that have grown in strength and numbers over the past decade, feeding on the wealth of accumulating experimental data [145,146]. A significant amount of time and resources have been invested into developing specialised databases and algorithms that predict potential interacting partners of particular lncRNA candidates, some of which have required supercomputer scale processing [147,148]. Exploring how some of these resources can help guide or supplement experimental approaches should therefore form a valuable addition to the secondary characterisation strategy for lncRNA candidates.

RNA:RNA interaction prediction stands to be the most well investigated in large part due to the strong overlap with the small RNA/miRNA field. For instance, lncRNAs acting as ceRNAs can be predicted through their interactions with miRNAs for which a multitude of databases and tools already exist to predict their general propensity to bind certain RNA sequences (Table 3); also see [149]. However, some lncRNA:RNA specific prediction tools have been developed too. Recent evaluation of a dozen such tools has shown that the real-life performance is still fairly average [53]. Tools such as ASSA or RIblast [150,151] that incorporate other sequence information, such as length and GC content and provide useful statistical outputs may be most relevant for real human datasets but cannot be solely relied on for confident inference.

RNA:Protein prediction requires a different approach. Notably, network and correlation based predictions [152] have gained popularity owing to the large increase in available expression data, allowing indirect inference of lncRNA:Protein relationships. RNA:RNA prediction can also benefit from this approach and should be used to complement predictions [153,154]. As such, most resources presented are databases integrating multiple sources of evidence from orthogonal experiments in repositories such as GEO [68], Array-Express [155], ENCODE [67], TCGA [64] and the SRA (Sequence Read Archive) [156] as well as pre-calculated prediction and annotation databases such as as LncRNADisease [94], MNDR (Mammal ncRNA-Disease Repository) [97], eDGAR (Database of Disease-Gene Associations with annotated Relationships among genes) [157], circRNADisease [114], RAIN (RNA-protein Association and Interaction Networks) [158], RAID (RNA Association Interaction Database) [159], NPInter [160] and RISE (RNA Interactome from Sequencing Experiments) [161] (Table 3). For a comprehensive review of tools and databases dedicated to miRNA specific predictions please refer to [149,162].

These tools and databases should help prioritise the types of hypotheses and experiments planned for experimental validation in the cancer model system of choice utilising several of the numerous techniques covered in the rest of this article.

Table 3. Databases integrating computational and experimental sources for predicting lncRNA interactions. Databases are listed detailing the types of interactions they cover. A link to the hosting website is provided followed by the latest known version as well as the most recent publication describing the database. Most databases rely on a primary experimental (EXP) or computational (CPU) source, which are briefly explained. Any additional sources are also summarised.

Database	Link	Interaction Type	Primary Source	Additional Sources
NPInter v4 (2019) [160]	http://bigdata.ibp.ac.cn/npinter4 (accessed on 8 March 2021)	miRNA-RNA; ncRNA-DNA; ncRNA-Protein; circRNA	EXP: Re-processing and integration of experimental data (**GEO; ENCODE; RISE**)	CPU: miRNA binding (**miRanda, TargetScan**); Disease association (**LncRNADisease, MNDR, eDGAR** and **circRNADisease**)EXP: Literature mining
lncRRIsearch (2019) [163]	http://rtools.cbrc.jp/LncRRIsearch/ (accessed on 8 March 2021)	lncRNA-mRNA	CPU: RIBlast	EXP: Tissue expression
DIANA-LncBase v3 (2020) [164]	https://diana.e-ce.uth.gr/lncbasev3 (accessed on 8 March 2021)	miRNA-lncRNA	EXP: Re-processing and integration of experimental data (miRNA, AGO2-CLIP-Seq and CLIP-Seq)	CPU: Correlation with lncRNA expression
SPONGEdb v1 (2021) [165]	https://exbio.wzw.tum.de/sponge/home (accessed on 8 March 2021)	miRNA-lncRNA	CPU: **DIANA-LncBase, TargetScan, miRcode, miRTarBase**	EXP: TCGA expression
LnCeVar v1 (2020) [166]	http://www.bio-bigdata.net/LnCeVar/ (accessed on 8 March 2021)	miRNA-lncRNA	EXP: SNP and mutation data from **TCGA, COSMIC, 1000 Genomes Project**	CPU: Integration from **miRanda, mirBase, miRTarBase, TargetScan**
miRSponge v1 (2015) [167]	http://bio-bigdata.hrbmu.edu.cn/miRSponge/ (accessed on 8 March 2021)	miRNA-lncRNA miRNA-circRNA	EXP: Manual curation from literature	CPU: Integration from **TarBase, miRTarBase, miRanda, miRecord**
starBase/ENCORI v2 (2014/2021) [79]	http://starbase.sysu.edu.cn/ (accessed on 8 March 2021)	miRNA-ncRNA; RBP-RNA;RNA-RNA	EXP: Re-processing and integration of experimental data (CLIP-Seq & variations)	CPU: Correlation of RBP somatic mutation with diseases EXP: Pan-Cancer networks from expression profiles (**TCGA**)

Table 3. *Cont.*

Database	Link	Interaction Type	Primary Source	Additional Sources
RAID v3/RNAInter (2020) [168]	http://www.rna-society.org/raid/ (accessed on 8 March 2021)	RNA-Protein; RNA-RNA; RNA-Histone; RNA-Drug	EXP/CPU: Integration of literature sources and 35 databases.	EXP: Methylation, localisation and editing data from other databases.
RISE (2018) [161]	http://rise.life.tsinghua.edu.cn/index.html (accessed on 8 March 2021)	RNA-RNA	EXP: Integration from sequencing based studies	CPU: Integration with several other databases (**RAIN**, **RAID**, **NPInter**)
LncRNA2Target v2 (2019) [169]	http://123.59.132.21/lncrna2target/ (accessed on 8 March 2021)	lncRNA-RNA	EXP: Manual extraction of interaction associations from literature	EXP: Re-processing of lncRNA perturbation RNA-Seq datasets
LncExpDB (2020) [170]	https://bigd.big.ac.cn/lncexpdb/interactions (accessed on 8 March 2021)	lncRNA-mRNA	CPU: Co-expression network analysis and prediction	EXP: Expression extracted from public repositories (**GEO**, **SRA** and **ArrayExpress**)
LncACTdb v2 (2019) [171]	http://www.bio-bigdata.net/LncACTdb (accessed on 8 March 2021)	miRNA-lncRNA-mRNAmiRNA-circRNA	EXP: Manual curation from literature	CPU: Predictions from networks and integration with Pan-Cancer data (**TCGA**)

3.2. Sequencing Compatible Approaches

With the advent of next-generation sequencing, multiple high-throughput techniques have been developed which allow for increased screening capabilities to discover novel lncRNA interactions in cancer. One of the primary mechanisms through which lncRNAs have been documented to dysregulate cancer post-transcription is through their participation in ribonucleoprotein (RNP) complexes [44,172]. Considering the importance of RBPs, we will first introduce the RIP and CLIP methods, which have been more recently adapted to RNA-Seq. We will then focus on how related methods have been specifically tailored for capturing RNA interactions involving the RNA-induced silencing complex (RISC). Finally, methods for exploring RISC independent RNA interactions will be presented.

3.2.1. Ribonucleoprotein Complex Interaction Detection

LncRNAs often interact with RNA-binding proteins (RBPs), such as the interaction between *HOTAIR* and *EZH2* [173]. Several knock-on effects can result, such as competing with other mRNAs/lncRNAs for RBP binding and/or increasing the ceRNA potential of the lncRNA. In the *MACC1-AS1* to *PTBP1* interaction, such effects have significant consequences on breast cancer tumorigenesis [174]. RNA immunoprecipitation (**RIP**) is one of the first techniques employed to identify such RNAs bound to specific RBPs. RIP involves cell or tissue lysis, followed by immunoprecipitation of native RNA-protein complexes with a specific antibody against the target protein. As these complexes are not stabilized by covalent crosslinking, extra precaution must be taken during washing to remove nonspecific RNA while maintaining the RNA–protein interactions. This limitation makes detection of RNAs with low binding affinity to the protein of interest difficult. In addition, unstably bound RBPs may dissociate from their RNA targets and re-associate with other RNAs under harsh conditions [175,176].

Nevertheless, the RIP techniques have been successfully used over the years revealing relevant interactions in the context of cancer. For example, RIP was used by Tripathi et al. to investigate the interaction between lncRNA metastasis-associated lung adenocarcinoma transcript 1 (*MALAT1*) and the serine/arginine (SR) splicing factors [177]. *MALAT1* is known to be overexpressed in breast, pancreas, lung, colon and prostate carcinomas [178], in addition, it is associated with metastasis and poor survival in non-small cell lung cancer patients [26]. SR splicing factors can influence the alternative splicing (AS) events of many pre-mRNAs in a concentration and phosphorylation-dependent manner, but its cellular mechanism was unknown [179–181]. In the study, *MALAT1* was found to interact with SRSF1 and regulate cellular levels of its phosphorylated forms, which modulated AS events downstream. Further exploration of *MALAT1* interactions via **RIP-Seq** by Wang et al. was similarly fruitful. Their study revealed that *MALAT1* also binds to *EZH2* [182]. *EZH2* is overexpressed in endometrium, prostate and breast cancers [183]. In prostate cancer patients, *EZH2* is associated with increased cell proliferation, invasiveness and metastasis [184,185]. As *EZH2* had been shown to interact with several lncRNAs, such as *HOTAIR* [173] and *PCAT-1* [186], it was unclear which lncRNA was important for *EZH2*-driven prostate cancer progression. Ultimately, knockdown experiments demonstrated that *EZH2-MALAT1* association played a significant role in cancer progression [182], thus representing a new alternative target for treating prostate cancer.

To overcome the low specificity of RIP, crosslinking and immunoprecipitation (**CLIP**) was developed by Ule et al. [187]. CLIP involves the usage of ultraviolet (UV) light to form covalent bonds between RBPs and their direct binding RNAs. An advantage in itself since UV does not crosslink proteins to each other, significantly improving its specificity. In **CLIP-Seq**, after crosslinking, RNA is fragmented, purified and prepared for sequencing [188]. This has led CLIP-Seq to be accepted as a gold standard for identification of endogenous RNA–protein interactions [189].

Since the development of CLIP-Seq, there have been major advancements in CLIP methods that further increase specificity. The first is the development of hybrid CLIP (**hiCLIP**), which enables identification of RNA duplexes bound to RBPs. This is achieved by ligating the two RNA strands with an additional RNA adaptor, following that, the RNA duplexes are immunoprecipitated and sequenced. This method was used to identify mRNA–mRNA and mRNA–lncRNA duplexes bound by Staufen 1 [190]. The second major improvement to CLIP is individual nucleotide CLIP (**iCLIP**), which maps RBP binding sites at nucleotide resolution. A limitation of CLIP is that cDNAs prematurely truncate before the crosslinked nucleotide [191]. However, iCLIP exploits this limitation through the addition of a second adaptor to the 3′ end of cDNA after reverse transcription via circularization [192]. This enables prematurely truncated cDNA at the crosslinked nucleotide to be amplified and therefore improves sensitivity.

3.2.2. RISC Dependent RNA Interactions

The active miRNA research field has led to the development and improvement of a number of methods for establishing which RNAs are being targeted by miRNAs. Thus, establishing whether candidate lncRNAs are involved in RNAi mediated regulation can provide valuable insight into their function. In the first case, lncRNAs targeted by a RISC that is loaded with a complementary miRNA or siRNA may act as decoys or competitive endogenous RNAs (ceRNAs). Circ-lncRNAs may be particularly ideal as ceRNAs due totheir increased stability [38,107]. Colloquially these lncRNAs or circ-lncRNAs are said to "sponge" away interference from targets with other cellular functions, such as mRNAs [193]. Alternatively, lncRNAs may act as the precursors to miRNAs or siRNAs—a further processing step that may be mediated by other RNA binding proteins, such as HuR. These opposing roles can be determined primarily by whether the lncRNA co-occupies the RISC with suspected RNA targets/loads by capturing the ribonucleic or protein part of the complex.

As a first approach, it is possible to identify RNA–RNA interactions on the basis of co-occupation of the RISC complexes isolated via RIP or CLIP based techniques introduced earlier. **Ago2-RIP-Seq** and **Ago-HITS-CLIP** (also called **Ago-CLIP-Seq**) in particular focus on applying RIP-Seq and CLIP-Seq respectively to AGO2, pulling down all miRNAs and possible targets in a single experiment [194,195]. Photoactivatable-Ribonucleoside-Enhanced Crosslinking and Immunoprecipitation (**PAR-CLIP**) is another popular variant of CLIP-Seq that uses photoactivatable nucleoside analogues, such as 4-thiouridine (4SU) to crosslink RISC proteins, such as AGO2 or TNRC6 to the labelled RNAs [121,196]. A particular result of this method is the T to C transitions that occur at the crosslinking sites that can be used to enhance downstream analyses. This technique has been widely cited and implemented including in cancer research [197].

An alternative approach involves utilising a modified ribonucleotide probe to bait and capture any complementary RNAs when they are loaded in the RISC complex. This can be helpful to identify which miRNA/siRNAs are being sponged by a lncRNA acting as ceRNA (competitive endogenous RNA). Additionally, if a lncRNA is suspected of being processed into a miRNA/siRNA, a probe mimicking the lncRNA-derived miRNA/siRNA can be prepared to enable target identification. The first method to apply this concept employed biotinylated miRNA mimic probes to capture their targets in vivo [198]. An in vitro version employing digoxigenin instead has also shown similar performance and is known as the "labelled miRNA pull-down" (**LAMP**) assay system [199]. Many elaborate modifications have been devised to enhance the probes with interesting properties to better capture RNA-RNA duplexes.

A major type of enhancement to the original biotinylated approach has been the inclusion of photoactivatable tags or analogues into the probes (similar to PAR-CLIP). For example, **miR-TRAP** incorporates a psoralen analogue allowing photoactivatable crosslinking to targets, and more stringent purification [200]. The original miR-TRAP method has also recently been paired with RNA-Seq in **PCP-Seq** [201]. PCP-Seq was validated in A549 cancer cell lines.

Other crosslinking technologies incorporated into probes also alter the tag used for isolation procedures. *"Photoclicking"* a process borrowed from bioorthogonal protein chemistry uses tetrazole-ene or dibenzocyclooctyne (DBCO) [202]. **DBCO-tagged** mimic probes in particular have been reported to confer increased miRNA-RISC loading affinity and can be isolated via azide-immobilised magnetic beads [203]. In the **PA-miRNA** method, the biotin tag normally used for isolation is attached via a photo-cleavable linker. It is unclear whether this provides a particular advantage in identifying complementary targets, but the modification is claimed to allow the probe to be used as a photoactivatable source of miRNA [204]. **TargetLink** is a tagless method that utilises a Locked Nucleic Acid (LNA)-based probe for capturing crosslinked miRNA-mRNA complexes and was tested on a human colorectal cancer cell line yielding 12 target genes for miR-21 [205].

Similar approaches can be further enhanced by combining with the technologies used for RNA baiting. Such is the case for **miR-CATCH** which targets a single mRNA (or lncRNA) using a biotinylated DNA probe and crosslinked RISC ribonucleoprotein complexes to detect all miRNAs targetors [206,207]. **miR-CLIP** instead focuses on a single miRNA-like probe containing psoralen and biotin groups to capture the "targetome" after subsequent Ago2 immunoprecipitation and streptavidin purification followed by RNA sequencing [208]. Both techniques have shown promise in cancer research, miR-CATCH has been applied to *MSLN* mRNA, which is overexpressed in Malignant Pleural Mesothelioma and miR-CLIP was validated in Hela cells revealing the lncRNA H19 as a target of miR-106a [209].

Nucleoside analogues, such as diazirine and aryldiazirine, have shown promising results as a means of crosslinking RNA-RNA molecules and post-crosslink tagging has been developed [210–212]. Crosslinking chemistry is an active area of research that promises to deliver many more options that may give rise to further variations of RNA pull down methods [213]. Most interestingly, diazirine has even been encoded as an unnatural amino acid, which may open up new interesting possibilities for protein mediated interaction capture [214].

3.2.3. RISC Independent RNA Interactions

Although RISC independent lncRNA–RNA interactions may be less well known, they have been shown to regulate important biological processes, such as somatic tissue differentiation via post-transcriptional mechanisms [215] and cancer cell growth [216]. Interestingly, the extent of base-pairing could determine if the LncRNA–mRNA interactions positively or negatively regulate gene expression. LncRNAs associated with mRNAs through partial base-pairing have been found to promote mRNA decay [217] while more complete base-pairing protects the mRNA from degradation [218].

Many RISC independent RNA–RNA interactions may involve the participation of an RBP other than AGO2. Therefore, a similar RIP or CLIP based approach targeting other known ribonucleoproteins may allow identification of other proximal RNAs interacting with the lncRNA of interest. Additionally, **CLASH** (cross-linking ligation and sequencing of hybrids), a modified version of iCLIP, is another technique that allows for identification of RNA–RNA interactions by using a tagged "bait" protein [219]. After UV crosslinking of RNA–protein interactions, the bait is pulled down and RNA is recovered and sequenced. When a particular RBP is not targeted **MARIO** (Mapping RNA interactome in vivo) allows EZ-link biotinylation of the protein [220]. A biotinylated RNA-linker is then ligated in a similar fashion to hiCLIP allowing RNA–RNA interactions mediated by other proteins to be captured and sequenced. However, without an RNA binding protein to mediate the interaction, other solutions are required.

Given that lncRNA expression tends to be lower than mRNAs, the levels of endogenous lncRNA must be considered otherwise there might be insufficient material being pulled down. One way to overcome this technical limitation is to perform **in vitro transcribed biotin tagged mimics** of the lncRNA prior to pull down. Not only does this ensure sufficient lncRNA for the pull down but can also improve specificity rather than relying on antisense DNA probe binding. This technique was used in characterising the function of antisense lncRNA of MACC1 in gastric cancer [221]. In that study, bioinformatics predictions suggested that MACC1-AS1 contained a binding site for MACC1 mRNA and the interaction between the two RNAs was later validated via qRT-PCR.

RIA-seq (RNA interactome analysis and sequencing) allows for mapping of transcriptome-wide RNA-RNA interactions before selectively probing for your lncRNAs of interest [215]. In brief, cells are fixed with 1% glutaraldehyde before lysis. The RNA are then sonicated to a size range of 100 to 500 nucleotides before addition of antisense DNA probes. The probes are biotinylated and target specific regions of the lncRNA of interest. Thereafter, streptavidin binding captures the beads–biotin-probes–RNA complexes. The RNA is then eluted and qRT–PCR is used to detect enriched transcripts. Alternatively, high-throughput sequencing can be used though sufficient read depth is required to detect interaction. This technique was used to discover a novel mechanism of lncRNA-mRNA interaction in colorectal cancer. The cytoplasmic lncRNA SNHG5 was found to interact with and stabilise their target mRNAs by protecting them from degradation by STAU1. As such, it promotes colorectal cancer cell survival [222]. However, the specificity of RIA-Seq depends largely on the probe design.

Finally, ribonucleoprotein agnostic methods exist to perform transcriptome wide identification of all RNA complexes without specific baits or probes. **PARIS** (Psoralen Analysis of RNA Interactions and Structures), for example, combines psoralen crosslinking with proximity ligation to identify interactions and structural information of all RNAs [223]. In brief, live cells are UV crosslinked and lysed before RNA is extracted and fragmented. PAGE gel electrophoresis is then used to purify RNA where only RNA duplexes are obtained. The RNA duplexes then undergo proximity ligation followed by photo-decrosslinking before the RNA is prepared for sequencing. This technique allows for identification of long-range RNA structures ranging from 200 to over 1000 nt [224]. Apart from detecting just intramolecular interactions and structures, PARIS has been reported to also identify and refine RNA–RNA interactions to near base pair resolution. In addition, unlike other techniques that require specific RNA baits, PARIS allows for identification of native base-pairing interactions through cross-linking of live cells.

Another similar technique is **LIGR-seq** (LIGation of interacting RNA followed by high-throughput sequencing) which uses a psoralen derivative aminomethyltrioxalen (AMT) that intercalates into the RNA for UV crosslinking [225,226]. circRNA ligase is used for proximity ligation of RNA before sequencing. Unlike the PARIS protocol, enrichment of RNA complexes occurs through RNase R digestion of uncrosslinked RNAs. **SPLASH** (Sequencing of Psoralen crosslinked, Ligated, and Selected Hybrids) might be seen as a more robust variation of PARIS and LIGR-seq as it utilises biotin-labelled psoralen for enrichment of crosslinked RNA using streptavidin beads [227].

RIC-Seq (RNA in situ conformation sequencing) has also recently entered this arena of whole RNA-interactome and secondary structure mapping [228]. Similar to SPLASH, in situ proximity ligation of RNA complexes is applied and biotin enables pulldown of crosslinked RNA. However, it substitutes psoralen with pCp thereby labelling the 3' end of RNA [229] instead of staggered pyrimidines on opposite strands [230]. This step would seem to give RIC-Seq an edge over psoralen-based techniques as it appears to more effectively enrich RNA complexes allowing for detection of lowly expressed RNA [228].

Fundamentally, all these RNA–RNA interactome methods apply proximity ligation with key differences at the RNA-complex isolation steps. Interpreting the results from these largescale experiments is challenging especially considering the current bioinformatic tools are still somewhat underdeveloped. Effectively, analyses borrow tools traditionally used for HiC. However, they have all shown promise in being able to identify lncRNA interactions. For example, *MALAT1* was found to interact with *NEAT1* through analysis of RIC-Seq data [228]. Nevertheless, the aforementioned methods (summarised in Table 4) are providing valuable results that are being compiled into databases such as RISE and can be re-analysed using updated analytical tools when available to continue improving our understanding of RNA–RNA interactomes and structure. Such results can ultimately be followed by more precise experiments to reveal the regulatory effects of the lncRNA interaction.

Table 4. Summary of sequencing approaches for facilitating the characterization of lncRNAs.

Method	Specifications	Limitations	Requirements (Time/Special Resources)
RIP/RIP-seq [182] (tagged/endogenous RBP mediated RNA co-occupancy)	Characterization of native RNA-protein complexes without crosslinking; antibody enrichment	Low specificity; dependent on antibody availability	3–4 d/IP compatible antibody; Autoradiograph facilities
CLIP/CLIP-seq [187] (tagged/endogenous RBP mediated RNA co-occupancy)	RNA-protein interaction sites via RNA-Protein UV crosslinking; antibody enrichment	5′ and 3′ sites of RNA tags affected by cleavage and ligation biases; dependent on antibody availability	5–8 d/IP compatible antibody; UV Crosslinker; Autoradiograph facilities
hiCLIP [190] (tagged/endogenous RBP mediated RNA co-occupancy and RNA-duplexes)	RNA-protein interaction sites and RNA duplexes via UV crosslinking; antibody enrichment	May only capture highly expressed RNA species; dependent on antibody availability	5 d/IP compatible antibody; UV Crosslinker; Autoradiograph facilities
iCLIP [192] (tagged/endogenous RBP mediated RNA co-occupancy)	RNA-protein interaction sites at nucleotide resolution via UV crosslinking; antibody enrichment	miRNA-target interaction strength; dependent on antibody availability	5 d/IP compatible antbody; UV Crosslinker; Autoradiograph facilities
PAR-CLIP [121] (tagged/endogenous RBP mediated RNA co-occupancy)	RNA-protein interaction sites at nucleotide resolution; enhanced UV cross-linking and analysis choices; antibody enrichment	cultured cells only; 4-SU can induce cellular stress; dependent on antibody availability	5 d/IP compatible ant body; UV Crosslinker; Autoradiograph facilities
Biotin-mimics/LAMP [199] (tagged miRNA mimic probing RNA targets)	One miRNA to many RNA interactions; Biotin enrichment	Delivered by transfection to cultured cells; Requires known miRNA sequence	2 d/Streptavidin magnetic beads
miR-TRAP/PCP-seq [200,201] (tagged miRNA mimic probing RNA targets)	One miRNA to many RNA interactions at nucleotide resolution; UVA crosslinking; Poly-A enrichment	Delivered by transfection to cultured cells; Requires known miRNA sequence	2–3 d/UV Crosslinker
DBCO-tagged mimics [203] (tagged miRNA mimic probing RNA targets)	One miRNA to many RNA interactions; increased loading affinity; Click enrichment	Requires known miRNA sequence	3 d/Azide-immobilized magnetic Beads
PA-miRNA [204] (tagged miRNA mimic probing RNA targets)	One miRNA to many RNA interactions; Photocleavable linker; Biotin enrichment	Delivered by transfection to cultured cells; Requires known miRNA sequence; linker is not easily acquired	5 d/Solid phase synthesis; HPLC; Mass spectrometry; UV Crosslinker; Streptavidin magnetic beads

Table 4. *Cont.*

Method	Specifications	Limitations	Requirements (Time/Special Resources)
TargetLink [205] (tagged miRNA mimic probing RNA targets)	One or more miRNAs to many RNA targets; LNA+Biotin enrichment	Requires KO control; Requires known miRNA sequence	6 d/UV Crosslinker; HPLC; Streptavidin magnetic beads
miR-CATCH [206] (tagged RNA mimic probing miRNA targetors)	One RNA to many miRNA interactions; RNA-RISC crosslinking by formaldehyde; Biotin enrichment	Delivered by transfection to cultured cells; Requires known RNA sequence	3–4 d/Dynamag-2; FastPrep-24; Hybridization Oven; Streptavidin magnetic beads
miR-CLIP [208] (tagged miRNA mimic probing RNA targets)	One miRNA to many RNA interactions; RNA-RNA crosslinking by psoralen; Biotin enrichment	Delivered by transfection to cultured cells; Requires known miRNA sequence; Probe needs testing	3–4 d/HPLC; UV Crosslinker; Streptavidin magnetic beads
CLASH [219] (tagged RBP mediated RNA-Protein/duplex capture)	RNA-protein interaction sites and RNA duplexes via UV crosslinking; IgG+Ni-NTA enrichment	Delivered by transfection to cultured cells; Tagged protein expression design may be challenging	4–5 d/UV Crosslinker; Autoradiography facilities
MARIO [220] (endogenous RBP mediated RNA-duplex capture)	Global RNA-RNA interactions mediated by RBPs; RNA-Protein UV crosslinking; 2-step biotin enrichment; proximity ligation	Limited to RBP mediated interactions	5 d/UV Crosslinker; Streptavidin magnetic beads
RIA-seq [215] (endogenous RNA-duplex capture)	One RNA to all RNA interactions; glutaraldehyde crosslinking; biotin enrichment	Limited to RBP mediated interactions; probe preparation may be challenging	5 d/Streptavidin magnetic beads
PARIS [223] (endogenous RNA-duplex capture)	All to all RNA interactions; psoralen crosslinking of RNAs; 2D enrichment of crosslinked duplexes; proximity ligation	Possible AMT side effects; 2D gel setup may be challenging	5 d/UV Crosslinker; SequaGel UreaGel System
LIGR-seq [225] (endogenous RNA-duplex capture)	All to all RNA interactions; psoralen crosslinking of RNAs; RNAseR enrichment of crosslinked duplexes; proximity ligation	Possible AMT side effects	4 d/UV Crosslinker; RNAseR
SPLASH [227] (endogenous RNA-duplex capture)	All to all RNA interactions; psoralen crosslinking of RNAs; biotin enrichment of crosslinked duplexes; proximity ligation	Possible AMT side effects	4 d/UV Crosslinker; Streptavidin magnetic beads
RIC-seq [228] (endogenous RBP mediated RNA-duplex capture)	Global RNA-RNA interactions mediated by RBPs; RNA-Protein formaldehyde crosslinking; biotin enrichment; in situ proximity ligation	Limited to RBP mediated interactions; cell permeabilization may need optimizing	5 d/Streptavidin magnetic beads

3.3. Other Approaches and Biochemical Assays

Whether potential candidate lncRNA and interacting partners have been predicted using bioinformatic tools or via high-throughput sequencing techniques, it is also possible to apply other low- or medium-throughput technologies to characterise or further validate possible interactions.

3.3.1. Protein Interaction Assays

Microarrays provide an alternative method to Next-Generation Sequencing (NGS) for lncRNA-protein interaction studies. This relatively inexpensive method is able to provide information in a couple of hours. However, it has limitations in identifying novel RNA targets. It could be used to quantify and identify either annotated RNA targets or RBPs in ribonucleoprotein (RNP) complexes. RNP immunoprecipitation–microarray (RIP-Chip) has been applied successfully in detecting several QKI-5-binding lncRNAs, especially *lnc10* that regulates the apoptosis of germ cells during their development [231]. In this method, the crosslinking reaction may be omitted during RIP as cell extracts will be used to identify RBPs. However, the crosslinking step can give results with high backgrounds and introduce sequence biases. Briefly, cell extracts are used for immunoprecipitation against the protein of interest and then washed extensively, following which the RNP is eluted and dissociated into RNA and protein [232]. Besides this, protein microarrays have also been widely used to detect RBPs that interact with certain lncRNA. Here, lncRNAs are transcribed in vitro and labelled with Cy5 dye, then labelled lncRNAs are incubated with a protein microarray [233]. Protein microarrays have been able to detect the interaction between *TINCR* RNA and STAU1 protein [215]. Aberrant expression of *TINCR* RNA is implicated in the progression of many cancers. *TINCR* RNA overexpression in epithelial ovarian cancer has been reported to correlate with tumour size, metastasis and survival rates in the patients. By silencing *TINCR*, FGF2 expression is downregulated and leads to the inhibition of epithelial ovarian cancer progression [234].

Dot-blot assay is widely used to study lncRNA–protein interaction and is especially useful in mapping the protein binding region in lncRNA. In this assay, lncRNAs of interest are biotinylated and transcribed in vitro, followed by in vitro RNA-protein binding via incubation of the biotinylated lncRNA with recombinant protein. The bound lncRNA is partially digested by RNase to allow only a small fragment attached to the protein. The lncRNA–protein complexes are subjected to proteinase K treatment to dissociate the complexes. Subsequently, the lncRNA is purified and hybridized to nylon or PVDF membranes spotted with 54–60 mer antisense DNA oligonucleotides tiled along the lncRNA of interest [235,236]. The hybridized membrane is washed and visualized by the detection of streptavidin-HRP signals. This assay has successfully identified the motif sequence of *BCAR4* bound by SNIP1 and PNUTS, which is located at positions 235–288 and 991–1044 in *BCAR4* [236]. In a tumour microenvironment study, positions 355–414 and 1298–1353 of lncRNA *CamK-A* are bound and protected by PNCK and IκBα, which is important in tumour progression [237].

Mass spectrometry (MS) is commonly used to characterize various proteins that are associated with lncRNAs, following pull-down of the lncRNA of interest. It can identify and quantify molecules in complex mixtures based on their mass and charge. However, the quantification accuracy may not be correct due to the difference in mass spectrometric responses. To overcome this issue, stable-isotope labelling has been applied before proceeding with MS. The stable-isotope labelling by amino acids in cell culture (SILAC) has been shown to simplify the quantification and remove false-positive results [238]. This labelling is performed by simply growing two cell populations in two different mediums containing either light or heavy amino acids. Then, the cells are mixed, and proteins extracted for MS analysis. This method has been used to identify several proteins that are specifically enriched and found to interact with *Xist* to mediate transcriptional silencing [239]. Moreover, aberrant expression of *Xist* is associated with tumour progression and metastasis in multiple cancers. Knockdown of *Xist* in colorectal cancer has been proven to inhibit cell proliferation, invasion, and epithelial-mesenchymal transition (EMT) [240]. Larger tumour size and advanced stage of tumour are significantly correlated with high expression of *Xist*. Hence, *Xist* expression is used to predict the prognosis and survival of colorectal cancer patients [241].

3.3.2. RNA Interaction Assays

Co-sedimentation assays can be used whereby RNA is extracted from cells and fractionated using sucrose or glycerol gradients. The RNAs found in the different fractions are examined by Northern blot. RNAs found in the same gradient fractions are thought to interact with each other [242,243] though it does not directly demonstrate interaction. A more robust experiment would be the electrophoretic mobility shift assay (EMSA), which involves studying interaction of RNA fragments by observing rate of migration of the samples during gel electrophoresis [244]. If the lncRNA–mRNA interacts, the complex would have a larger molecular mass compared to separate strands of RNA. Therefore, the complex would migrate slower on the gel compared to non-paired RNAs. Samples can be extracted from cells or synthesised in vitro. Synthesising of RNA fragments could potentially demonstrate interaction between specific regions of the lncRNA-mRNA complexes. However, these techniques can only screen for a given set of molecules.

Ribonuclease protection assays (RPA) can also be used to detect these sense-antisense RNA duplexes. RPA involves isolation of total RNA followed by RNase and DNase digestion [245–247]. Duplexed RNA should be protected from digestion and will be detected by PCR and gel electrophoresis or qRT-PCR. This technique was used to demonstrate interaction between PDCD4-AS1 and PDCD4 mRNA in breast cancer [248].

Microarrays as mentioned earlier have been used to identify the alternative splicing (AS) events regulated by MALAT1 for example. PolyA+ RNA isolated from MALAT1-antisense oligo treated and control HeLa cells were isolated and prepared into labelled cDNA. This was hybridized to a custom AS microarray. The GenASAP algorithm was then used to estimate the percent exon inclusion. Semiquantitative RT-PCR using primers specific for exons flanking the AS events was performed to validate the microarray predictions. This assay revealed that MALAT1 depleted cells have changes in AS of B-MYB and MGEA6 pre-mRNAs [177].

4. Closing Remarks

Throughout this review, we have introduced both computational resources and experimental methods to perform primary and secondary characterisations of lncRNAs to ascertain their potential roles in post-transcriptional regulation and cancer. Primary characterisation establishes basic ground truths relating expression, localisation and relative importance in a model system of choice under normal or perturbed conditions. Secondary characterisation particularly focuses on identifying the interacting RNAs and RBPs that the lncRNAs may influence and importantly deregulate in cancer states due to aberrant expression or non-coding mutations affecting their binding. The RIP and CLIP technologies in particular have been well adapted to next-generation sequencing allowing these newer methods to become reference options for performing high-throughput screening of RNA/Protein or RNA/RNA interactions. Furthermore, a host of improvements to the cross-linking biochemistry have been incorporated and enabled progressive advances in specificity and reliability. It will be important to continue generating these types of experiments and complementing the growing databases dedicated to cataloguing and integrating this information with other valuable sources as presented in Section 2.1, which are already useful starting points for orienting experimental strategies for more novel lncRNAs. The appearance of disease and cancer specialised lncRNA databases and metadatabases will prove highly valuable in integrating the diverse interaction sources and placing them in relevant cancer contexts to identify interesting regulatory patterns or correlations in diseased states.

Some of the findings concerning lncRNA functions in cancer may have direct applications for therapies involving Antisense oligonucleotides (ASOs) for example. ASOs are DNA:RNA chimeras that direct RNase H to degrade target RNAs [249] such as target lncRNAs associated with cancer. During preclinical trials, they were able to target *MALAT1* in vivo, resulting in reduced metastasis [250]. Additionally, ASOs can operate through other mechanisms, such as steric blocking of TF binding and modulating splicing [251]. Unfortunately, RNA molecules like lncRNAs are able to form multiple conformations given their intrinsic flexibility [252]. This makes predicting their structure a challenge which could impede the success of targeted regulation of their expression.

As non-coding RNA continues to take on importance in influencing fundamental processes such as post-transcriptional regulation it will be interesting to integrate this knowledge with findings relating to other novel post-transcriptional regulatory mechanisms such as RNA modifications. This field has also benefited from the adoption of CLIP technologies to perform epitranscriptomic studies on some of the hundreds of modifications that are likely to affect RNA stability, structure, localisation and interactions—lncRNAs included [253–255]. All of these CLIP-based sequencing methods will continue evolving with the maturation of third-generation sequencing, which is already enabling native RNA sequencing including RNA modification detection and structural footprinting [256,257]. In the near future, it may well be possible to capture RNA interactomes, methylomes and structuromes in single experiments to reveal a more complete landscape of the post-transcriptional regulatory mechanisms susceptible to exploitation by cancers.

Funding: This work is supported by the **National Research Foundation (NRF) Singapore**, under its Singapore NRF Fellowship (NRF-NRFF2018-04). In addition, J.M. Carter, D.A. Ang and N. Sim are funded by the Nanyang Assistant Professorship (NAP) Start-up-grant to Y.L., from the Nanyang Technological University (NTU), Singapore.

Institutional Review Board Statement: Not applicable.

Informed Consent Statement: Not applicable.

Data Availability Statement: No new data were created or analyzed in this study. Data sharing is not applicable to this article.

Acknowledgments: The authors thank present and past colleagues for inspiring discussions and Aswin Doekhie for critical reading of the manuscript.

Conflicts of Interest: The authors declare no conflict of interest.

References

1. Djebali, S.; Davis, C.A.; Merkel, A.; Dobin, A.; Lassmann, T.; Mortazavi, A.; Tanzer, A.; Lagarde, J.; Lin, W.; Schlesinger, F.; et al. Landscape of transcription in human cells. *Nature* **2012**, *489*, 101–108. [CrossRef]
2. 6 Non-coding RNA characterization. *Nature* **2019**. [CrossRef]
3. Fabbri, M.; Girnita, L.; Varani, G.; Calin, G.A. Decrypting noncoding RNA interactions, structures, and functional networks. *Genome Res.* **2019**, *29*, 1377–1388. [CrossRef] [PubMed]
4. Yi, S.; Lin, S.; Li, Y.; Zhao, W.; Mills, G.B.; Sahni, N. Functional variomics and network perturbation: Connecting genotype to phenotype in cancer. *Nat. Rev. Genet.* **2017**, *18*, 395–410. [CrossRef]
5. Bradner, J.E.; Hnisz, D.; Young, R.A. Transcriptional Addiction in Cancer. *Cell* **2017**, *168*, 629–643. [CrossRef] [PubMed]
6. Jewer, M.; Findlay, S.D.; Postovit, L.-M. Post-transcriptional regulation in cancer progression: Microenvironmental control of alternative splicing and translation. *J. Cell Commun. Signal.* **2012**, *6*, 233–248. [CrossRef] [PubMed]
7. Vaklavas, C.; Blume, S.W.; Grizzle, W.E. Translational Dysregulation in Cancer: Molecular Insights and Potential Clinical Applications in Biomarker Development. *Front. Oncol.* **2017**, *7*, 158. [CrossRef]
8. Han, Z.-J.; Feng, Y.-H.; Gu, B.-H.; Li, Y.-M.; Chen, H. The post-translational modification, SUMOylation, and cancer (Review). *Int. J. Oncol.* **2018**, *52*, 1081–1094. [CrossRef] [PubMed]
9. Lynch, M. Rate, molecular spectrum, and consequences of human mutation. *Proc. Natl. Acad. Sci. USA* **2010**, *107*, 961–968. [CrossRef] [PubMed]
10. Araya, C.L.; Cenik, C.; Reuter, J.A.; Kiss, G.; Pande, V.S.; Snyder, M.P.; Greenleaf, W.J. Identification of significantly mutated regions across cancer types highlights a rich landscape of functional molecular alterations. *Nat. Genet.* **2016**, *48*, 117–125. [CrossRef]

11. Khurana, E.; Fu, Y.; Chakravarty, D.; Demichelis, F.; Rubin, M.A.; Gerstein, M. Role of non-coding sequence variants in cancer. *Nat. Rev. Genet.* **2016**, *17*, 93–108. [CrossRef]
12. Zhou, S.; Treloar, A.E.; Lupien, M. Emergence of the Noncoding Cancer Genome: A Target of Genetic and Epigenetic Alterations. *Cancer Discov.* **2016**, *6*, 1215–1229. [CrossRef] [PubMed]
13. Pratt, H.; Weng, Z. Decoding the non-coding genome: Opportunities and challenges of genomic and epigenomic consortium data. *Curr. Opin. Syst. Biol.* **2018**, *11*, 82–90. [CrossRef]
14. He, L.; Hannon, G.J. MicroRNAs: Small RNAs with a big role in gene regulation. *Nat. Rev. Genet.* **2004**, *5*, 522–531. [CrossRef] [PubMed]
15. Esteller, M. Non-coding RNAs in human disease. *Nat. Rev. Genet.* **2011**, *12*, 861–874. [CrossRef]
16. Gloss, B.S.; Dinger, M.E. The specificity of long noncoding RNA expression. *Biochim. Biophys. Acta* **2016**, *1859*, 16–22. [CrossRef]
17. Haigis, K.M.; Cichowski, K.; Elledge, S.J. Tissue-specificity in cancer: The rule, not the exception. *Science* **2019**, *363*, 1150–1151. [CrossRef]
18. Zhang, Y.; Zhang, X.-O.; Chen, T.; Xiang, J.-F.; Yin, Q.-F.; Xing, Y.-H.; Zhu, S.; Yang, L.; Chen, L.-L. Circular intronic long noncoding RNAs. *Mol. Cell* **2013**, *51*, 792–806. [CrossRef]
19. Suzuki, H.; Zuo, Y.; Wang, J.; Zhang, M.Q.; Malhotra, A.; Mayeda, A. Characterization of RNase R-digested cellular RNA source that consists of lariat and circular RNAs from pre-mRNA splicing. *Nucleic Acids Res.* **2006**, *34*, e63. [CrossRef]
20. Derrien, T.; Johnson, R.; Bussotti, G.; Tanzer, A.; Djebali, S.; Tilgner, H.; Guernec, G.; Martin, D.; Merkel, A.; Knowles, D.G.; et al. The GENCODE v7 catalog of human long noncoding RNAs: Analysis of their gene structure, evolution, and expression. *Genome Res.* **2012**, *22*, 1775–1789. [CrossRef]
21. St; St. Laurent, G.; Wahlestedt, C.; Kapranov, P. The Landscape of long noncoding RNA classification. *Trends Genet.* **2015**, *31*, 239–251. [CrossRef]
22. Wenric, S.; ElGuendi, S.; Caberg, J.-H.; Bezzaou, W.; Fasquelle, C.; Charloteaux, B.; Karim, L.; Hennuy, B.; Frères, P.; Collignon, J.; et al. Transcriptome-wide analysis of natural antisense transcripts shows their potential role in breast cancer. *Sci. Rep.* **2017**, *7*, 17452. [CrossRef]
23. Sanchez de Groot, N.; Armaos, A.; Graña-Montes, R.; Alriquet, M.; Calloni, G.; Vabulas, R.M.; Tartaglia, G.G. RNA structure drives interaction with proteins. *Nat. Commun.* **2019**, *10*, 3246. [CrossRef]
24. Gutschner, T.; Diederichs, S. The hallmarks of cancer: A long non-coding RNA point of view. *RNA Biol.* **2012**, *9*, 703–719. [CrossRef]
25. Fatica, A.; Bozzoni, I. Long non-coding RNAs: New players in cell differentiation and development. *Nat. Rev. Genet.* **2014**, *15*, 7–21. [CrossRef] [PubMed]
26. Ji, P.; Diederichs, S.; Wang, W.; Böing, S.; Metzger, R.; Schneider, P.M.; Tidow, N.; Brandt, B.; Buerger, H.; Bulk, E.; et al. MALAT-1, a novel noncoding RNA, and thymosin beta4 predict metastasis and survival in early-stage non-small cell lung cancer. *Oncogene* **2003**, *22*, 8031–8041. [CrossRef]
27. Matouk, I.J.; DeGroot, N.; Mezan, S.; Ayesh, S.; Abu-lail, R.; Hochberg, A.; Galun, E. The H19 non-coding RNA is essential for human tumor growth. *PLoS ONE* **2007**, *2*, e845. [CrossRef] [PubMed]
28. Yildirim, E.; Kirby, J.E.; Brown, D.E.; Mercier, F.E.; Sadreyev, R.I.; Scadden, D.T.; Lee, J.T. Xist RNA is a potent suppressor of hematologic cancer in mice. *Cell* **2013**, *152*, 727–742. [CrossRef]
29. Sirchia, S.M.; Tabano, S.; Monti, L.; Recalcati, M.P.; Gariboldi, M.; Grati, F.R.; Porta, G.; Finelli, P.; Radice, P.; Miozzo, M. Misbehaviour of XIST RNA in breast cancer cells. *PLoS ONE* **2009**, *4*, e5559. [CrossRef]
30. Wang, H.; Li, Q.; Tang, S.; Li, M.; Feng, A.; Qin, L.; Liu, Z.; Wang, X. The role of long noncoding RNA HOTAIR in the acquired multidrug resistance to imatinib in chronic myeloid leukemia cells. *Hematology* **2017**, *22*, 208–216. [CrossRef] [PubMed]
31. Xue, X.; Yang, Y.A.; Zhang, A.; Fong, K.-W.; Kim, J.; Song, B.; Li, S.; Zhao, J.C.; Yu, J. LncRNA HOTAIR enhances ER signaling and confers tamoxifen resistance in breast cancer. *Oncogene* **2016**, *35*, 2746–2755. [CrossRef]
32. Jarroux, J.; Morillon, A.; Pinskaya, M. History, Discovery, and Classification of lncRNAs. *Adv. Exp. Med. Biol.* **2017**, *1008*, 1–46. [CrossRef]
33. Wang, W.-T.; Han, C.; Sun, Y.-M.; Chen, T.-Q.; Chen, Y.-Q. Noncoding RNAs in cancer therapy resistance and targeted drug development. *J. Hematol. Oncol.* **2019**, *12*, 55. [CrossRef]
34. Chi, Y.; Wang, D.; Wang, J.; Yu, W.; Yang, J. Long Non-Coding RNA in the Pathogenesis of Cancers. *Cells* **2019**, *8*, 1015. [CrossRef] [PubMed]
35. Yang, Q.; Du, W.W.; Wu, N.; Yang, W.; Awan, F.M.; Fang, L.; Ma, J.; Li, X.; Zeng, Y.; Yang, Z.; et al. A circular RNA promotes tumorigenesis by inducing c-myc nuclear translocation. *Cell Death Differ.* **2017**, *24*, 1609–1620. [CrossRef] [PubMed]
36. Su, M.; Xiao, Y.; Ma, J.; Tang, Y.; Tian, B.; Zhang, Y.; Li, X.; Wu, Z.; Yang, D.; Zhou, Y.; et al. Circular RNAs in Cancer: Emerging functions in hallmarks, stemness, resistance and roles as potential biomarkers. *Mol. Cancer* **2019**, *18*, 90. [CrossRef]
37. Qi, P.; Du, X. The long non-coding RNAs, a new cancer diagnostic and therapeutic gold mine. *Mod. Pathol.* **2013**, *26*, 155–165. [CrossRef] [PubMed]
38. Salmena, L.; Poliseno, L.; Tay, Y.; Kats, L.; Pandolfi, P.P. A ceRNA hypothesis: The Rosetta Stone of a hidden RNA language? *Cell* **2011**, *146*, 353–358. [CrossRef] [PubMed]
39. Wang, K.C.; Chang, H.Y. Molecular mechanisms of long noncoding RNAs. *Mol. Cell* **2011**, *43*, 904–914. [CrossRef]

40. Zhang, M.; Matyunina, L.V.; Walker, L.D.; Chen, W.; Xiao, H.; Benigno, B.B.; Wu, R.; McDonald, J.F. Evidence for the importance of post-transcriptional regulatory changes in ovarian cancer progression and the contribution of miRNAs. *Sci. Rep.* **2017**, *7*, 8171. [CrossRef]

41. O'Brien, J.; Hayder, H.; Zayed, Y.; Peng, C. Overview of MicroRNA Biogenesis, Mechanisms of Actions, and Circulation. *Front. Endocrinol.* **2018**, *9*, 402. [CrossRef]

42. Yoon, J.-H.; Abdelmohsen, K.; Gorospe, M. Functional interactions among microRNAs and long noncoding RNAs. *Semin. Cell Dev. Biol.* **2014**, *34*, 9–14. [CrossRef]

43. Huang, B.; Song, J.H.; Cheng, Y.; Abraham, J.M.; Ibrahim, S.; Sun, Z.; Ke, X.; Meltzer, S.J. Long non-coding antisense RNA KRT7-AS is activated in gastric cancers and supports cancer cell progression by increasing KRT7 expression. *Oncogene* **2016**, *35*, 4927–4936. [CrossRef] [PubMed]

44. He, R.-Z.; Luo, D.-X.; Mo, Y.-Y. Emerging roles of lncRNAs in the post-transcriptional regulation in cancer. *Genes Dis.* **2019**, *6*, 6–15. [CrossRef] [PubMed]

45. Sakamoto, Y.; Sereewattanawoot, S.; Suzuki, A. A new era of long-read sequencing for cancer genomics. *J. Hum. Genet.* **2020**, *65*, 3–10. [CrossRef] [PubMed]

46. Gardini, A. Global Run-On Sequencing (GRO-Seq). *Methods Mol. Biol.* **2017**, *1468*, 111–120. [CrossRef] [PubMed]

47. Mahat, D.B.; Kwak, H.; Booth, G.T.; Jonkers, I.H.; Danko, C.G.; Patel, R.K.; Waters, C.T.; Munson, K.; Core, L.J.; Lis, J.T. Base-pair-resolution genome-wide mapping of active RNA polymerases using precision nuclear run-on (PRO-seq). *Nat. Protoc.* **2016**, *11*, 1455–1476. [CrossRef]

48. Paulsen, M.T.; Veloso, A.; Prasad, J.; Bedi, K.; Ljungman, E.A.; Magnuson, B.; Wilson, T.E.; Ljungman, M. Use of Bru-Seq and BruChase-Seq for genome-wide assessment of the synthesis and stability of RNA. *Methods* **2014**, *67*, 45–54. [CrossRef]

49. Plessy, C.; Bertin, N.; Takahashi, H.; Simone, R.; Salimullah, M.; Lassmann, T.; Vitezic, M.; Severin, J.; Olivarius, S.; Lazarevic, D.; et al. Linking promoters to functional transcripts in small samples with nanoCAGE and CAGEscan. *Nat. Methods* **2010**, *7*, 528–534. [CrossRef]

50. Workman, R.E.; Tang, A.D.; Tang, P.S.; Jain, M.; Tyson, J.R.; Razaghi, R.; Zuzarte, P.C.; Gilpatrick, T.; Payne, A.; Quick, J.; et al. Nanopore native RNA sequencing of a human poly(A) transcriptome. *Nat. Methods* **2019**, *16*, 1297–1305. [CrossRef]

51. Raha, D.; Hong, M.; Snyder, M. ChIP-Seq: A method for global identification of regulatory elements in the genome. *Curr. Protoc. Mol. Biol.* **2010**, *21*. [CrossRef]

52. Corces, M.R.; Trevino, A.E.; Hamilton, E.G.; Greenside, P.G.; Sinnott-Armstrong, N.A.; Vesuna, S.; Satpathy, A.T.; Rubin, A.J.; Montine, K.S.; Wu, B.; et al. An improved ATAC-seq protocol reduces background and enables interrogation of frozen tissues. *Nat. Methods* **2017**, *14*, 959–962. [CrossRef]

53. Antonov, I.V.; Mazurov, E.; Borodovsky, M.; Medvedeva, Y.A. Prediction of lncRNAs and their interactions with nucleic acids: Benchmarking bioinformatics tools. *Brief. Bioinform.* **2019**, *20*, 551–564. [CrossRef]

54. Li, J.; Zhang, X.; Liu, C. The computational approaches of lncRNA identification based on coding potential: Status quo and challenges. *Comput. Struct. Biotechnol. J.* **2020**, *18*, 3666–3677. [CrossRef]

55. Jeck, W.R.; Sharpless, N.E. Detecting and characterizing circular RNAs. *Nat. Biotechnol.* **2014**, *32*, 453–461. [CrossRef] [PubMed]

56. Zeng, X.; Lin, W.; Guo, M.; Zou, Q. A comprehensive overview and evaluation of circular RNA detection tools. *PLoS Comput. Biol.* **2017**, *13*, e1005420. [CrossRef] [PubMed]

57. Szabo, L.; Salzman, J. Detecting circular RNAs: Bioinformatic and experimental challenges. *Nat. Rev. Genet.* **2016**, *17*, 679–692. [CrossRef]

58. Szabo, L.; Morey, R.; Palpant, N.J.; Wang, P.L.; Afari, N.; Jiang, C.; Parast, M.M.; Murry, C.E.; Laurent, L.C.; Salzman, J. Statistically based splicing detection reveals neural enrichment and tissue-specific induction of circular RNA during human fetal development. *Genome Biol.* **2015**, *16*, 126. [CrossRef] [PubMed]

59. Gao, Y.; Wang, J.; Zhao, F. CIRI: An efficient and unbiased algorithm for de novo circular RNA identification. *Genome Biol.* **2015**, *16*, 4. [CrossRef]

60. Gao, Y.; Zhang, J.; Zhao, F. Circular RNA identification based on multiple seed matching. *Brief. Bioinform.* **2018**, *19*, 803–810. [CrossRef]

61. Zou, D.; Ma, L.; Yu, J.; Zhang, Z. Biological databases for human research. *Genom. Proteom. Bioinform.* **2015**, *13*, 55–63. [CrossRef]

62. Landrum, M.J.; Chitipiralla, S.; Brown, G.R.; Chen, C.; Gu, B.; Hart, J.; Hoffman, D.; Jang, W.; Kaur, K.; Liu, C.; et al. ClinVar: Improvements to accessing data. *Nucleic Acids Res.* **2020**, *48*, D835–D844. [CrossRef] [PubMed]

63. Tate, J.G.; Bamford, S.; Jubb, H.C.; Sondka, Z.; Beare, D.M.; Bindal, N.; Boutselakis, H.; Cole, C.G.; Creatore, C.; Dawson, E.; et al. COSMIC: The Catalogue of Somatic Mutations in Cancer. *Nucleic Acids Res.* **2019**, *47*, D941–D947. [CrossRef] [PubMed]

64. Cancer Genome Atlas Research Network; Weinstein, J.N.; Collisson, E.A.; Mills, G.B.; Shaw, K.R.M.; Ozenberger, B.A.; Ellrott, K.; Shmulevich, I.; Sander, C.; Stuart, J.M. The Cancer Genome Atlas Pan-Cancer analysis project. *Nat. Genet.* **2013**, *45*, 1113–1120. [CrossRef]

65. Clarke, L.; Fairley, S.; Zheng-Bradley, X.; Streeter, I.; Perry, E.; Lowy, E.; Tassé, A.-M.; Flicek, P. The international Genome sample resource (IGSR): A worldwide collection of genome variation incorporating the 1000 Genomes Project data. *Nucleic Acids Res.* **2017**, *45*, D854–D859. [CrossRef] [PubMed]

66. Frankish, A.; Diekhans, M.; Ferreira, A.-M.; Johnson, R.; Jungreis, I.; Loveland, J.; Mudge, J.M.; Sisu, C.; Wright, J.; Armstrong, J.; et al. GENCODE reference annotation for the human and mouse genomes. *Nucleic Acids Res.* **2019**, *47*, D766–D773. [CrossRef] [PubMed]

67. ENCODE Project Consortium; Moore, J.E.; Purcaro, M.J.; Pratt, H.E.; Epstein, C.B.; Shoresh, N.; Adrian, J.; Kawli, T.; Davis, C.A.; Dobin, A.; et al. Expanded encyclopaedias of DNA elements in the human and mouse genomes. *Nature* **2020**, *583*, 699–710. [CrossRef]

68. Geo NCBI-GEO. Available online: https://www.ncbi.nlm.nih.gov/geo/ (accessed on 29 January 2021).

69. Sherry, S.T.; Ward, M.; Sirotkin, K. dbSNP-database for single nucleotide polymorphisms and other classes of minor genetic variation. *Genome Res.* **1999**, *9*, 677–679.

70. The UniProt Consortium UniProt: The universal protein knowledgebase. *Nucleic Acids Res.* **2017**, *45*, D158–D169. [CrossRef]

71. The Human Protein Atlas. Available online: https://www.proteinatlas.org/ (accessed on 29 January 2021).

72. GTEx Portal. Available online: https://gtexportal.org/home/ (accessed on 29 January 2021).

73. Lizio, M.; Abugessaisa, I.; Noguchi, S.; Kondo, A.; Hasegawa, A.; Hon, C.C.; de Hoon, M.; Severin, J.; Oki, S.; Hayashizaki, Y.; et al. Update of the FANTOM web resource: Expansion to provide additional transcriptome atlases. *Nucleic Acids Res.* **2019**, *47*, D752–D758. [CrossRef] [PubMed]

74. Broad Institute Cancer Cell Line Encyclopedia (CCLE). Available online: https://portals.broadinstitute.org/ccle (accessed on 29 January 2021).

75. Schriml, L.M.; Mitraka, E.; Munro, J.; Tauber, B.; Schor, M.; Nickle, L.; Felix, V.; Jeng, L.; Bearer, C.; Lichenstein, R.; et al. Human Disease Ontology 2018 update: Classification, content and workflow expansion. *Nucleic Acids Res.* **2019**, *47*, D955–D962. [CrossRef]

76. Gene Ontology Consortium the Gene Ontology resource: Enriching a GOld mine. *Nucleic Acids Res.* **2021**, *49*, D325–D334. [CrossRef]

77. NCBI-MeSH. Available online: https://www.ncbi.nlm.nih.gov/mesh/ (accessed on 29 January 2021).

78. GenomeOC Therapeutically Applicable Research to Generate Effective Treatments. Available online: https://ocg.cancer.gov/programs/target (accessed on 29 January 2021).

79. Li, J.-H.; Liu, S.; Zhou, H.; Qu, L.-H.; Yang, J.-H. starBase v2.0: Decoding miRNA-ceRNA, miRNA-ncRNA and protein-RNA interaction networks from large-scale CLIP-Seq data. *Nucleic Acids Res.* **2014**, *42*, D92–D97. [CrossRef]

80. Agarwal, V.; Bell, G.W.; Nam, J.-W.; Bartel, D.P. Predicting effective microRNA target sites in mammalian mRNAs. *Elife* **2015**, *4*. [CrossRef]

81. Blin, K.; Dieterich, C.; Wurmus, R.; Rajewsky, N.; Landthaler, M.; Akalin, A. DoRiNA 2.0–upgrading the doRiNA database of RNA interactions in post-transcriptional regulation. *Nucleic Acids Res.* **2015**, *43*, D160–D167. [CrossRef]

82. Jeggari, A.; Marks, D.S.; Larsson, E. miRcode: A map of putative microRNA target sites in the long non-coding transcriptome. *Bioinformatics* **2012**, *28*, 2062–2063. [CrossRef]

83. Huang, H.-Y.; Lin, Y.-C.-D.; Li, J.; Huang, K.-Y.; Shrestha, S.; Hong, H.-C.; Tang, Y.; Chen, Y.-G.; Jin, C.-N.; Yu, Y.; et al. miRTarBase 2020: Updates to the experimentally validated microRNA-target interaction database. *Nucleic Acids Res.* **2020**, *48*, D148–D154. [CrossRef]

84. Huang, Z.; Shi, J.; Gao, Y.; Cui, C.; Zhang, S.; Li, J.; Zhou, Y.; Cui, Q. HMDD v3.0: A database for experimentally supported human microRNA-disease associations. *Nucleic Acids Res.* **2019**, *47*, D1013–D1017. [CrossRef] [PubMed]

85. Wang, D.; Gu, J.; Wang, T.; Ding, Z. OncomiRDB: A database for the experimentally verified oncogenic and tumor-suppressive microRNAs. *Bioinformatics* **2014**, *30*, 2237–2238. [CrossRef] [PubMed]

86. Yang, Z.; Wu, L.; Wang, A.; Tang, W.; Zhao, Y.; Zhao, H.; Teschendorff, A.E. dbDEMC 2.0: Updated database of differentially expressed miRNAs in human cancers. *Nucleic Acids Res.* **2017**, *45*, D812–D818. [CrossRef]

87. Xiao, F.; Zuo, Z.; Cai, G.; Kang, S.; Gao, X.; Li, T. miRecords: An integrated resource for microRNA-target interactions. *Nucleic Acids Res.* **2009**, *37*, D105–D110. [CrossRef]

88. Zhao, L.; Wang, J.; Li, Y.; Song, T.; Wu, Y.; Fang, S.; Bu, D.; Li, H.; Sun, L.; Pei, D.; et al. NONCODEV6: An updated database dedicated to long non-coding RNA annotation in both animals and plants. *Nucleic Acids Res.* **2021**, *49*, D165–D171. [CrossRef]

89. Volders, P.-J.; Anckaert, J.; Verheggen, K.; Nuytens, J.; Martens, L.; Mestdagh, P.; Vandesompele, J. LNCipedia 5: Towards a reference set of human long non-coding RNAs. *Nucleic Acids Res.* **2019**, *47*, D135–D139. [CrossRef]

90. Mas-Ponte, D.; Carlevaro-Fita, J.; Palumbo, E.; Hermoso Pulido, T.; Guigo, R.; Johnson, R. LncATLAS database for subcellular localization of long noncoding RNAs. *RNA* **2017**, *23*, 1080–1087. [CrossRef] [PubMed]

91. Ma, L.; Li, A.; Zou, D.; Xu, X.; Xia, L.; Yu, J.; Bajic, V.B.; Zhang, Z. LncRNAWiki: Harnessing community knowledge in collaborative curation of human long non-coding RNAs. *Nucleic Acids Res.* **2015**, *43*, D187–D192. [CrossRef] [PubMed]

92. Ma, L.; Cao, J.; Liu, L.; Du, Q.; Li, Z.; Zou, D.; Bajic, V.B.; Zhang, Z. LncBook: A curated knowledgebase of human long non-coding RNAs. *Nucleic Acids Res.* **2019**, *47*, D128–D134. [CrossRef] [PubMed]

93. Gao, Y.; Shang, S.; Guo, S.; Li, X.; Zhou, H.; Liu, H.; Sun, Y.; Wang, J.; Wang, P.; Zhi, H.; et al. Lnc2Cancer 3.0: An updated resource for experimentally supported lncRNA/circRNA cancer associations and web tools based on RNA-seq and scRNA-seq data. *Nucleic Acids Res.* **2021**, *49*, D1251–D1258. [CrossRef] [PubMed]

94. Bao, Z.; Yang, Z.; Huang, Z.; Zhou, Y.; Cui, Q.; Dong, D. LncRNADisease 2.0: An updated database of long non-coding RNA-associated diseases. *Nucleic Acids Res.* **2019**, *47*, D1034–D1037. [CrossRef]

95. Li, Y.; Li, L.; Wang, Z.; Pan, T.; Sahni, N.; Jin, X.; Wang, G.; Li, J.; Zheng, X.; Zhang, Y.; et al. LncMAP: Pan-cancer atlas of long noncoding RNA-mediated transcriptional network perturbations. *Nucleic Acids Res.* **2018**, *46*, 1113–1123. [CrossRef]

96. Li, J.; Han, L.; Roebuck, P.; Diao, L.; Liu, L.; Yuan, Y.; Weinstein, J.N.; Liang, H. TANRIC: An Interactive Open Platform to Explore the Function of lncRNAs in Cancer. *Cancer Res.* **2015**, *75*, 3728–3737. [CrossRef]

97. Ning, L.; Cui, T.; Zheng, B.; Wang, N.; Luo, J.; Yang, B.; Du, M.; Cheng, J.; Dou, Y.; Wang, D. MNDR v3.0: Mammal ncRNA-disease repository with increased coverage and annotation. *Nucleic Acids Res.* **2021**, *49*, D160–D164. [CrossRef]

98. Miao, Y.-R.; Liu, W.; Zhang, Q.; Guo, A.-Y. lncRNASNP2: An updated database of functional SNPs and mutations in human and mouse lncRNAs. *Nucleic Acids Res.* **2018**, *46*, D276–D280. [CrossRef]

99. Chan, W.-L.; Huang, H.-D.; Chang, J.-G. lncRNAMap: A map of putative regulatory functions in the long non-coding transcriptome. *Comput. Biol. Chem.* **2014**, *50*, 41–49. [CrossRef]

100. Zhao, H.; Shi, J.; Zhang, Y.; Xie, A.; Yu, L.; Zhang, C.; Lei, J.; Xu, H.; Leng, Z.; Li, T.; et al. LncTarD: A manually-curated database of experimentally-supported functional lncRNA-target regulations in human diseases. *Nucleic Acids Res.* **2020**, *48*, D118–D126. [CrossRef]

101. Zhou, B.; Zhao, H.; Yu, J.; Guo, C.; Dou, X.; Song, F.; Hu, G.; Cao, Z.; Qu, Y.; Yang, Y.; et al. EVLncRNAs: A manually curated database for long non-coding RNAs validated by low-throughput experiments. *Nucleic Acids Res.* **2018**, *46*, D100–D105. [CrossRef]

102. Lv, D.; Xu, K.; Jin, X.; Li, J.; Shi, Y.; Zhang, M.; Jin, X.; Li, Y.; Xu, J.; Li, X. LncSpA: LncRNA Spatial Atlas of Expression across Normal and Cancer Tissues. *Cancer Res.* **2020**, *80*, 2067–2071. [CrossRef]

103. Wu, W.; Ji, P.; Zhao, F. CircAtlas: An integrated resource of one million highly accurate circular RNAs from 1070 vertebrate transcriptomes. *Genome Biol.* **2020**, *21*, 101. [CrossRef]

104. Chen, X.; Han, P.; Zhou, T.; Guo, X.; Song, X.; Li, Y. circRNADb: A comprehensive database for human circular RNAs with protein-coding annotations. *Sci. Rep.* **2016**, *6*, 34985. [CrossRef]

105. Meng, X.; Hu, D.; Zhang, P.; Chen, Q.; Chen, M. CircFunBase: A database for functional circular RNAs. *Database* **2019**, *2019*. [CrossRef]

106. Glažar, P.; Papavasileiou, P.; Rajewsky, N. circBase: A database for circular RNAs. *RNA* **2014**, *20*, 1666–1670. [CrossRef]

107. Memczak, S.; Jens, M.; Elefsinioti, A.; Torti, F.; Krueger, J.; Rybak, A.; Maier, L.; Mackowiak, S.D.; Gregersen, L.H.; Munschauer, M.; et al. Circular RNAs are a large class of animal RNAs with regulatory potency. *Nature* **2013**, *495*, 333–338. [CrossRef]

108. Salzman, J.; Chen, R.E.; Olsen, M.N.; Wang, P.L.; Brown, P.O. Cell-type specific features of circular RNA expression. *PLoS Genet.* **2013**, *9*, e1003777. [CrossRef]

109. Jeck, W.R.; Sorrentino, J.A.; Wang, K.; Slevin, M.K.; Burd, C.E.; Liu, J.; Marzluff, W.F.; Sharpless, N.E. Circular RNAs are abundant, conserved, and associated with ALU repeats. *RNA* **2013**, *19*, 141–157. [CrossRef] [PubMed]

110. Rybak-Wolf, A.; Stottmeister, C.; Glažar, P.; Jens, M.; Pino, N.; Giusti, S.; Hanan, M.; Behm, M.; Bartok, O.; Ashwal-Fluss, R.; et al. Circular RNAs in the Mammalian Brain Are Highly Abundant, Conserved, and Dynamically Expressed. *Mol. Cell* **2015**, *58*, 870–885. [CrossRef]

111. Maass, P.G.; Glažar, P.; Memczak, S.; Dittmar, G.; Hollfinger, I.; Schreyer, L.; Sauer, A.V.; Toka, O.; Aiuti, A.; Luft, F.C.; et al. A map of human circular RNAs in clinically relevant tissues. *J. Mol. Med.* **2017**, *95*, 1179–1189. [CrossRef] [PubMed]

112. Liu, M.; Wang, Q.; Shen, J.; Yang, B.B.; Ding, X. Circbank: A comprehensive database for circRNA with standard nomenclature. *RNA Biol.* **2019**, *16*, 899–905. [CrossRef] [PubMed]

113. Dong, R.; Ma, X.-K.; Li, G.-W.; Yang, L. CIRCpedia v2: An Updated Database for Comprehensive Circular RNA Annotation and Expression Comparison. *Genom. Proteom. Bioinform.* **2018**, *16*, 226–233. [CrossRef]

114. Zhao, Z.; Wang, K.; Wu, F.; Wang, W.; Zhang, K.; Hu, H.; Liu, Y.; Jiang, T. circRNA disease: A manually curated database of experimentally supported circRNA-disease associations. *Cell Death Dis.* **2018**, *9*, 475. [CrossRef]

115. Fan, C.; Lei, X.; Fang, Z.; Jiang, Q.; Wu, F.-X. CircR2Disease: A manually curated database for experimentally supported circular RNAs associated with various diseases. *Database* **2018**, *2018*. [CrossRef]

116. Xia, S.; Feng, J.; Lei, L.; Hu, J.; Xia, L.; Wang, J.; Xiang, Y.; Liu, L.; Zhong, S.; Han, L.; et al. Comprehensive characterization of tissue-specific circular RNAs in the human and mouse genomes. *Brief. Bioinform.* **2017**, *18*, 984–992. [CrossRef]

117. Rophina, M.; Sharma, D.; Poojary, M.; Scaria, V. Circad: A comprehensive manually curated resource of circular RNA associated with diseases. *Database* **2020**, *2020*. [CrossRef]

118. Zhao, M.; Qu, H. circVAR database: Genome-wide archive of genetic variants for human circular RNAs. *BMC Genom.* **2020**, *21*, 750. [CrossRef]

119. Xia, S.; Feng, J.; Chen, K.; Ma, Y.; Gong, J.; Cai, F.; Jin, Y.; Gao, Y.; Xia, L.; Chang, H.; et al. CSCD: A database for cancer-specific circular RNAs. *Nucleic Acids Res.* **2018**, *46*, D925–D929. [CrossRef] [PubMed]

120. Ghosal, S.; Das, S.; Sen, R.; Basak, P.; Chakrabarti, J. Circ2Traits: A comprehensive database for circular RNA potentially associated with disease and traits. *Front. Genet.* **2013**, *4*, 283. [CrossRef]

121. Hafner, M.; Landthaler, M.; Burger, L.; Khorshid, M.; Hausser, J.; Berninger, P.; Rothballer, A.; Ascano, M., Jr.; Jungkamp, A.-C.; Munschauer, M.; et al. Transcriptome-wide identification of RNA-binding protein and microRNA target sites by PAR-CLIP. *Cell* **2010**, *141*, 129–141. [CrossRef]

122. Yao, D.; Zhang, L.; Zheng, M.; Sun, X.; Lu, Y.; Liu, P. Circ2Disease: A manually curated database of experimentally validated circRNAs in human disease. *Sci. Rep.* **2018**, *8*, 11018. [CrossRef]

123. Dudekula, D.B.; Panda, A.C.; Grammatikakis, I.; De, S.; Abdelmohsen, K.; Gorospe, M. CircInteractome: A web tool for exploring circular RNAs and their interacting proteins and microRNAs. *RNA Biol.* **2016**, *13*, 34–42. [CrossRef] [PubMed]
124. Kevil, C.G.; Walsh, L.; Laroux, F.S.; Kalogeris, T.; Grisham, M.B.; Alexander, J.S. An improved, rapid Northern protocol. *Biochem. Biophys. Res. Commun.* **1997**, *238*, 277–279. [CrossRef] [PubMed]
125. Hu, X.; Feng, Y.; Hu, Z.; Zhang, Y.; Yuan, C.-X.; Xu, X.; Zhang, L. Detection of Long Noncoding RNA Expression by Nonradioactive Northern Blots. In *Long Non-Coding RNAs: Methods and Protocols*; Feng, Y., Zhang, L., Eds.; Springer: New York, NY, USA, 2016; pp. 177–188. ISBN 9781493933785.
126. Streit, S.; Michalski, C.W.; Erkan, M.; Kleeff, J.; Friess, H. Northern blot analysis for detection and quantification of RNA in pancreatic cancer cells and tissues. *Nat. Protoc.* **2009**, *4*, 37–43. [CrossRef]
127. Zang, S.; Lin, R.-J. Northwestern Blot Analysis: Detecting RNA–Protein Interaction After Gel Separation of Protein Mixture. In *RNA-Protein Complexes and Interactions: Methods and Protocols*; Lin, R.-J., Ed.; Springer: New York, NY, USA, 2016; pp. 111–125. ISBN 9781493935918.
128. Bustin, S.A.; Benes, V.; Garson, J.A.; Hellemans, J.; Huggett, J.; Kubista, M.; Mueller, R.; Nolan, T.; Pfaffl, M.W.; Shipley, G.L.; et al. The MIQE guidelines: Minimum information for publication of quantitative real-time PCR experiments. *Clin. Chem.* **2009**, *55*, 611–622. [CrossRef]
129. Jo, J.; Choi, S.; Oh, J.; Lee, S.-G.; Choi, S.Y.; Kim, K.K.; Park, C. Conventionally used reference genes are not outstanding for normalization of gene expression in human cancer research. *BMC Bioinform.* **2019**, *20*, 245. [CrossRef] [PubMed]
130. Olivarius, S.; Plessy, C.; Carninci, P. High-throughput verification of transcriptional starting sites by Deep-RACE. *Biotechniques* **2009**, *46*, 130–132. [CrossRef] [PubMed]
131. Lagarde, J.; Uszczynska-Ratajczak, B.; Santoyo-Lopez, J.; Gonzalez, J.M.; Tapanari, E.; Mudge, J.M.; Steward, C.A.; Wilming, L.; Tanzer, A.; Howald, C.; et al. Extension of human lncRNA transcripts by RACE coupled with long-read high-throughput sequencing (RACE-Seq). *Nat. Commun.* **2016**, *7*, 12339. [CrossRef] [PubMed]
132. Coassin, S.R.; Orjalo, A.V.; Semaan, S.J.; Johansson, H.E. Simultaneous Detection of Nuclear and Cytoplasmic RNA Variants Utilizing Stellaris® RNA Fluorescence In Situ Hybridization in Adherent Cells. In *In Situ Hybridization Protocols*; Nielsen, B.S., Ed.; Springer: New York, NY, USA, 2014; pp. 189–199. ISBN 9781493914593.
133. Orjalo, A.V.; Johansson, H.E. Stellaris® RNA Fluorescence In Situ Hybridization for the Simultaneous Detection of Immature and Mature Long Noncod-ing RNAs in Adherent Cells. In *Long Non-Coding RNAs: Methods and Protocols*; Feng, Y., Zhang, L., Eds.; Springer: New York, NY, USA, 2016; pp. 119–134. ISBN 9781493933785.
134. Hacisuleyman, E.; Goff, L.A.; Trapnell, C.; Williams, A.; Henao-Mejia, J.; Sun, L.; McClanahan, P.; Hendrickson, D.G.; Sauvageau, M.; Kelley, D.R.; et al. Topological organization of multichromosomal regions by the long intergenic noncoding RNA Firre. *Nat. Struct. Mol. Biol.* **2014**, *21*, 198–206. [CrossRef] [PubMed]
135. Maamar, H.; Cabili, M.N.; Rinn, J.; Raj, A. linc-HOXA1 is a noncoding RNA that represses Hoxa1 transcription in cis. *Genes Dev.* **2013**, *27*, 1260–1271. [CrossRef]
136. Dunagin, M.; Cabili, M.N.; Rinn, J.; Raj, A. Visualization of lncRNA by Single-Molecule Fluorescence In Situ Hybridization. In *Nuclear Bodies and Noncoding RNAs: Methods and Protocols*; Nakagawa, S., Hirose, T., Eds.; Springer: New York, NY, USA, 2015; pp. 3–19. ISBN 9781493922536.
137. Mortimer, S.A.; Weeks, K.M. A fast-acting reagent for accurate analysis of RNA secondary and tertiary structure by SHAPE chemistry. *J. Am. Chem. Soc.* **2007**, *129*, 4144–4145. [CrossRef] [PubMed]
138. Merino, E.J.; Wilkinson, K.A.; Coughlan, J.L.; Weeks, K.M. RNA structure analysis at single nucleotide resolution by selective 2′-hydroxyl acylation and primer extension (SHAPE). *J. Am. Chem. Soc.* **2005**, *127*, 4223–4231. [CrossRef] [PubMed]
139. Smola, M.J.; Rice, G.M.; Busan, S.; Siegfried, N.A.; Weeks, K.M. Selective 2′-hydroxyl acylation analyzed by primer extension and mutational profiling (SHAPE-MaP) for direct, versatile and accurate RNA structure analysis. *Nat. Protoc.* **2015**, *10*, 1643–1669. [CrossRef]
140. Schmidt, K.; Weidmann, C.A.; Hilimire, T.A.; Yee, E.; Hatfield, B.M.; Schneekloth, J.S., Jr.; Weeks, K.M.; Novina, C.D. Targeting the Oncogenic Long Non-coding RNA SLNCR1 by Blocking Its Sequence-Specific Binding to the Androgen Receptor. *Cell Rep.* **2020**, *30*, 541–554.e5. [CrossRef]
141. Shields, E.J.; Petracovici, A.F.; Bonasio, R. lncRedibly versatile: Biochemical and biological functions of long noncoding RNAs. *Biochem. J.* **2019**, *476*, 1083–1104. [CrossRef]
142. Mishra, K.; Kanduri, C. Understanding Long Noncoding RNA and Chromatin Interactions: What We Know So Far. *Noncoding RNA* **2019**, *5*. [CrossRef] [PubMed]
143. Zhang, Y.; Long, Y.; Kwoh, C.K. Deep learning based DNA:RNA triplex forming potential prediction. *BMC Bioinform.* **2020**, *21*, 522. [CrossRef] [PubMed]
144. Kuo, C.-C.; Hänzelmann, S.; Sentürk Cetin, N.; Frank, S.; Zajzon, B.; Derks, J.-P.; Akhade, V.S.; Ahuja, G.; Kanduri, C.; Grummt, I.; et al. Detection of RNA-DNA binding sites in long noncoding RNAs. *Nucleic Acids Res.* **2019**, *47*, e32. [CrossRef] [PubMed]
145. Alam, T.; Al-Absi, H.R.H.; Schmeier, S. Deep Learning in LncRNAome: Contribution, Challenges, and Perspectives. *Noncoding RNA* **2020**, *6*. [CrossRef]
146. Iwakiri, J.; Hamada, M.; Asai, K. Bioinformatics tools for lncRNA research. *Biochim. Biophys. Acta* **2016**, *1859*, 23–30. [CrossRef] [PubMed]

147. Terai, G.; Iwakiri, J.; Kameda, T.; Hamada, M.; Asai, K. Comprehensive prediction of lncRNA-RNA interactions in human transcriptome. *BMC Genom.* **2016**, *17*, 12. [CrossRef] [PubMed]
148. Iwakiri, J.; Terai, G.; Hamada, M. Computational prediction of lncRNA-mRNA interactionsby integrating tissue specificity in human transcriptome. *Biol. Direct* **2017**, *12*, 15. [CrossRef]
149. Mar-Aguilar, F.; Rodríguez-Padilla, C.; Reséndez-Pérez, D. Web-based tools for microRNAs involved in human cancer (Review). *Oncol. Lett.* **2016**, *11*, 3563–3570. [CrossRef]
150. Antonov, I.; Marakhonov, A.; Zamkova, M.; Medvedeva, Y. ASSA: Fast identification of statistically significant interactions between long RNAs. *J. Bioinform. Comput. Biol.* **2018**, *16*, 1840001. [CrossRef]
151. Fukunaga, T.; Hamada, M. RIblast: An ultrafast RNA-RNA interaction prediction system based on a seed-and-extension approach. *Bioinformatics* **2017**, *33*, 2666–2674. [CrossRef]
152. Zhang, J.; Le, T.D.; Liu, L.; Li, J. Inferring and analyzing module-specific lncRNA-mRNA causal regulatory networks in human cancer. *Brief. Bioinform.* **2019**, *20*, 1403–1419. [CrossRef]
153. Pyfrom, S.C.; Luo, H.; Payton, J.E. PLAIDOH: A novel method for functional prediction of long non-coding RNAs identifies cancer-specific LncRNA activities. *BMC Genom.* **2019**, *20*, 137. [CrossRef]
154. Gawronski, A.R.; Uhl, M.; Zhang, Y.; Lin, Y.-Y.; Niknafs, Y.S.; Ramnarine, V.R.; Malik, R.; Feng, F.; Chinnaiyan, A.M.; Collins, C.C.; et al. MechRNA: Prediction of lncRNA mechanisms from RNA-RNA and RNA-protein interactions. *Bioinformatics* **2018**, *34*, 3101–3110. [CrossRef]
155. Athar, A.; Füllgrabe, A.; George, N.; Iqbal, H.; Huerta, L.; Ali, A.; Snow, C.; Fonseca, N.A.; Petryszak, R.; Papatheodorou, I.; et al. ArrayExpress update—From bulk to single-cell expression data. *Nucleic Acids Res.* **2019**, *47*, D711–D715. [CrossRef]
156. NCBI-SRA. Available online: https://www.ncbi.nlm.nih.gov/sra (accessed on 29 January 2021).
157. Babbi, G.; Martelli, P.L.; Profiti, G.; Bovo, S.; Savojardo, C.; Casadio, R. eDGAR: A database of Disease-Gene Associations with annotated Relationships among genes. *BMC Genom.* **2017**, *18*, 554. [CrossRef]
158. Junge, A.; Refsgaard, J.C.; Garde, C.; Pan, X.; Santos, A.; Alkan, F.; Anthon, C.; von Mering, C.; Workman, C.T.; Jensen, L.J.; et al. RAIN: RNA-protein Association and Interaction Networks. *Database* **2017**, *2017*. [CrossRef] [PubMed]
159. Yi, Y.; Zhao, Y.; Li, C.; Zhang, L.; Huang, H.; Li, Y.; Liu, L.; Hou, P.; Cui, T.; Tan, P.; et al. RAID v2.0: An updated resource of RNA-associated interactions across organisms. *Nucleic Acids Res.* **2017**, *45*, D115–D118. [CrossRef] [PubMed]
160. Teng, X.; Chen, X.; Xue, H.; Tang, Y.; Zhang, P.; Kang, Q.; Hao, Y.; Chen, R.; Zhao, Y.; He, S. NPInter v4.0: An integrated database of ncRNA interactions. *Nucleic Acids Res.* **2020**, *48*, D160–D165. [CrossRef] [PubMed]
161. Gong, J.; Shao, D.; Xu, K.; Lu, Z.; Lu, Z.J.; Yang, Y.T.; Zhang, Q.C. RISE: A database of RNA interactome from sequencing experiments. *Nucleic Acids Res.* **2018**, *46*, D194–D201. [CrossRef] [PubMed]
162. Chen, L.; Heikkinen, L.; Wang, C.; Yang, Y.; Sun, H.; Wong, G. Trends in the development of miRNA bioinformatics tools. *Brief. Bioinform.* **2019**, *20*, 1836–1852. [CrossRef] [PubMed]
163. Fukunaga, T.; Iwakiri, J.; Ono, Y.; Hamada, M. LncRRIsearch: A Web Server for lncRNA-RNA Interaction Prediction Integrated With Tissue-Specific Expression and Subcellular Localization Data. *Front. Genet.* **2019**, *10*, 462. [CrossRef] [PubMed]
164. Karagkouni, D.; Paraskevopoulou, M.D.; Tastsoglou, S.; Skoufos, G.; Karavangeli, A.; Pierros, V.; Zacharopoulou, E.; Hatzigeor-giou, A.G. DIANA-LncBase v3: Indexing experimentally supported miRNA targets on non-coding transcripts. *Nucleic Acids Res.* **2020**, *48*, D101–D110. [CrossRef] [PubMed]
165. Hoffmann, M.; Pachl, E.; Hartung, M.; Stiegler, V.; Baumbach, J.; Schulz, M.H.; List, M. SPONGEdb: A pan-cancer resource for competing endogenous RNA interactions. *NAR Cancer* **2021**, *3*. [CrossRef]
166. Wang, P.; Li, X.; Gao, Y.; Guo, Q.; Ning, S.; Zhang, Y.; Shang, S.; Wang, J.; Wang, Y.; Zhi, H.; et al. LnCeVar: A comprehensive database of genomic variations that disturb ceRNA network regulation. *Nucleic Acids Res.* **2020**, *48*, D111–D117. [CrossRef] [PubMed]
167. Wang, P.; Zhi, H.; Zhang, Y.; Liu, Y.; Zhang, J.; Gao, Y.; Guo, M.; Ning, S.; Li, X. miRSponge: A manually curated database for experimentally supported miRNA sponges and ceRNAs. *Database* **2015**, *2015*. [CrossRef]
168. Lin, Y.; Liu, T.; Cui, T.; Wang, Z.; Zhang, Y.; Tan, P.; Huang, Y.; Yu, J.; Wang, D. RNAInter in 2020: RNA interactome repository with increased coverage and annotation. *Nucleic Acids Res.* **2020**, *48*, D189–D197. [CrossRef] [PubMed]
169. Cheng, L.; Wang, P.; Tian, R.; Wang, S.; Guo, Q.; Luo, M.; Zhou, W.; Liu, G.; Jiang, H.; Jiang, Q. LncRNA2Target v2.0: A comprehensive database for target genes of lncRNAs in human and mouse. *Nucleic Acids Res.* **2019**, *47*, D140–D144. [CrossRef] [PubMed]
170. Li, Z.; Liu, L.; Jiang, S.; Li, Q.; Feng, C.; Du, Q.; Zou, D.; Xiao, J.; Zhang, Z.; Ma, L. LncExpDB: An expression database of human long non-coding RNAs. *Nucleic Acids Res.* **2021**, *49*, D962–D968. [CrossRef]
171. Wang, P.; Li, X.; Gao, Y.; Guo, Q.; Wang, Y.; Fang, Y.; Ma, X.; Zhi, H.; Zhou, D.; Shen, W.; et al. LncACTdb 2.0: An updated database of experimentally supported ceRNA interactions curated from low- and high-throughput experiments. *Nucleic Acids Res.* **2019**, *47*, D121–D127. [CrossRef] [PubMed]
172. Liu, L.; Zhang, Y.; Lu, J. The roles of long noncoding RNAs in breast cancer metastasis. *Cell Death Dis.* **2020**, *11*, 749. [CrossRef]
173. Gupta, R.A.; Shah, N.; Wang, K.C.; Kim, J.; Horlings, H.M.; Wong, D.J.; Tsai, M.-C.; Hung, T.; Argani, P.; Rinn, J.L.; et al. Long non-coding RNA HOTAIR reprograms chromatin state to promote cancer metastasis. *Nature* **2010**, *464*, 1071–1076. [CrossRef] [PubMed]

174. Zhang, X.; Zhou, Y.; Chen, S.; Li, W.; Chen, W.; Gu, W. LncRNA MACC1-AS1 sponges multiple miRNAs and RNA-binding protein PTBP1. *Oncogenesis* **2019**, *8*, 73. [CrossRef]
175. Mili, S.; Steitz, J.A. Evidence for reassociation of RNA-binding proteins after cell lysis: Implications for the interpretation of immunoprecipitation analyses. *RNA* **2004**, *10*, 1692–1694. [CrossRef] [PubMed]
176. Riley, K.J.; Yario, T.A.; Steitz, J.A. Association of Argonaute proteins and microRNAs can occur after cell lysis. *RNA* **2012**, *18*, 1581–1585. [CrossRef] [PubMed]
177. Tripathi, V.; Ellis, J.D.; Shen, Z.; Song, D.Y.; Pan, Q.; Watt, A.T.; Freier, S.M.; Bennett, C.F.; Sharma, A.; Bubulya, P.A.; et al. The nuclear-retained noncoding RNA MALAT1 regulates alternative splicing by modulating SR splicing factor phosphorylation. *Mol. Cell* **2010**, *39*, 925–938. [CrossRef] [PubMed]
178. Lin, R.; Maeda, S.; Liu, C.; Karin, M.; Edgington, T.S. A large noncoding RNA is a marker for murine hepatocellular carcinomas and a spectrum of human carcinomas. *Oncogene* **2007**, *26*, 851–858. [CrossRef] [PubMed]
179. Stamm, S. Regulation of Alternative Splicing by Reversible Protein Phosphorylation. *J. Biol. Chem.* **2008**, *283*, 1223–1227. [CrossRef]
180. Calarco, J.A.; Superina, S.; O'Hanlon, D.; Gabut, M.; Raj, B.; Pan, Q.; Skalska, U.; Clarke, L.; Gelinas, D.; van der Kooy, D.; et al. Regulation of vertebrate nervous system alternative splicing and development by an SR-related protein. *Cell* **2009**, *138*, 898–910. [CrossRef]
181. Bourgeois, C.F.; Lejeune, F.; Stévenin, J. Broad Specificity of SR (Serine/Arginine) Proteins in the Regulation of Alternative Splicing of Pre-Messenger RNA. In *Progress in Nucleic Acid Research and Molecular Biology*; Academic Press: Cambridge, MA, USA, 2004; Volume 78, pp. 37–88.
182. Wang, D.; Ding, L.; Wang, L.; Zhao, Y.; Sun, Z.; Karnes, R.J.; Zhang, J.; Huang, H. LncRNA MALAT1 enhances oncogenic activities of EZH2 in castration-resistant prostate cancer. *Oncotarget* **2015**, *6*, 41045–41055. [CrossRef]
183. Bachmann, I.M.; Halvorsen, O.J.; Collett, K.; Stefansson, I.M.; Straume, O.; Haukaas, S.A.; Salvesen, H.B.; Otte, A.P.; Akslen, L.A. EZH2 expression is associated with high proliferation rate and aggressive tumor subgroups in cutaneous melanoma and cancers of the endometrium, prostate, and breast. *J. Clin. Oncol.* **2006**, *24*, 268–273. [CrossRef]
184. Bryant, R.J.; Cross, N.A.; Eaton, C.L.; Hamdy, F.C.; Cunliffe, V.T. EZH2 promotes proliferation and invasiveness of prostate cancer cells. *Prostate* **2007**, *67*, 547–556. [CrossRef] [PubMed]
185. Saramäki, O.R.; Tammela, T.L.J.; Martikainen, P.M.; Vessella, R.L.; Visakorpi, T. The gene for polycomb group protein enhancer of zeste homolog 2 (EZH2) is amplified in late-stage prostate cancer. *Genes Chromosomes Cancer* **2006**, *45*, 639–645. [CrossRef] [PubMed]
186. Prensner, J.R.; Iyer, M.K.; Balbin, O.A.; Dhanasekaran, S.M.; Cao, Q.; Brenner, J.C.; Laxman, B.; Asangani, I.A.; Grasso, C.S.; Kominsky, H.D.; et al. Transcriptome sequencing across a prostate cancer cohort identifies PCAT-1, an unannotated lincRNA implicated in disease progression. *Nat. Biotechnol.* **2011**, *29*, 742–749. [CrossRef] [PubMed]
187. Ule, J.; Jensen, K.B.; Ruggiu, M.; Mele, A.; Ule, A.; Darnell, R.B. CLIP identifies Nova-regulated RNA networks in the brain. *Science* **2003**, *302*, 1212–1215. [CrossRef]
188. Haberman, N.; Huppertz, I.; Attig, J.; König, J.; Wang, Z.; Hauer, C.; Hentze, M.W.; Kulozik, A.E.; Le Hir, H.; Curk, T.; et al. Insights into the design and interpretation of iCLIP experiments. *Genome Biol.* **2017**, *18*, 7. [CrossRef]
189. Conlon, E.G.; Manley, J.L. RNA-binding proteins in neurodegeneration: Mechanisms in aggregate. *Genes Dev.* **2017**, *31*, 1509–1528. [CrossRef]
190. Sugimoto, Y.; Vigilante, A.; Darbo, E.; Zirra, A.; Militti, C.; D'Ambrogio, A.; Luscombe, N.M.; Ule, J. hiCLIP reveals the in vivo atlas of mRNA secondary structures recognized by Staufen 1. *Nature* **2015**, *519*, 491–494. [CrossRef]
191. Urlaub, H.; Hartmuth, K.; Lührmann, R. A two-tracked approach to analyze RNA-protein crosslinking sites in native, nonlabeled small nuclear ribonucleoprotein particles. *Methods* **2002**, *26*, 170–181. [CrossRef]
192. König, J.; Zarnack, K.; Rot, G.; Curk, T.; Kayikci, M.; Zupan, B.; Turner, D.J.; Luscombe, N.M.; Ule, J. iCLIP reveals the function of hnRNP particles in splicing at individual nucleotide resolution. *Nat. Struct. Mol. Biol.* **2010**, *17*, 909–915. [CrossRef]
193. Zheng, Z.-Q.; Li, Z.-X.; Zhou, G.-Q.; Lin, L.; Zhang, L.-L.; Lv, J.-W.; Huang, X.-D.; Liu, R.-Q.; Chen, F.; He, X.-J.; et al. Long Noncoding RNA FAM225A Promotes Nasopharyngeal Carcinoma Tumorigenesis and Metastasis by Acting as ceRNA to Sponge miR-590-3p/miR-1275 and Upregulate ITGB3. *Cancer Res.* **2019**, *79*, 4612–4626. [CrossRef]
194. Petri, R.; Jakobsson, J. Identifying miRNA Targets Using AGO-RIPseq. In *mRNA Decay: Methods and Protocols*; Lamandé, S.R., Ed.; Springer: New York, NY, USA, 2018; pp. 131–140. ISBN 9781493975402.
195. Chi, S.W.; Zang, J.B.; Mele, A.; Darnell, R.B. Argonaute HITS-CLIP decodes microRNA-mRNA interaction maps. *Nature* **2009**, *460*, 479–486. [CrossRef]
196. Pollum, M.; Jockusch, S.; Crespo-Hernández, C.E. Increase in the photoreactivity of uracil derivatives by doubling thionation. *Phys. Chem. Chem. Phys.* **2015**, *17*, 27851–27861. [CrossRef] [PubMed]
197. Hamilton, M.P.; Rajapakshe, K.I.; Bader, D.A.; Cerne, J.Z.; Smith, E.A.; Coarfa, C.; Hartig, S.M.; McGuire, S.E. The Landscape of microRNA Targeting in Prostate Cancer Defined by AGO-PAR-CLIP. *Neoplasia* **2016**, *18*, 356–370. [CrossRef]
198. Orom, U.A.; Lund, A.H. Isolation of microRNA targets using biotinylated synthetic microRNAs. *Methods* **2007**, *43*, 162–165. [CrossRef] [PubMed]
199. Hsu, R.-J.; Yang, H.-J.; Tsai, H.-J. Labeled microRNA pull-down assay system: An experimental approach for high-throughput identification of microRNA-target mRNAs. *Nucleic Acids Res.* **2009**, *37*, e77. [CrossRef] [PubMed]

200. Baigude, H.; Li, Z.; Zhou, Y.; Rana, T.M. miR-TRAP: A benchtop chemical biology strategy to identify microRNA targets. *Angew. Chem. Int. Ed.* **2012**, *51*, 5880–5883. [CrossRef]

201. Su, Z.; Ganbold, T.; Baigude, H. Analysis and Identification of Tumorigenic Targets of MicroRNA in Cancer Cells by Photoreactive Chemical Probes. *Int. J. Mol. Sci.* **2020**, *21*, 1545. [CrossRef] [PubMed]

202. Li, J.; Huang, L.; Xiao, X.; Chen, Y.; Wang, X.; Zhou, Z.; Zhang, C.; Zhang, Y. Photoclickable MicroRNA for the Intracellular Target Identification of MicroRNAs. *J. Am. Chem. Soc.* **2016**, *138*, 15943–15949. [CrossRef]

203. Zhang, P.; Fu, H.; Du, S.; Wang, F.; Yang, J.; Cai, W.; Liu, D. Click RNA for Rapid Capture and Identification of Intracellular MicroRNA Targets. *Anal. Chem.* **2019**, *91*, 15740–15747. [CrossRef]

204. Chen, L.; Sun, Y.; Li, J.; Zhang, Y. A photoactivatable microRNA probe for identification of microRNA targets and light-controlled suppression of microRNA target expression. *Chem. Commun.* **2020**, *56*, 627–630. [CrossRef]

205. Xu, Y.; Chen, Y.; Li, D.; Liu, Q.; Xuan, Z.; Li, W.-H. TargetLink, a new method for identifying the endogenous target set of a specific microRNA in intact living cells. *RNA Biol.* **2017**, *14*, 259–274. [CrossRef]

206. Vencken, S.; Hassan, T.; McElvaney, N.G.; Smith, S.G.J.; Greene, C.M. miR-CATCH: MicroRNA Capture Affinity Technology. In *RNA Interference: Challenges and Therapeutic Opportunities*; Sioud, M., Ed.; Springer: New York, NY, USA, 2015; pp. 365–373. ISBN 9781493915385.

207. Hassan, T.; Smith, S.G.J.; Gaughan, K.; Oglesby, I.K.; O'Neill, S.; McElvaney, N.G.; Greene, C.M. Isolation and identification of cell-specific microRNAs targeting a messenger RNA using a biotinylated anti-sense oligonucleotide capture affinity technique. *Nucleic Acids Res.* **2013**, *41*, e71. [CrossRef]

208. Imig, J.; Brunschweiger, A.; Brümmer, A.; Guennewig, B.; Mittal, N.; Kishore, S.; Tsikrika, P.; Gerber, A.P.; Zavolan, M.; Hall, J. miR-CLIP capture of a miRNA targetome uncovers a lincRNA H19-miR-106a interaction. *Nat. Chem. Biol.* **2015**, *11*, 107–114. [CrossRef]

209. De Santi, C.; Vencken, S.; Blake, J.; Haase, B.; Benes, V.; Gemignani, F.; Landi, S.; Greene, C.M. Identification of MiR-21-5p as a Functional Regulator of Mesothelin Expression Using MicroRNA Capture Affinity Coupled with Next Generation Sequencing. *PLoS ONE* **2017**, *12*, e0170999. [CrossRef] [PubMed]

210. Nakamoto, K.; Ueno, Y. Diazirine-containing RNA photo-cross-linking probes for capturing microRNA targets. *J. Org. Chem.* **2014**, *79*, 2463–2472. [CrossRef] [PubMed]

211. Nakamoto, K.; Minami, K.; Akao, Y.; Ueno, Y. Labeling of target mRNAs using a photo-reactive microRNA probe. *Chem. Commun.* **2016**, *52*, 6720–6722. [CrossRef]

212. Nakamoto, K.; Akao, Y.; Ueno, Y. Diazirine-containing tag-free RNA probes for efficient RISC-loading and photoaffinity labeling of microRNA targets. *Bioorg. Med. Chem. Lett.* **2018**, *28*, 2906–2909. [CrossRef] [PubMed]

213. Ashwood, B.; Pollum, M.; Crespo-Hernández, C.E. Photochemical and Photodynamical Properties of Sulfur-Substituted Nucleic Acid Bases. *Photochem. Photobiol.* **2019**, *95*, 33–58. [CrossRef] [PubMed]

214. Dziuba, D.; Hoffmann, J.-E.; Hentze, M.W.; Schultz, C. A Genetically Encoded Diazirine Analogue for RNA-Protein Photo-crosslinking. *Chembiochem* **2020**, *21*, 88–93. [CrossRef]

215. Kretz, M.; Siprashvili, Z.; Chu, C.; Webster, D.E.; Zehnder, A.; Qu, K.; Lee, C.S.; Flockhart, R.J.; Groff, A.F.; Chow, J.; et al. Control of somatic tissue differentiation by the long non-coding RNA TINCR. *Nature* **2013**, *493*, 231–235. [CrossRef]

216. Abdelmohsen, K.; Panda, A.C.; Kang, M.-J.; Guo, R.; Kim, J.; Grammatikakis, I.; Yoon, J.-H.; Dudekula, D.B.; Noh, J.H.; Yang, X.; et al. 7SL RNA represses p53 translation by competing with HuR. *Nucleic Acids Res.* **2014**, *42*, 10099–10111. [CrossRef] [PubMed]

217. Gong, C.; Maquat, L.E. lncRNAs transactivate STAU1-mediated mRNA decay by duplexing with 3′ UTRs via Alu elements. *Nature* **2011**, *470*, 284–288. [CrossRef] [PubMed]

218. Faghihi, M.A.; Modarresi, F.; Khalil, A.M.; Wood, D.E.; Sahagan, B.G.; Morgan, T.E.; Finch, C.E.; St Laurent, G., 3rd; Kenny, P.J.; Wahlestedt, C. Expression of a noncoding RNA is elevated in Alzheimer's disease and drives rapid feed-forward regulation of beta-secretase. *Nat. Med.* **2008**, *14*, 723–730. [CrossRef]

219. Helwak, A.; Tollervey, D. Mapping the miRNA interactome by cross-linking ligation and sequencing of hybrids (CLASH). *Nat. Protoc.* **2014**, *9*, 711–728. [CrossRef]

220. Nguyen, T.C.; Cao, X.; Yu, P.; Xiao, S.; Lu, J.; Biase, F.H.; Sridhar, B.; Huang, N.; Zhang, K.; Zhong, S. Mapping RNA-RNA interactome and RNA structure in vivo by MARIO. *Nat. Commun.* **2016**, *7*, 12023. [CrossRef]

221. Zhao, Y.; Liu, Y.; Lin, L.; Huang, Q.; He, W.; Zhang, S.; Dong, S.; Wen, Z.; Rao, J.; Liao, W.; et al. The lncRNA MACC1-AS1 promotes gastric cancer cell metabolic plasticity via AMPK/Lin28 mediated mRNA stability of MACC1. *Mol. Cancer* **2018**, *17*, 69. [CrossRef] [PubMed]

222. Damas, N.D.; Marcatti, M.; Côme, C.; Christensen, L.L.; Nielsen, M.M.; Baumgartner, R.; Gylling, H.M.; Maglieri, G.; Rundsten, C.F.; Seemann, S.E.; et al. SNHG5 promotes colorectal cancer cell survival by counteracting STAU1-mediated mRNA destabilization. *Nat. Commun.* **2016**, *7*, 13875. [CrossRef] [PubMed]

223. Lu, Z.; Gong, J.; Zhang, Q.C. PARIS: Psoralen Analysis of RNA Interactions and Structures with High Throughput and Resolution. *Methods Mol. Biol.* **2018**, *1649*, 59–84. [CrossRef]

224. Lu, Z.; Zhang, Q.C.; Lee, B.; Flynn, R.A.; Smith, M.A.; Robinson, J.T.; Davidovich, C.; Gooding, A.R.; Goodrich, K.J.; Mattick, J.S.; et al. RNA Duplex Map in Living Cells Reveals Higher-Order Transcriptome Structure. *Cell* **2016**, *165*, 1267–1279. [CrossRef] [PubMed]

225. Sharma, E.; Sterne-Weiler, T.; O'Hanlon, D.; Blencowe, B.J. Global Mapping of Human RNA-RNA Interactions. *Mol. Cell* **2016**, *62*, 618–626. [CrossRef] [PubMed]

226. Calvet, J.P.; Pederson, T. Photochemical cross-linking of secondary structure in HeLa cell heterogeneous nuclear RNA in situ 1. *Nucleic Acids Res.* **1979**, *6*, 1993–2001. [CrossRef]

227. Aw, J.G.A.; Shen, Y.; Wilm, A.; Sun, M.; Lim, X.N.; Boon, K.-L.; Tapsin, S.; Chan, Y.-S.; Tan, C.-P.; Sim, A.Y.L.; et al. In Vivo Mapping of Eukaryotic RNA Interactomes Reveals Principles of Higher-Order Organization and Regulation. *Mol. Cell* **2016**, *62*, 603–617. [CrossRef]

228. Cai, Z.; Cao, C.; Ji, L.; Ye, R.; Wang, D.; Xia, C.; Wang, S.; Du, Z.; Hu, N.; Yu, X.; et al. RIC-seq for global in situ profiling of RNA-RNA spatial interactions. *Nature* **2020**, *582*, 432–437. [CrossRef] [PubMed]

229. Huang, Z.; Szostak, J.W. A simple method for 3'-labeling of RNA. *Nucleic Acids Res.* **1996**, *24*, 4360–4361. [CrossRef] [PubMed]

230. Cimino, G.D.; Gamper, H.B.; Isaacs, S.T.; Hearst, J.E. Psoralens as photoactive probes of nucleic acid structure and function: Organic chemistry, photochemistry, and biochemistry. *Annu. Rev. Biochem.* **1985**, *54*, 1151–1193. [CrossRef] [PubMed]

231. Li, K.; Zhong, S.; Luo, Y.; Zou, D.; Li, M.; Li, Y.; Lu, Y.; Miao, S.; Wang, L.; Song, W. A long noncoding RNA binding to QKI-5 regulates germ cell apoptosis via p38 MAPK signaling pathway. *Cell Death Dis.* **2019**, *10*, 699. [CrossRef]

232. Keene, J.D.; Komisarow, J.M.; Friedersdorf, M.B. RIP-Chip: The isolation and identification of mRNAs, microRNAs and protein components of ribonucleoprotein complexes from cell extracts. *Nat. Protoc.* **2006**, *1*, 302–307. [CrossRef]

233. Siprashvili, Z.; Webster, D.E.; Kretz, M.; Johnston, D.; Rinn, J.L.; Chang, H.Y.; Khavari, P.A. Identification of proteins binding coding and non-coding human RNAs using protein microarrays. *BMC Genom.* **2012**, *13*, 633. [CrossRef]

234. Li, R.; Wang, Y.; Xu, Y.; He, X.; Li, Y. Silencing the long noncoding RNA, TINCR, a molecular sponge of miR-335, inhibits the malignant phenotype of epithelial ovarian cancer via FGF2 suppression. *Int. J. Oncol.* **2019**, *55*, 1110–1124. [CrossRef] [PubMed]

235. Tu, J.; Tian, G.; Cheung, H.-H.; Wei, W.; Lee, T.-L. Gas5 is an essential lncRNA regulator for self-renewal and pluripotency of mouse embryonic stem cells and induced pluripotent stem cells. *Stem Cell Res. Ther.* **2018**, *9*, 71. [CrossRef]

236. Xing, Z.; Lin, A.; Li, C.; Liang, K.; Wang, S.; Liu, Y.; Park, P.K.; Qin, L.; Wei, Y.; Hawke, D.H.; et al. lncRNA directs cooperative epigenetic regulation downstream of chemokine signals. *Cell* **2014**, *159*, 1110–1125. [CrossRef] [PubMed]

237. Sang, L.-J.; Ju, H.-Q.; Liu, G.-P.; Tian, T.; Ma, G.-L.; Lu, Y.-X.; Liu, Z.-X.; Pan, R.-L.; Li, R.-H.; Piao, H.-L.; et al. LncRNA CamK-A Regulates Ca2+-Signaling-Mediated Tumor Microenvironment Remodeling. *Mol. Cell* **2018**, *72*, 71–83. [CrossRef]

238. Mann, M. Functional and quantitative proteomics using SILAC. *Nat. Rev. Mol. Cell Biol.* **2006**, *7*, 952–958. [CrossRef] [PubMed]

239. McHugh, C.A.; Chen, C.-K.; Chow, A.; Surka, C.F.; Tran, C.; McDonel, P.; Pandya-Jones, A.; Blanco, M.; Burghard, C.; Moradian, A.; et al. The Xist lncRNA interacts directly with SHARP to silence transcription through HDAC3. *Nature* **2015**, *521*, 232–236. [CrossRef] [PubMed]

240. Chen, D.-L.; Chen, L.-Z.; Lu, Y.-X.; Zhang, D.-S.; Zeng, Z.-L.; Pan, Z.-Z.; Huang, P.; Wang, F.-H.; Li, Y.-H.; Ju, H.-Q.; et al. Long noncoding RNA XIST expedites metastasis and modulates epithelial-mesenchymal transition in colorectal cancer. *Cell Death Dis.* **2017**, *8*, e3011. [CrossRef] [PubMed]

241. Zhang, X.-T.; Pan, S.-X.; Wang, A.-H.; Kong, Q.-Y.; Jiang, K.-T.; Yu, Z.-B. Long Non-Coding RNA (lncRNA) X-Inactive Specific Transcript (XIST) Plays a Critical Role in Predicting Clinical Prognosis and Progression of Colorectal Cancer. *Med. Sci. Monit.* **2019**, *25*, 6429–6435. [CrossRef]

242. Liang, X.-H.; Fournier, M.J. The helicase Has1p is required for snoRNA release from pre-rRNA. *Mol. Cell. Biol.* **2006**, *26*, 7437–7450. [CrossRef]

243. Fayet-Lebaron, E.; Atzorn, V.; Henry, Y.; Kiss, T. 18S rRNA processing requires base pairings of snR30 H/ACA snoRNA to eukaryote-specific 18S sequences. *EMBO J.* **2009**, *28*, 1260–1270. [CrossRef]

244. Bak, G.; Han, K.; Kim, K.-S.; Lee, Y. Electrophoretic mobility shift assay of RNA-RNA complexes. *Methods Mol. Biol.* **2015**, *1240*, 153–163. [CrossRef] [PubMed]

245. Li, K.; Blum, Y.; Verma, A.; Liu, Z.; Pramanik, K.; Leigh, N.R.; Chun, C.Z.; Samant, G.V.; Zhao, B.; Garnaas, M.K.; et al. A noncoding antisense RNA in tie-1 locus regulates tie-1 function in vivo. *Blood* **2010**, *115*, 133–139. [CrossRef]

246. Gilman, M. Ribonuclease protection assay. *Curr. Protoc. Mol. Biol.* **2001**, *4*. [CrossRef] [PubMed]

247. Melton, D.A.; Krieg, P.A.; Rebagliati, M.R.; Maniatis, T.; Zinn, K.; Green, M.R. Efficient in vitro synthesis of biologically active RNA and RNA hybridization probes from plasmids containing a bacteriophage SP6 promoter. *Nucleic Acids Res.* **1984**, *12*, 7035–7056. [CrossRef] [PubMed]

248. Jadaliha, M.; Gholamalamdari, O.; Tang, W.; Zhang, Y.; Petracovici, A.; Hao, Q.; Tariq, A.; Kim, T.G.; Holton, S.E.; Singh, D.K.; et al. A natural antisense lncRNA controls breast cancer progression by promoting tumor suppressor gene mRNA stability. *PLoS Genet.* **2018**, *14*, e1007802. [CrossRef] [PubMed]

249. Ideue, T.; Hino, K.; Kitao, S.; Yokoi, T.; Hirose, T. Efficient oligonucleotide-mediated degradation of nuclear noncoding RNAs in mammalian cultured cells. *RNA* **2009**, *15*, 1578–1587. [CrossRef]

250. Arun, G.; Diermeier, S.; Akerman, M.; Chang, K.-C.; Wilkinson, J.E.; Hearn, S.; Kim, Y.; MacLeod, A.R.; Krainer, A.R.; Norton, L.; et al. Differentiation of mammary tumors and reduction in metastasis upon Malat1 lncRNA loss. *Genes Dev.* **2016**, *30*, 34–51. [CrossRef]

251. Bennett, C.F.; Baker, B.F.; Pham, N.; Swayze, E.; Geary, R.S. Pharmacology of Antisense Drugs. *Annu. Rev. Pharmacol. Toxicol.* **2017**, *57*, 81–105. [CrossRef]

252. Dethoff, E.A.; Chugh, J.; Mustoe, A.M.; Al-Hashimi, H.M. Functional complexity and regulation through RNA dynamics. *Nature* **2012**, *482*, 322–330. [CrossRef]
253. Hussain, S.; Sajini, A.A.; Blanco, S.; Dietmann, S.; Lombard, P.; Sugimoto, Y.; Paramor, M.; Gleeson, J.G.; Odom, D.T.; Ule, J.; et al. NSun2-mediated cytosine-5 methylation of vault noncoding RNA determines its processing into regulatory small RNAs. *Cell Rep.* **2013**, *4*, 255–261. [CrossRef]
254. Carter, J.-M.; Emmett, W.; Mozos, I.R.; Kotter, A.; Helm, M.; Ule, J.; Hussain, S. FICC-Seq: A method for enzyme-specified profiling of methyl-5-uridine in cellular RNA. *Nucleic Acids Res.* **2019**, *47*, e113. [CrossRef]
255. Linder, B.; Grozhik, A.V.; Olarerin-George, A.O.; Meydan, C.; Mason, C.E.; Jaffrey, S.R. Single-nucleotide-resolution mapping of m6A and m6Am throughout the transcriptome. *Nat. Methods* **2015**, *12*, 767–772. [CrossRef]
256. Aw, J.G.A.; Lim, S.W.; Wang, J.X.; Lambert, F.R.P.; Tan, W.T.; Shen, Y.; Zhang, Y.; Kaewsapsak, P.; Li, C.; Ng, S.B.; et al. Determination of isoform-specific RNA structure with nanopore long reads. *Nat. Biotechnol.* **2020**. [CrossRef]
257. Liu, H.; Begik, O.; Lucas, M.C.; Ramirez, J.M.; Mason, C.E.; Wiener, D.; Schwartz, S.; Mattick, J.S.; Smith, M.A.; Novoa, E.M. Accurate detection of m6A RNA modifications in native RNA sequences. *Nat. Commun.* **2019**, *10*, 4079. [CrossRef] [PubMed]

Review

Long Non-Coding RNA-Ribonucleoprotein Networks in the Post-Transcriptional Control of Gene Expression

Paola Briata * and Roberto Gherzi *

Gene Expression Regulation Laboratory, Ospedale Policlinico San Martino IRCCS, 16132 Genova, Italy
* Correspondence: paola.briata@hsanmartino.it (P.B.); rgherzi@ucsd.edu (R.G.)

Received: 31 August 2020; Accepted: 16 September 2020; Published: 17 September 2020

Abstract: Although mammals possess roughly the same number of protein-coding genes as worms, it is evident that the non-coding transcriptome content has become far broader and more sophisticated during evolution. Indeed, the vital regulatory importance of both short and long non-coding RNAs (lncRNAs) has been demonstrated during the last two decades. RNA binding proteins (RBPs) represent approximately 7.5% of all proteins and regulate the fate and function of a huge number of transcripts thus contributing to ensure cellular homeostasis. Transcriptomic and proteomic studies revealed that RBP-based complexes often include lncRNAs. This review will describe examples of how lncRNA-RBP networks can virtually control all the post-transcriptional events in the cell.

Keywords: long non-coding RNA 1; RNA binding protein 2; post-transcriptional regulation

---

1. Introduction

Mammalian genomes are pervasively transcribed even though, in humans, only 19,000 proteins are coded for by less than 2% of the genome and, in the last two decades, it has become clear that the vast majority of the genome is transcribed as non-coding RNAs (ncRNAs) [1]. Long non-coding RNAs (lncRNAs), a largely underexplored class of ncRNAs arbitrarily classified as >200 nucleotides long, account for most of this pervasive transcription and more and more lncRNAs have been demonstrated to be functional molecules rather than transcriptional noise [1,2]. They are expressed in many different cell types and tissues at different levels, display strong cell- and tissue- specific expression, and are often characterized by poor conservation among species, at least at the primary sequence level [1,2]. Besides lncRNAs that display genomic features in common with protein coding-genes, others can be assigned to the following categories: (*i*) lncRNAs that are intergenic to protein-coding genes (lincRNAs); (*ii*) natural antisense transcripts (AS); and (*iii*) intronic lncRNAs [1,2]. In general, lncRNAs exhibit a surprisingly wide range of sizes, structural arrangements and functions and can be detected in the nucleus and/or the cytoplasm of expressing cells. All these features endow them with diverse and enormous functional potential even though they have also presented experimental challenges for their analysis [1,2].

Like proteins, lncRNAs exert their roles in all cell functions operating through different mechanisms. Their versatile features depend on several reasons but mainly on their subcellular localization and the adoption of specific structural modules with interacting partners, a process that may undergo dynamic changes in response to local cellular environments [3]. lncRNAs have been shown to be involved in diverse fundamental cellular processes such as proliferation and apoptosis, development and differentiation, X chromosome inactivation, and genomic imprinting [3]. They have also been implicated in human diseases such as coronary artery disease, amyotrophic lateral sclerosis, and Alzheimer's disease [4–6] as well as in cancer with either oncogenic or tumor suppression functions [7]. LncRNAs can mediate their effects in *cis* or in *trans* by directly binding to DNA, RNA or proteins and can (*i*) influence the function of transcriptional complexes; (*ii*) modulate chromatin structures; (*iii*) regulate genome organization through interaction with nuclear matrix proteins; (*iv*) function as scaffolds to form ribonucleoprotein (RNP) complexes; (*v*) act as decoys for proteins and micro-RNAs (miRNAs) [2,3]. Thus, lncRNA-mediated control of gene expression may take place at transcriptional and/or post-transcriptional levels [2,3].

In general, lncRNAs interact with RNA-binding proteins (RBPs) that are conventionally viewed as proteins that bind to RNA through one or multiple RNA-binding domains and then change the fate or function of the bound RNAs [8]. A wide range of RBPs has been discovered and investigated over the years and proved to regulate gene expression at many levels but these are generally viewed as key players in post-transcriptional events [9,10]. The combination of the versatility of their RNA-binding domains with their structural flexibility enables RBPs to be involved in virtually all the post-transcriptional regulatory layers in the cell and to control the metabolism of a large array of transcripts [9,10]. RBPs establish highly dynamic interactions with other proteins, as well as with coding and non-coding RNAs, creating functional RNPs that regulate pre-mRNA splicing and polyadenylation, mRNA export, stability, localization and translation [9,10].

Excellent reviews are available on the roles of lncRNAs in transcriptional regulation and genomic organization. This review will focus on different levels of post-transcriptional control exerted by lncRNA/RBP interactions (*i*) polyadenylation and pre-mRNA splicing, (*ii*) mRNA export, (*iii*) mRNA decay, (*iv*) translation, (*v*) protein stability, (*vi*) miRNA maturation from precursors. We will not consider post-transcriptional effects dependent on base pairing between lncRNAs and other RNA species that do not involve RBPs.

2. LncRNAs, RBPs, and Regulation of pre-mRNA Processing

In order to produce a mature mRNA that can be efficiently translated into a protein, pre-mRNAs need extensive processing that can be recapitulated in (*i*) addition of cap structures at their 5′-end (capping), (*ii*) addition of stretches of A nucleotides at their 3′end (polyadenylation), and (*iii*) removal of introns with joining of exons (splicing). In certain circumstances, splicing and polyadenylation reactions can be modulated in order to originate two or more mRNA isoforms from a single pre-mRNA with processes defined as alternative polyadenylation (AP) and alternative splicing (AS) that concern more than 90% of intron-containing genes in humans [11,12]. The initial post-transcriptional modifications of pre-mRNA molecules—5′-end capping, splicing, and 3′-end formation by cleavage/polyadenylation—occur co-transcriptionally in the nucleus [13]. Indeed, seminal experiments performed in the early 2000s revealed that coupling early modifications of pre-mRNA with polymerase II-dependent transcription accelerates, by several orders of magnitude, the process of mRNA maturation [13]. Therefore, one could properly refer to these events as co- and post-transcriptional modification of nascent mRNAs. In recent years, a number of reports indicated that lncRNAs directly regulate AS events by utilizing three distinct modes: (*i*) the interaction with specific splicing factors (SFs) as well as with other SF-associated RBPs; (*ii*) the formation of RNA-RNA duplexes with pre-mRNA molecules [2,3], and (*iii*) the induction of chromatin remodeling that indirectly favors the AS of specific genes [2,3]. We will discuss here only the first mode of regulation.

Studies performed by the Chess laboratory in 2007 revealed that two abundant, predominantly nuclear lncRNAs, *MALAT1* (Metastasis Associated Lung Adenocarcinoma Transcript 1) and *NEAT1* (Nuclear Enriched Abundant Transcript 1), are associated with nuclear domains enriched in pre-mRNA splicing factors that are located in the interchromatin regions of the nucleoplasm of mammalian cells (speckles and paraspeckles) [14].

MALAT1 co-localizes with several transcription factors as well as pre-mRNA processing factors and plays a critical role in coordinating transcriptional and post-transcriptional gene regulation [15]. Numerous RBPs (hnRNPH1, hnRNPK, hnRNPA1, hnRNPL, and PCBP1, just to mention a few) are required to ensure *MALAT1* proper localization to nuclear speckles [15]. Further, *MALAT1* has been described to interact with component of the pre-mRNA splicing complex (RNPS1, SRRM1, and AQR) as well as with a number of RBPs involved in specific pre-mRNA AS events (SRSF1, SRSF2, SRSF3, SON, hnRNPC, hnRNPH1, hnRNPL among others) [16,17]. Overall, *MALAT1* localizes to hundreds of genomic sites belonging to active genes, modulates the recruitment of splicing factors to a large number of actively transcribing loci, and its silencing severely affects pre-mRNA splicing in cultured cells [17–21]. Further, Prasanth and coworkers reported that *MALAT1* is able to modulate the phosphorylation status of the SF SRSF1 further reinforcing the notion that the lncRNA exerts a biological role as a coordinator of pre-mRNA splicing [17] (see also Section 5).

Kingston and coworkers have demonstrated that *MALAT1* colocalizes to many of its chromatin binding sites with another abundant lncRNA, *NEAT1*, even though the two lncRNAs display overall distinct binding patterns thus suggesting that they exert partly overlapping functions [20]. Interestingly, proteomic experiments revealed that both *MALAT1* and *NEAT1* interact with a common set of proteins that include the splicing factor ESRP2 and the scaffold protein SAFB2 that is involved in the regulated phosphorylation of SRSF1 by the kinase SAPK1 [20]. *NEAT1* is an exquisitely nuclear lncRNA and an essential structural component of paraspeckles that include the splicing factors SFPQ and NONO and control different aspects of gene expression [22]. Similar to *MALAT1*, also *NEAT1* recently proved to play an important role in modulating AS events. Shelkovnikova laboratory, taking advantage of a *Neat1* knockout mouse model, demonstrated that the lncRNA controls the AS of a group of genes important for neuronal proliferation and differentiation, cell–cell interactions in the central nervous system (CNS), synaptogenesis, and axon guidance [23]. Interestingly, *Neat1* also controls the AS of a group of RBPs including hnRNPA2B1, hnRNPH1, hnRNPD, hnRNPK, SRSF5, and SRSF7 [23]. *Neat1* knockout mice display a phenotype characterized by deficit in social interaction and rhythmic patterns of CNS activity [23]. Further evidence of the role of *Neat1* in regulating AS derived from a recent study that demonstrated the interaction of the lncRNA with the multifunctional RBP KHSRP. *Neat1*-KHSRP complex controls the process of metastatization of soft-tissue sarcomas by regulating AS events [24].

Another lncRNA localized to a nuclear compartment enriched in pre-mRNA splicing factors, is *Miat* (Myocardial Infarction Associated Transcript, a.k.a. *Gomafu*) that has been reported by Mattick and coworkers to be implicated in the pathogenesis of schizophrenia, a debilitating mental disorder affecting about 1% of the world population [25]. Authors demonstrated that *Miat* can regulate neuronal activity-dependent AS likely by acting as a scaffold for splicing factors (including SF1, SRSF1, and QK1) [25]. *Miat* transient downregulation that occurs upon neuronal depolarization allows the release of the splicing factors thus affecting AS events in neuronal cells [25].

A mass spectrometry-based analysis of molecular partners of *PANDAR* (Promoter Of *CDKN1A* Antisense DNA Damage Activated RNA)—a lncRNA involved in the regulation of proliferation and senescence whose overexpression has been observed in several human cancers and correlates with poor survival rate—allowed the identification of an unanticipated function of this lncRNA in modulating AS. Hennig and coworkers demonstrated that *PANDAR* interaction with PTBP1, a factor implicated in the regulation of AS events, results in modulated AS of *BCL2L1* pre-mRNA that encodes a potent inhibitor of cell death [26]. Authors hypothesize that *PANDAR* exerts a decoy function [26]. PTBP1 also interacts with *Pnky*, a neural-specific, nuclear lncRNA and modulates the expression and the AS of an overlapping set of transcripts [27]. Double knockdown experiments performed in neuronal stem cells indicate that the RBP and the lncRNA function in the same pathway [27].

The interaction of *LINC01133* with the SF SRSF6 proved to contribute to the ability of the lncRNA to modulate the Epithelial to Mesenchymal Transition (EMT) in colorectal cancers [28]. *LINC01133* is an abundant lncRNA whose expression is down-regulated upon colon cancer cell treatment with TGFβ, a potent inducer of EMT [28]. *LINC01133*-mediated inhibition of the SRSF6 function appears to be required for the lncRNA-mediated inhibition of EMT [28]. This observation supports the notion previously reported by our laboratory that TGFβ induces EMT by modulating the activity of RBPs involved in AS regulation [29].

By investigating the functions of *DSCAM-AS1* (Down Syndrome Cell Adhesion Molecule antisense 1)—a lncRNA overexpressed in invasive breast cancers—De Bortoli and coworkers reported that the lncRNA, besides affecting global gene expression and producing changes in the AS of its targets, influences polyadenylation by regulating the alternative 3′ UTR usage of 360 genes [30]. These changes in the early steps of the post-transcriptional regulation of gene expression appear to depend on the interaction between *DSCAM-AS1* and the nucleoplasm-enriched RBP hnRNPL [30].

3. LncRNAs, RBPs, and Regulation of mRNA Nuclear Export

Mature (capped, spliced, polyadenylated) mRNAs rapidly associate with RBPs and, together with various other RNA species (rRNA, tRNA, miRNA precursors, lncRNA), are transported from the nucleus to the cytoplasm through the nuclear pore complex (NPC) in the context of RNPs [31]. Despite the fact that mammalian cells synthesize a multitude of distinct mRNAs and that the composition of each individual RNP is unique and extremely dynamic throughout its life, export of the vast majority of mRNAs utilizes a single export receptor, the heterodimeric export receptor NXF1-NXT1 that mediates translocation through the NPC [31]. The export receptor is displaced at the cytoplasmic side of the NPC to release the RNPs into the cytoplasm. Directionality of the transport is controlled by distinct sets of DEAD-box ATPases that regulate RNPs association to and dissociation from the NXF1-NXT1 complex [31,32]. Importantly, mRNA nuclear export can undergo intense regulation by a variety of stimuli [32] that can also contribute to drug-induced eradication of cancer cells [33].

Recently, Prasanth and coworkers demonstrated that the overexpression of a predominantly nuclear lncRNA (*ROCR*, a.k.a. *LINC02095*) promotes breast cancer proliferation by facilitating the expression of the oncogenic transcription factor SOX9 [34]. *ROCR* favors both transcription and nuclear export of *SOX9* mRNA and its silencing in breast cancer cells reduces the cytoplasmic levels of *SOX9* mRNA [34]. Interestingly, SOX9 displays strong nuclear localization in highly invasive triple-negative breast cancer cells as opposed to other breast cancer subtypes [34]. Although nuclear retention of *SOX9* mRNA in cells depleted of *ROCR* is demonstrated, authors do not provide information on how the lncRNA affects the process of mRNA export and on the identity of the RBP(s) that, interacting with *ROCR*, contributes to its function.

Chromosome translocations may result in the exchange of DNA sequences between genes. Many such gene fusions are strong driver mutations in neoplasia and have provided fundamental insights into the pathogenetic mechanisms of certain tumors [35]. Chimeric mRNAs resulting from genomic rearrangements need to be translocated to the cytoplasm in order to be translated into the resulting oncogenic proteins [35]. Wang and coworkers recently reported on the involvement of the *MALAT1* in the regulation of nuclear export of chimeric mRNAs encoding the oncogenic fusion proteins PML-RARA, MLL-AF9, MLL-ENL, and AML1-ETO [36]. These authors show that nuclear export of the chimeric mRNAs depends on the *MALAT1* expression levels [36]. They propose a complex regulatory mechanism that involves the methylation of mRNAs to form N6-methyladenosine (m6A). m6A modification of mRNA accounts for the most abundant mRNA internal modification and has emerged as a widespread regulatory mechanism that controls gene expression in diverse physiological processes [37]. RBPs able to catalyze the m6A modification (writers), to recognize the m6A modification (readers), and to abrogate this specific modification (erasers) have been identified and characterized in recent years [37]. m6A has been reported to enhance mRNA export from the nucleus through the interaction of the m6A-modified mRNAs with the "reader" RBPs YTHDC1 and SRSF3 that function as adaptors for the NXF1-dependent mRNA export pathway [37]. Wang and coworkers provide evidence that *MALAT1*, upon interaction with oncogenic fusion proteins in nuclear speckles, promotes the interaction between the fusion proteins and the m6A methyltransferase cofactor METTL14 thus controlling the chimeric mRNA-exporting process through the m6A reader YTHDC1 [36]. The results of this study suggest the possibility that other lncRNAs, besides *MALAT1*, could provide a platform for the association of m6A "readers" with m6A-modified specific mRNAs to influence their nuclear export.

4. LncRNAs, RBPs, and Regulation of mRNA Decay

It is well known that the abundance of an mRNA is a function not only of its synthesis, processing, and nuclear export, but also of its degradation rate in the cytoplasm [38]. mRNA decay is an essential step in gene expression as it can rapidly set the levels of transcripts that undergo translation. A multitude of RBPs and/or non-coding RNAs can bind to specific elements of a certain mRNA and dictate its degradation rates via their ability to recruit (or exclude) the mRNA degradation machineries which perform the complex events of deadenylation, decapping and degradation of the RNA body [38]. Several cues can activate signal transduction pathways and modify the general mRNA decay machinery through their interaction with specific RBPs and this affects the mRNA decay rate and abundance [38]. We will describe and discuss here below examples of lncRNAs that contribute to the regulation of mRNA decay through their interaction with RBPs and, in turn, modulate important cellular functions and crucial pathological events.

An important example of lncRNA-RBP network operating in the cytoplasm and modulating the relevant cell function of maintaining genomic stability in human cells is based on the lncRNA *NORAD* [39,40]. *NORAD* (non-coding RNA activated by DNA damage) is highly conserved, broadly and abundantly expressed in mammalian cells and tissues, and induced after DNA damage [39,40]. Importantly, inactivation of *NORAD* triggers dramatic aneuploidy in previously karyotypically stable cell lines. In a search for *NORAD*-interacting proteins, Mendell and co-workers found that this lncRNA functions as a multivalent binding platform for the PUMILIO (PUM) family of RBPs, with the capacity to sequester a significant fraction of the cellular pool of PUM1 and PUM2 and, in turn, to limit their ability to repress target mRNAs [39]. RBPs of the PUM family bind with high specificity to sequences in the 3′ UTRs of target mRNAs and stimulate deadenylation and decapping, resulting in accelerated turnover and decreased translation [41]. Among PUM targets are a large set of factors that are critical for mitosis, DNA repair as well as DNA replication and their excessive repression in the absence of *NORAD* perturbs accurate chromosome segregation and can induce tetraploidization [39–41]. These findings have revealed a lncRNA-dependent mechanism that regulates a highly dosage-sensitive family of RBPs, uncovering a post-transcriptional regulatory axis that maintains genomic stability in mammalian cells and contributes to an emerging concept that a major class of lncRNAs function as molecular decoys. More recently, *NORAD*, whose sequence is characterized by several repetitive units, has been studied in order to identify additional interacting partners [42]. Ulitsky and coworkers found the RBP KHDRBS1 (a.k.a. SAM68) binds to *NORAD* and is required for *NORAD* function in antagonizing PUM [42]. This provides a paradigm for how repeated elements in lncRNAs synergistically contribute to complex tasks and for how a lncRNA can interact with multiple RBPs in order to operate a specific function.

Another lncRNA endowed with several distinct functions is *H19* [43]. In a systematic search to detect regulatory RNA species interacting with the RBP KHSRP in multipotent mesenchymal C2C12 cells, we identified, among others, *H19* [44]. We demonstrated that KHSRP directly interacts with *H19* in the cytoplasm of proliferating undifferentiated C2C12 cells and that this interaction favors the decay-promoting function of KHSRP on labile transcripts, such as *Myog*, through recruitment of the Exosome complex [44]. AKT activation during C2C12 differentiation induces KHSRP dissociation from *H19* and, as a consequence, *Myog* mRNA is stabilized whereas KHSRP is able to shuttle to nuclei where it promotes maturation of myogenic miRNAs from precursors, thus favoring myogenic differentiation (see also Section 6) [44]. In a sense, *H19* can be viewed as a modulator of two important and distinct post-transcriptional regulatory steps that lead to myogenic differentiation.

Recently, we identified a lncRNA expressed in epithelial tissues which we termed *Epr* (Epithelial cell Program Regulator, a.k.a. *BC030874*). *Epr* is rapidly downregulated by TGF-β and its sustained expression largely reshapes the transcriptome, favors the acquisition of epithelial traits, and reduces cell proliferation in cultured mammary gland cells as well as in an animal model of orthotopic transplantation [45]. Mechanistically, *Epr* interacts with chromatin and regulates the transcription of several genes [46] including the cyclin-dependent kinase inhibitor *Cdkn1a*. Interestingly, *Epr* changes *Cdkn1a* gene expression by affecting both its transcription and mRNA decay through its association with the transcription factor SMAD3 and the RBP KHSRP, respectively [45]. KHSRP is predominantly an mRNA decay promoting factor in this cellular context and the interaction with *Epr* blocks its ability to induce decay of *Cdkn1a* mRNA.

The lncRNA *LERFS* (Lowly Expressed in Rheumatoid Fibroblast-like Synoviocytes) is expressed at low levels in fibroblast-like synoviocytes (FLSs) derived from patients suffering for rheumatoid arthritis (RA) and regulates the migration, invasion, and proliferation of FLSs through interaction with the RBP SYNCRIP (a.k.a. hnRNPQ) [47]. Under healthy conditions, the *LERFS*-SYNCRIP complex, by binding to the mRNA of *RHOA*, *RAC1*, and *CDC42*—the small GTPase proteins that control the motility and proliferation of FLSs—, decreases the stability and/or translation of the target mRNAs and downregulates their protein levels [47]. In RA FLSs, decreased *LERFS* levels induce a reduction of the *LERFS*-SYNCRIP complex and this, in turn, reduces the binding of SYNCRIP to the target mRNAs thus increasing their stability or translation [47]. More specifically, *LERFS* and SYNCRIP regulate the stability and the translation of *RAC1* mRNA but regulate only the mRNA translation of *RHOA* and *CDC42* (see also Section 4) [47]. In general, these findings suggest that a decrease in synovial *LERFS* may contribute to the synovial aggression and joint destruction that are features of RA and targeting *LERFS* may have therapeutic potential in patients suffering for RA.

The lncRNA *UCA1* (Urothelial Carcinoma-Associated 1) has been found as a target of the CAPERα/TBX3 transcriptional repressor complex which is required to prevent premature senescence of primary cells, to regulate the activity of core senescence pathways in mouse embryos, and to control cell proliferation by repressing the transcription of *CDKN2A* gene (a.k.a. p16INK) and the RB pathway [48]. *UCA1* is a direct transcriptional target of CAPERα/TBX3 repression and its overexpression is sufficient to induce senescence [48]. In proliferating cells, hnRNPA1 binds and destabilizes *CDKN2A* mRNA whereas during senescence, *UCA1* sequesters hnRNPA1 and this, in turn, stabilizes *CDKN2A* mRNA [48]. Dissociation of the CAPERα/TBX3 co-repressor during oncogenic stress activates *UCA1* which, therefore, can be considered a tumor suppressor. See Section 4 for *UCA1*-dependent translational regulation and its opposite outcome in tumorigenesis.

Akiyama and colleagues demonstrated that *MYU* (MYC-Upregulated, a.k.a. *VPS9D1-AS1*) is a lncRNA transcriptionally induced by MYC upon its activation by the WNT signaling [49]. *MYU* is upregulated in most colon cancers and required for the tumorigenicity of colon cancer cells. Mechanistically, *MYU* associates with the RBP hnRNPK to stabilize *CDK6* mRNA and thereby promotes the G1-S transition of the cell cycle [49]. The authors also propose that hnRNPK and *MYU* hinder the inhibitory effect of miR-16 on *CDK6* mRNA [49]. Importantly, the WNT/MYC/*MYU*-mediated upregulation of CDK6 is essential for cell cycle progression and clonogenicity of colon cancer cells [49].

Another lncRNA playing a role in tumorigenesis is *LINC-ROR* (Regulator of Reprogramming) whose knockout in colon cancer cells suppresses cell proliferation and tumor growth. *LINC-ROR* plays an oncogenic role in part through regulation of *MYC* mRNA expression [50]. The lncRNA interacts with the RBPs PTBP1 (a.k.a. hnRNPI) and hnRNPD (a.k.a. AUF1) and is required for PTBP1 binding to *MYC* mRNA, while the interaction of *LINC-ROR* with hnRNPD inhibits its binding to *MYC* mRNA. As a result, *MYC* mRNA stability is increased and this leads to enhanced cell proliferation and tumorigenesis [50]. See also Section 4 for *LINC-ROR* functions in translation.

Cao and coworkers demonstrated that miR-1 promotes IFNG- (a.k.a IFN-γ) activated innate response in macrophages during *Listeria monocytogenes* infection through increasing the expression of *Stat1* mRNA [51]. From a mechanistic point of view, miR-1 targets the lncRNA *Sros1* (Suppressive non-coding RNA of STAT1) for degradation [51]. In noninfected macrophages *Sros1* blocks the interaction of *Stat1* mRNA with the RBP CAPRIN1 while the *Listeria monocytogenes*-induced degradation of *Sros1* releases CAPRIN1 that is made available to bind and stabilize the *Stat1* mRNA thus leading to increased STAT1 protein levels [51]. This ultimately strengthens IFNG signaling in the macrophages and promotes an innate immune response to intracellular bacterial infection.

5. LncRNAs, RBPs, and Translation Regulation

Translation is a multistep process comprising initiation, elongation, termination and ribosome recycling [52]. During initiation, the ribosome is recruited to the mRNA and scans the 5′ untranslated region of the transcript for the presence of the translation start codon. Under most conditions, initiation is the rate-limiting step of translation and therefore it is tightly regulated. Several key signaling pathways, including mammalian/mechanistic target of rapamycin (mTOR), mitogen activated protein kinases (MAPKs), and integrated stress response (ISR) pathways, converge on the initiation step to control the rate of protein synthesis in response to a variety of stimuli [52]. Control of mRNA translation plays a pivotal role in the regulation of gene expression in embryonic and adult tissues and defects in the translation process are deleterious for development and physiology [52]. During recent years, several lncRNAs have been identified as regulators of distinct steps of their target mRNA translation.

The lncRNA *TRERNA1* (Translational Regulatory, a.k.a. *treRNA*) was identified through genome-wide computational analysis [53]. *TRERNA1* is upregulated in breast cancer primary and lymph node metastasis samples and its expression stimulates tumor invasion in vitro and metastasis in vivo [53]. Authors found that *TRERNA1* downregulates the expression of the epithelial marker CDH1 (a.k.a. E-cadherin) by suppressing the translation of its mRNA and identified a novel RNP complex—consisting of the RBPs hnRNPK, FXR1, and FXR2 as well as the splicing factors PUF60 and SF3B3—that is required for *TRERNA1* function [53]. In more detail, PUF60-SF3B3 dimer interacts with hnRNP K, FXR1, and FXR2 to form a *TRERNA1*-containing RNP complex that, in turn, binds to eIF4G1 affecting translation [53].

Mo and coworkers have found that *LINC-ROR* is transcriptionally induced by TP53 (a.k.a. p53) and, at the same time, is a strong negative regulator of TP53-mediated cell cycle arrest and apoptosis [54]. Unlike MDM2 that causes TP53 degradation through the ubiquitin–proteasome pathway, *LINC-ROR* suppresses TP53 translation through direct interaction with the phosphorylated form of the RBP PTBP1 (a.k.a. hnRNPI) in the cytoplasm [54]. This suggests that the *LINC-ROR*-PTBP1-TP53 axis may constitute an additional surveillance network for the cell to better respond to various stresses (see also Section 3 for the role of *LINC-ROR* in mRNA decay control). The same group demonstrated that PTBP1 can also form a functional RNP with the lncRNA *UCA1* and increase the *UCA1* RNA stability [55]. In addition, in this case the phosphorylated form of PTBP1, predominantly in the cytoplasm, is responsible for the interaction with *UCA1* [55]. The interaction of *UCA1* with PTBP1 suppresses the protein level of CDKN1B (a.k.a. p27KIP1) by competitive inhibition, although the precise mechanism is still unclear. Authors demonstrate that the complex comprising *UCA1* and PTBP1, has an oncogenic role in breast cancer both in vitro and in vivo [55]. See Section 3 for *UCA1*-dependent regulation of mRNA stability and its opposite outcome in tumorigenesis.

LncMyoD (a.k.a. *1700025L06Rik*) is a lncRNA whose primary sequence is not well conserved between human and mouse models while its locus, gene structure, and function are preserved [56]. *LncMyoD* is transcribed next to the *Myod* gene and is directly activated by MYOD during myoblast differentiation. Knockdown of *LncMyoD* strongly inhibits terminal muscle differentiation, mainly due to an unsuccessful exit from the cell cycle [56]. Authors demonstrate that *LncMyoD* directly binds to the RBP IGF2BP2 (a.k.a IMP2) and negatively regulates IGF2BP2-mediated translation of genes able to modulate proliferation such as NRAS and MYC and this contributes to the failure of myoblast terminal differentiation [56].

Bozzoni and co-workers describe another regulatory circuitry controlled by a muscle-specific cytoplasmic lncRNA, *Lnc-Smart* (Skeletal Muscle Regulator of Translation, a.k.a. *Gm14635*), which is essential for proper differentiation of murine myogenic precursors [57]. By direct base pairing with a G-quadruplex region present in the *Mlx-γ* mRNA, *Lnc-Smart* prevents the translation of the mRNA by counteracting the activity of the RBP DHX36 endowed with RNA helicase function [57]. The time-restricted, specific effect of *Lnc-Smart* on the translation of *Mlx-γ* isoform modulates also the general subcellular localization of total MLX proteins (isoforms α and β), impacting on their transcriptional output and promoting proper myogenesis and mature myotube formation [57]. In more detail, *Lnc-Smart* depletion leads to alteration of the differentiation program with defects in myoblast fusion while its overexpression produces an apoptotic phenotype. Authors propose that *lnc-SMaRT* needs to be precisely controlled in time and quantity in order to fine-tune the balance between differentiation and apoptosis to ensure proper myogenesis [57].

The lncRNA *BCYRN1* (Brain Cytoplasmic RNA, a.k.a. *BC200*) regulates RNA metabolism in neural cells by modulating local translation in the postsynaptic dendritic microdomain by interacting with components of the translational machinery, such as eIF4A, eIF4B, and PABPC1 [58]. Lee and coworkers identified the RBPs hnRNPE1 and hnRNPE2 as *BCYRN1*-interacting proteins using a yeast three-hybrid screening. hnRNPE1 and hnRNPE2 bind to *BCYRN1* and can rescue the *BCYRN1*-dependent inhibition of translation by competing with eIF4A for binding to the lncRNA in an in vitro system [58].

6. LncRNAs, RBPs, and Post-Translational Modifications

Post-translational modifications occur in almost every protein during or after its translation and represent an extremely powerful tool operated by the cell in order to regulate the activity, stability, localization, interactions or folding of proteins by inducing their covalent linkage to new functional chemical groups, such as phosphate, acetyl, methyl, carbohydrate and ubiquitin [59]. Different post-translational modifications lead to distinct effects on target proteins and result in disparate biological consequences, from survival to apoptosis, from proliferation to differentiation, from activation to quiescence [59].

FUS (Fused in Sarcoma) is a multifunctional RBP that plays essential roles in post- transcriptional gene expression and possesses the ability to contribute to RNP granule formation via an RNA-dependent self-association [60]. FUS ability to interact with multiple RNA species accounts for its multiple functions. FUS (*i*) binds to nascent pre-mRNAs and acts as a molecular mediator between RNA polymerase II and RNAU1 small nuclear RNA-containing RNP thereby coupling transcription and splicing, (*ii*) binds to its own pre-mRNA and autoregulates its expression, and (*iii*) promotes homologous recombination during DNA double-strand break repair [60]. Numerous mutations in the *FUS* gene have been identified in patients suffering for two severe neurodegenerative disorders, amyotrophic lateral sclerosis and frontotemporal lobar degeneration [60]. Although the molecular mechanisms of FUS-dependent neurotoxicity are poorly understood, high concentrations of the RBP within RNA granules have been proposed to promote the formation of irreversible pathological aggregates [60]. Two recent papers point to lncRNA-dependent post-translational modifications of FUS as critical mechanisms affecting the cellular concentration and activity of the RBP and, in turn, its cellular functions. Nagai and coworkers reported that silencing of the *Drosophila* lncRNA *hsrω* converts FUS from a mono-to di-methylated arginine status via upregulation of the arginine methyltransferase 5 (PRMT5) [61]. PRMT5-dependent modification of FUS promotes its proteasomal degradation, thus leading to a strong downregulation of its cellular levels. Although in this case FUS regulation by the lncRNA is indirect, it is also interesting to note that *hsrω* interacts with and organizes a number of RBPs including TARDBP, hnRNPAB and hnRNPA2B1 and FUS itself [61]. Further, authors show that an increase in FUS causes a downregulation of PRMT5 expression leading to an autoregulatory accumulation of FUS, thus increasing the complexity of this regulatory mechanism [61].

Wu and coworkers investigated the functions of the lncRNA *RMST* (RhabdoMyosarcoma-associated Transcript) that has been characterized as a tumor suppressor in triple-negative breast cancers as well as a regulator of neuronal differentiation and brain development [62]. Authors reported that FUS and *RMST* directly interact and *RMST* enhances FUS SUMOylation [62] but fails to provide a mechanistic explanation for the *RMST*-dependent FUS SUMOylation. *RMST*-induced SUMOylation is required for the interaction between FUS and hnRNPD that is able to affect the stability of ATG4D protein, a factor involved in the biogenesis of autophagosomes, vesicles that contain cellular material intended to be degraded by autophagy [62]. Altogether, these data suggest that *RMST*-dependent SUMOylation of FUS promotes the hnRNPD-mediated stabilization of ATG4D and potentially impacts on the autophagic process [62].

The lncRNA *OCC1* (Overexpressed in Colon Carcinoma-1) plays a tumor suppressive role in colorectal cancer [63]. *OCC1* knockdown promotes cell growth both in vitro and in vivo, which is largely due to its ability to inhibit G0 to G1 and G1 to S phase cell cycle transitions [63]. *OCC1* exerts its function by destabilizing ELAVL1 (a.k.a. HuR) an RBP that, by interacting with the 3' untranslated regions of its target mRNAs, can stabilize thousands of transcripts [64]. *OCC1* enhances the binding of an ubiquitin E3 ligase to ELAVL1 and renders the RBP susceptible to ubiquitination and degradation, thereby reducing the levels of ELAVL1 and, in turn, of its target mRNAs, including the mRNAs associated with cancer cell growth [63]. This report confirms the original observation that ELAVL1 undergoes regulated ubiquitination and proteasome degradation [64] and represents an example of a lncRNA that indirectly regulates the stability of a group of mRNAs through modulation of the post-translational modification of an RBP [63].

As anticipated in Section 1, levels of *MALAT1* affect the ratio between dephosphorylated and phosphorylated SF SRSF1 with a not completely defined mechanism [17].

7. LncRNAs, RBPs, and Maturation of microRNAs from Precursors

A flood of studies published in the last 20 years have demonstrated that microRNAs (miRNAs) regulate the entire spectrum of cellular functions and a number of reports clearly demonstrated that miRNA biogenesis is an important regulatory step that controls the cellular levels of miRNAs and, consequently, their functions [65]. The biogenesis of miRNAs involves two distinct enzymatic reactions carried out by distinct multiprotein complexes located in different cellular compartments [65]. First, primary miRNAs (pri-miRNAs) are processed to precursor miRNAs (pre-miRNAs) through the intervention of the DROSHA-containing complex in the nucleus. Next, through the interaction with XPO5 (a.k.a. exportin-5) and RAN, the pre-miRNA is transported into the cytoplasm where it undergoes a second round of processing catalyzed by the DICER1-containing protein complex. Finally, one strand of the resulting short (21–25 nt) RNA duplex, that corresponds to the mature miRNA, is loaded into the RISC (RNA Induced Silencing Complex) to exert its mRNA targeting functions [65]. Numerous studies have demonstrated that specific RBPs associate with the enzymatic complexes responsible for miRNA maturation to provide specificity and/or to regulate their activity [65].

Groundbreaking investigations conducted in 2015 by Filipowicz laboratory demonstrated that, during the course of postnatal development of retinal photoreceptors, the accumulation of mature miR-183/96/182 is delayed compared with pri-miR-183/96/182 [66]. Authors identified the lncRNA *Rncr4* (named after Retinal Non-Coding RNA 4) that is expressed in maturing photoreceptors as a factor activating pri-miR-183/96/182 maturation [66]. *Rncr4* modulates the activity of the DEAD-box RNA helicase/ATPase DDX3X, an RBP that exerts a potent inhibition on pri-miR-183/96/182 maturation in early phases of postnatal photoreceptor development [66]. Authors observe that the photoreceptor-specific DDX3X silencing results in a significant decrease in pri-miRNAs and a strong increase in mature miR-183/96/182 levels in photoreceptors when compared with controls [66]. MiR-183/96/182 control the expression of CRB1 that is a component of the molecular scaffold involved in the formation and integrity of tight junctions between retinal glia and photoreceptors that controls proper development of polarity in the eye [66]. Altogether the study reveals that *Rncr4*-regulated timing of miR-183/96/182 maturation from precursors is an essential step for obtaining the even distribution of cells across retinal layers.

More recently, Portman and coworkers utilized a different model of organ development—sexual maturation in *Caenorabditis Elegans (C. Elegans)*—to prove the involvement of lncRNA-regulated miRNA maturation from precursors during development [67]. The *C. Elegans* RBP LIN-28, similarly to mammalian LIN28, is a negative regulator of the maturation of let-7 miRNA family members from their pri-miRNAs and Portman and coworkers demonstrated that the lncRNA *lep-5* inhibits LIN-28 function thus promoting the maturation of let-7 that, in turn, controls the onset of sexual maturation in the nervous system of roundworms [67]. Mechanistically, *lep-5* functions as an RNA scaffold, forming a tripartite complex with LEP-2 (whose mammalian homolog is MKRN1 an E3 ubiquitin ligase that promotes the ubiquitination and proteasomal degradation of target proteins) and LIN-28 to promote LIN-28 degradation [67]. The well-known conservation of regulatory mechanisms across species allowed Portman and coworkers to hypothesize that an unidentified *lep-5*-like lncRNA may exist in mammals and play a key role in sexual maturation [67].

The heterodimeric complex formed by the two RBPs NONO and SFQP (a.k.a. PSF) has been defined as a prototypical multipurpose molecular scaffold that dynamically mediates a wide range of protein–protein and protein–nucleic acid interactions [68]. Indeed, the NONO-SFQP complex (*i*) controls pre-mRNA splicing and polyadenylation processes [68], (*ii*) plays a role in nuclear retention of defective RNAs—when associated with the nuclear matrix protein MATR3—, and (*iii*) promotes DNA double-strand break repair via the canonical non-homologous end joining pathway [68]. Fu and coworkers reported an additional function for the NONO-SFQP complex by demonstrating its ability to bind to a large number of pri-miRNAs and to globally enhance pri-miRNA processing into pre-miRNAs by the DROSHA complex [69]. The NONO-SFQP heterodimer is involved in paraspeckle formation and integrity and, therefore, it is not surprising that it interacts with the paraspeckle-enriched lncRNA *Neat1*. The authors also prove that *Neat1* specifically links NONO-SFQP heterodimer with the DROSHA complex thus modulating its enzymatic activity [69].

As we have discussed in Section 3, the lncRNA *H19* is endowed with remarkably distinct regulatory properties. Wu and coworkers recently reported that *H19* suppresses the expression of PTBP1 in cholestatic mouse livers [70]. Authors have observed that PTBP1 and *H19* interact under normal conditions but fail to provide information about the mechanism by which *H19* controls PTBP1 expression [70]. It would be interesting to investigate whether *H19* exerts a scaffold function by bridging together a putative ubiquitin ligase with PTBP1 in order to promote its degradation similarly to what *lep-5* does with LIN-28 in *C. Elegans* (see above, [67]). Authors report a suppressive effect of PTBP1 on the maturation of let-7 family members from their pre-miRNA precursors and suggest that *H19*-dependent PTBP1 downregulation ultimately leads to enhanced levels of let-7 family members in cholestatic mouse livers [70].

Our laboratory has reported that *H19* is indirectly implicated in the processing of a specific subset of miRNAs, the so-called myogenic-miRNAs, whose enhanced expression contributes to myogenesis and muscle regeneration [44]. Indeed, during myogenic differentiation of multipotent mesenchymal C2C12 cells, AKT-dependent phosphorylation of the RBP KHSRP induces its dismissal from the cytoplasm (where it is associated with *H19* to promote decay of labile mRNAs including *Myog*, see Section 3) and its translocation to cell nuclei where KHSRP is repurposed to induce myogenic-pri-miRNAs maturation [44].

8. Take-Home Message

It is evident from the above Sections that the networks based on lncRNA-RBP interactions represent highly versatile tools to post-transcriptionally regulate gene expression. We have discussed examples of specific lncRNAs that, through interactions with distinct sets of RBPs, regulate complex layers of post-transcriptional control (Summarized in Figure 1 and Table 1).

LncRNAs usually display a cell- or tissue-restricted expression while RBPs are more broadly expressed. Thus, a lncRNA can provide a cell- and/or tissue-specific function to an RBP. Further, since the expression levels of lncRNAs can be modulated by extracellular signals and RBP functions can be post-translationally modulated by the same and/or different pathways, the functional outcome of lncRNA-RBP complexes can be tightly controlled in a time- and space-specific manner. This results in a huge regulatory potential.

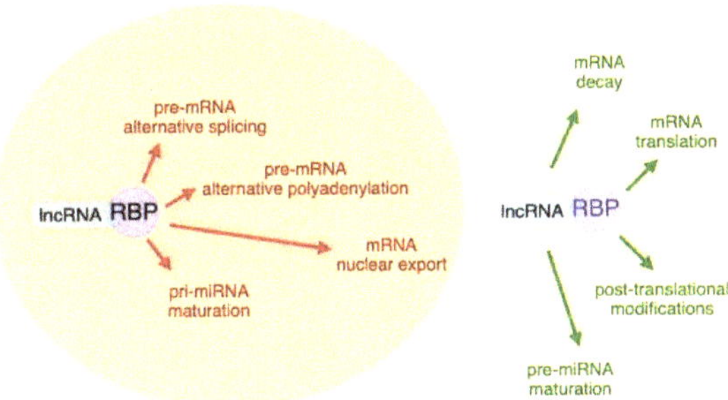

Figure 1. Nuclear and cytoplasmic functions of long non-coding RNA (lncRNA)-RNA binding protein (RBP) networks.

Table 1. Summary of the lncRNA-RBP networks described in this review. The ENSEMBL accession number is provided in parentheses. In the case of *Drosophila* and *C. Elegans* lncRNAs, the accession numbers to FlyBase and WormBase, respectively, are provided in parentheses.

LncRNA	RBP	Function	Mechanism of Action	Ref.	Cell Outcome
MALAT1 (ENSG00000251562)	Multiple splicing regulators	Alternative Splicing	Scaffold	[16–21]	Various
	YTHDC1, SRSF3	Nuclear export	Scaffold	[37]	Oncogenesis
	SRSF1	Alternative Splicing	Scaffold	[17]	Unknown
NEAT1 (hsa ENSG00000245532) (mmu ENSMUSG00000092274)	Multiple splicing regulators	Alternative Splicing	Scaffold	[20,22–24]	Various
	NONO, SFQP	Pri-miRNA processing	Scaffold	[70]	Myoblast differentiation
Miat (ENSMUSG00000097767)	Multiple splicing regulators	Alternative Splicing	Scaffold	[25]	Control of neuronal depolarization
PANDAR (ENSG00000281450)	PTBP1	Alternative Splicing	Decoy	[26]	Apoptosis
Pnky (ENSMUSG00000107859)	PTBP1	Alternative Splicing	Unknown	[27]	Neurogenesis
LINC01133 (ENSG00000224259	SRSF6	Alternative Splicing	Decoy	[28]	Epithelial to mesenchymal transition
DSCAM-AS1 (ENSG00000235123)	hnRNPL	Alternative Splicing, Alternative polyadenylation	Scaffold	[30]	Cancer progression
ROCR (ENSG00000228639)	unknown	Nuclear export	Unknown	[34]	Cancer progression
NORAD (ENSG00000260032)	PUMILIO	mRNA decay	Decoy	[39–42]	Genome stability
H19 (hsa ENSG00000130600) (mmu ENSMUSG00000000031)	KHSRP	mRNA decay	Scaffold	[44]	Myoblast differentiation
	Phospho-KHSRP	Pri-miRNA processing	Release of scaffold function	[44]	Myoblast differentiation
	PTBP1	Pre-miRNA processing	Indirect regulation?	[71]	Liver disease

Table 1. *Cont.*

LncRNA	RBP	Function	Mechanism of Action	Ref.	Cell Outcome
Epr (ENSMUSG00000074300)	KHSRP	mRNA decay	Decoy	[45]	Cell proliferation
LERFS (ENSG00000234665)	SYNCRIP	mRNA decay/mRNA translation	Scaffold	[47]	Synoviocyte proliferation and motility
UCA1 (ENSG00000214049)	hnRNPA1	mRNA decay	Decoy	[48]	Cell proliferation, senescence
	PTBP1	mRNA translation	Decoy	[55]	Cell proliferation
MY (ENSG00000261373)	hnRNPK	mRNA decay	Scaffold	[49]	Cell proliferation
LINC-ROR (ENSG00000258609)	PTBP1, hnRNPD	mRNA decay	Scaffold Decoy	[50]	Cell proliferation
	Phospho-PTBP1	mRNA translation	Decoy	[54]	Cell proliferation, apoptosis
LncMyoD (ENSMUST00000209655)	IGF2BP2	mRNA translation	Decoy	[56]	Myoblast differentiation
LncSMaRT (ENSMUSG00000087591)	DHX36	mRNA translation	Decoy	[57]	Myoblast differentiation
BCYRN1 (ENSG00000236824)	hnRNPE1/E2	mRNA translation	Decoy	[58]	Post-synaptic translation
HSRω (Drosophila) (FlyBase ID FBgn0001234)	FUS	Post-translation modification	Indirect regulation	[61]	Neurotoxicity
RMST (ENSG00000255794)	FUS	Post-translation modification	Scaffold?	[62]	Autophagy
OCC1 (ENSG000002351629	ELAVL1	Post-translation modification	Scaffold	[63]	Cell proliferation
Rncr4 (ENSMUSG00000103108)	DDX3X	Pri-miRNA processing	Decoy	[66]	Photoreceptor development
Lep-5 (C. Elegans) (WormBase H36L18.2)	LIN-28	Pri-miRNA processing	Scaffold	[67]	Sexual development

It is known that many lncRNAs function as molecular decoys and we have reviewed examples of abundant lncRNAs that exert part of their biological functions through this mechanism (e.g., *MALAT1, NEAT1, H19, NORAD*). However, the generally low abundance of many lncRNAs can generate debate on the stoichiometry of their interaction with the usually abundant RBPs. More and more evidence points to the functional relevance of specialized membrane-free subcellular compartments where high abundance of lncRNAs may not be required because their local concentration might be the limiting step. Indeed, ncRNAs have been viewed as potential mediators of liquid–liquid phase separation through their ability to operate as molecular scaffolds for the binding of RBPs, thus regulating the sizes and the dynamics of membrane-free organelles that carry out biological processes [71]. Phase separation is an emerging paradigm for understanding spatial and temporal regulation of a variety of cellular processes and additional studies will be needed to clarify its role in the post-transcriptional regulatory layer of gene expression [72].

In conclusion, the complexity of lncRNA-RBP functional networks is often increased by the experimental evidence that some post-transcriptional modifications of gene expression occur co-transcriptionally and by the ability of some lncRNAs to exert both transcriptional and post-transcriptional functions in a coordinated way. Recently developed technologies aimed at analyzing—in the context of distinct cell compartments—macromolecular complexes including lncRNAs, chromatin, and RBPs in an "almost-native" status, will allow researchers to portray, at a better resolution, the elaborate scenario of the interactions that we have described.

Author Contributions: P.B. and R.G. equally contributed to all the steps of this review preparation. Both authors have read and agreed to the published version of the manuscript.

Funding: This work was supported by the Fondazione AIRC per la Ricerca sul Cancro (IG 21541).

Acknowledgments: L. Brondolo (Gene Expression Regulation Laboratory, Ospedale Policlinico San Martino) contributed to the early phase of the bibliographic search.

Conflicts of Interest: The authors declare no conflict of interest.

References

1. Deveson, I.W.; Hardwick, S.A.; Mercer, T.R.; Mattick, J.S. The Dimensions, Dynamics, and Relevance of the Mammalian Noncoding Transcriptome. *Trends Genet.* **2017**, *33*, 464–478. [CrossRef]
2. Quinn, J.J.; Chang, H.Y. Unique features of long non-coding RNA biogenesis and function. *Nat. Rev. Genet.* **2016**, *17*, 47–62. [CrossRef]
3. Yao, R.W.; Wang, Y.; Chen, L.L. Cellular functions of long noncoding RNAs. *Nat. Cell Biol.* **2019**, *21*, 542–551. [CrossRef]
4. Cardona-Monzonís, A.; García-Giménez, J.; Mena-Mollá, S.; Pareja-Galeano, H.; de la Guía-Galipienso, F.; Lippi, G.; Pallardò, F.V.; Sanchis-Gomar, F. Non-coding RNAs and Coronary Artery Disease. *Adv. Exp. Med. Biol.* **2020**, *1229*, 273–285. [CrossRef]
5. Gagliardi, S.; Pandini, C.; Garofalo, M.; Bordoni, M.; Pansarasa, O.; Cereda, C. Long non coding RNAs and ALS: Still much to do. *Noncoding RNA Res.* **2018**, *3*, 226–231. [CrossRef]
6. Faghihi, M.A.; Modarresi, F.; Khalil, A.M.; Wood, D.E.; Sahagan, B.G.; Morgan, T.E.; Finch, C.E.; St Laurent, G., 3rd; Kenny, P.J.; Wahlestedt, C. Expression of a noncoding RNA is elevated in Alzheimer's disease and drives rapid feed-forward regulation of beta-secretase. *Nat. Med.* **2008**, *14*, 723–730. [CrossRef]
7. Slack, F.J.; Chinnaiyan, A.M. The Role of Non-coding RNAs in Oncology. *Cell* **2019**, *179*, 1033–1055. [CrossRef]
8. Dangelmaier, E.; Lal, A. Adaptor proteins in long noncoding RNA biology. *Biochim. Biophys. Acta Gene Regul. Mech.* **2020**, *1863*, 194370. [CrossRef]
9. Glisovic, T.; Bachorik, J.L.; Yong, J.; Dreyfuss, G. RNA-binding proteins and post-transcriptional gene regulation. *FEBS Lett.* **2008**, *582*, 1977–1986. [CrossRef]
10. Hentze, M.W.; Castello, A.; Schwarzl, T.; Preiss, T. A brave new world of RNA-binding proteins. *Nat. Rev. Mol. Cell Biol.* **2018**, *19*, 327–341. [CrossRef]
11. Gruber, A.J.; Zavolan, M. Alternative cleavage and polyadenylation in health and disease. *Nat. Rev. Genet.* **2019**, *20*, 599–614. [CrossRef]
12. Matera, A.G.; Wang, Z. A day in the life of the spliceosome. *Nat. Rev. Mol. Cell Biol.* **2014**, *15*, 108–121. [CrossRef]
13. Bentley, D.L. Rules of engagement: Co-transcriptional recruitment of pre-mRNA processing factors. *Curr. Opin. Cell Biol.* **2005**, *17*, 251–256. [CrossRef]
14. Hutchinson, J.N.; Ensminger, A.W.; Clemson, C.M.; Lynch, C.R.; Lawrence, J.B.; Chess, A. A screen for nuclear transcripts identifies two linked noncoding RNAs associated with SC35 splicing domains. *BMC Genomics* **2007**, *8*, 39. [CrossRef]
15. Spector, D.L.; Lamond, A.I. Nuclear speckles. *Cold Spring Harb. Perspect. Biol.* **2011**, *3*. [CrossRef]
16. Miyagawa, R.; Tano, K.; Mizuno, R.; Nakamura, Y.; Ijiri, K.; Rakwal, R.; Shibato, J.; Masuo, Y.; Mayeda, A.; Hirose, T.; et al. Identification of cis- and trans-acting factors involved in the localization of MALAT-1 noncoding RNA to nuclear speckles. *RNA* **2012**, *18*, 738–751. [CrossRef]
17. Tripathi, V.; Ellis, J.D.; Shen, Z.; Song, D.Y.; Pan, Q.; Watt, A.T.; Freier, S.M.; Bennett, C.F.; Sharma, A.; Bubulya, P.A.; et al. The nuclear-retained noncoding RNA MALAT1 regulates alternative splicing by modulating SR splicing factor phosphorylation. *Mol. Cell* **2010**, *39*, 925–938. [CrossRef]
18. Mao, Y.S.; Zhang, B.; Spector, D.L. Biogenesis and function of nuclear bodies. *Trends Genet.* **2011**, *27*, 295–306. [CrossRef]
19. Malakar, P.; Shilo, A.; Mogilevsky, A.; Stein, I.; Pikarsky, E.; Nevo, Y.; Benyamini, H.; Elgavish, S.; Zong, X.; Prasanth, K.V.; et al. Long Noncoding RNA MALAT1 Promotes Hepatocellular Carcinoma Development by SRSF1 Upregulation and mTOR Activation. *Cancer Res.* **2017**, *77*, 1155–1167. [CrossRef]

20. West, J.A.; Davis, C.P.; Sunwoo, H.; Simon, M.D.; Sadreyev, R.I.; Wang, P.I.; Tolstorukov, M.Y.; Kingston, R.E. The long noncoding RNAs NEAT1 and MALAT1 bind active chromatin sites. *Mol. Cell* **2014**, *55*, 791–802. [CrossRef]

21. Engreitz, J.M.; Sirokman, K.; McDonel, P.; Shishkin, A.A.; Surka, C.; Russell, P.; Grossman, S.R.; Chow, A.Y.; Guttman, M.; Lander, E.S. RNA-RNA interactions enable specific targeting of noncoding RNAs to nascent Pre-mRNAs and chromatin sites. *Cell* **2014**, *159*, 188–199. [CrossRef]

22. Fox, A.H.; Lamond, A.I. Paraspeckles. *Cold Spring Harb. Perspect. Biol.* **2010**, *2*, a000687. [CrossRef] [PubMed]

23. Kukharsky, M.S.; Ninkina, N.N.; An, H.; Telezhkin, V.; Wei, W.; Meritens, C.R.; Cooper-Knock, J.; Nakagawa, S.; Hirose, T.; Buchman, V.L.; et al. Long non-coding RNA Neat1 regulates adaptive behavioural response to stress in mice. *Transl. Psychiatry* **2020**, *10*, 171. [CrossRef]

24. Huang, J.; Sachdeva, M.; Xu, E.; Robinson, T.J.; Luo, L.; Ma, Y.; Williams, N.T.; Lopez, O.; Cervia, L.D.; Yuan, F.; et al. The Long Noncoding RNA NEAT1 Promotes Sarcoma Metastasis by Regulating RNA Splicing Pathways. *Mol. Cancer Res.* **2020**. [CrossRef]

25. Barry, G.; Briggs, J.A.; Vanichkina, D.P.; Poth, E.M.; Beveridge, N.J.; Ratnu, V.S.; Nayler, S.P.; Nones, K.; Hu, J.; Bredy, T.W.; et al. The long non-coding RNA Gomafu is acutely regulated in response to neuronal activation and involved in schizophrenia-associated alternative splicing. *Mol. Psychiatry* **2014**, *19*, 486–494. [CrossRef]

26. Pospiech, N.; Cibis, H.; Dietrich, L.; Müller, F.; Bange, T.; Hennig, S. Identification of novel PANDAR protein interaction partners involved in splicing regulation. *Sci. Rep.* **2018**, *8*, 2798. [CrossRef]

27. Ramos, A.D.; Andersen, R.E.; Liu, S.J.; Nowakowski, T.J.; Hong, S.J.; Gertz, C.; Salinas, R.D.; Zarabi, H.; Kriegstein, A.R.; Lim, D.A. The long noncoding RNA Pnky regulates neuronal differentiation of embryonic and postnatal neural stem cells. *Cell Stem Cell* **2015**, *16*, 439–447. [CrossRef] [PubMed]

28. Kong, J.; Sun, W.; Li, C.; Wan, L.; Wang, S.; Wu, Y.; Xu, E.; Zhang, H.; Lai, M. Long non-coding RNA LINC01133 inhibits epithelial-mesenchymal transition and metastasis in colorectal cancer by interacting with SRSF6. *Cancer Lett.* **2016**, *380*, 476–484. [CrossRef]

29. Puppo, M.; Bucci, G.; Rossi, M.; Giovarelli, M.; Bordo, D.; Moshiri, A.; Gorlero, F.; Gherzi, R.; Briata, P. miRNA-mediated KHSRP Silencing rewires distinct post-transcriptional programs during TGF-β-Induced Epithelial-to-Mesenchymal Transition. *Cell Rep.* **2016**, *16*, 967–978. [CrossRef]

30. Elhasnaoui, J.; Miano, V.; Ferrero, G.; Doria, E.; Leon, A.E.; Fabricio, A.S.C.; Annaratone, L.; Castellano, I.; Sapino, A.; De Bortoli, M. DSCAM-AS1-Driven Proliferation of Breast Cancer Cells Involves Regulation of Alternative Exon Splicing and 3′-End Usage. *Cancers* **2020**, *12*, 1453. [CrossRef]

31. Xie, Y.; Ren, Y. Mechanisms of nuclear mRNA export: A structural perspective. *Traffic* **2019**, *20*, 829–840. [CrossRef] [PubMed]

32. Heinrich, S.; Derrer, C.P.; Lari, A.; Weis, K.; Montpetit, B. Temporal and spatial regulation of mRNA export: Single particle RNA-imaging provides new tools and insights. *Bioessays* **2017**, *39*. [CrossRef] [PubMed]

33. Harris, M.N.; Ozpolat, B.; Abdi, F.; Gu, S.; Legler, A.; Mawuenyega, K.G.; Tirado-Gomez, M.; Lopez-Berestein, G.; Chen, X. Comparative proteomic analysis of all-trans-retinoic acid treatment reveals systematic posttranscriptional control mechanisms in acute promyelocytic leukemia. *Blood* **2004**, *104*, 1314–1323. [CrossRef] [PubMed]

34. Tariq, A.; Hao, Q.; Sun, Q.; Singh, D.K.; Jadaliha, M.; Zhang, Y.; Chetlangia, N.; Ma, J.; Holton, S.E.; Bhargava, R.; et al. LncRNA-mediated regulation of SOX9 expression in basal subtype breast cancer cells. *RNA* **2020**, *26*, 175–185. [CrossRef]

35. Mertens, F.; Johansson, B.; Fioretos, T.; Mitelman, F. The emerging complexity of gene fusions in cancer. *Nat. Rev. Cancer* **2015**, *15*, 371–381. [CrossRef]

36. Chen, Z.H.; Chen, T.Q.; Zeng, Z.C.; Wang, D.; Han, C.; Sun, Y.M.; Huang, W.; Sun, L.Y.; Fang, K.; Chen, Y.Q.; et al. Nuclear export of chimeric mRNAs depends on an lncRNA-triggered autoregulatory loop in blood malignancies. *Cell Death Dis.* **2020**, *11*, 566. [CrossRef]

37. Yi, Y.C.; Chen, X.Y.; Zhang, J.; Zhu, J.S. Novel insights into the interplay between m6A modification and noncoding RNAs in cancer. *Mol. Cancer* **2020**, *19*, 121. [CrossRef]

38. Wu, X.; Brewer, G. The regulation of mRNA stability in mammalian cells: 2.0. *Gene* **2012**, *500*, 10–21. [CrossRef]

39. Lee, S.; Kopp, F.; Chang, T.C.; Sataluri, A.; Chen, B.; Sivakumar, S.; Yu, H.; Xie, Y.; Mendell, J.T. Noncoding RNA NORAD Regulates Genomic Stability by Sequestering PUMILIO Proteins. *Cell* **2016**, *164*, 69–80. [CrossRef]

40. Tichon, A.; Gil, N.; Lubelsky, Y.; Havkin Solomon, T.; Lemze, D.; Itzkovitz, S.; Stern-Ginossar, N.; Ulitsky, I. A conserved abundant cytoplasmic long noncoding RNA modulates repression by Pumilio proteins in human cells. *Nat. Commun.* **2016**, *7*, 12209. [CrossRef]

41. Goldstrohm, A.C.; Hall, T.M.T.; McKenney, K.M. Post-transcriptional Regulatory Functions of Mammalian Pumilio Proteins. *Trends Genet.* **2018**, *34*, 972–990. [CrossRef]

42. Tichon, A.; Perry, R.B.; Stojic, L.; Ulitsky, I. SAM68 is required for regulation of Pumilio by the NORAD long noncoding RNA. *Genes Dev.* **2018**, *32*, 70–78. [CrossRef] [PubMed]

43. Alipoor, B.; Parvar, S.N.; Sabati, Z.; Ghaedi, H.; Ghasemi, H. An updated review of the H19 lncRNA in human cancer: Molecular mechanism and diagnostic and therapeutic importance. *Mol. Biol. Rep.* **2020**, *47*, 6357–6374. [CrossRef] [PubMed]

44. Giovarelli, M.; Bucci, G.; Ramos, A.; Bordo, D.; Wilusz, C.J.; Chen, C.-Y.; Puppo, M.; Briata, P.; Gherzi, R. H19 long noncoding RNA controls the mRNA decay promoting function of KSRP. *Proc. Natl. Acad. Sci. USA* **2014**, *111*, E5023–E5028. [CrossRef]

45. Rossi, M.; Bucci, G.; Rizzotto, D.; Bordo, D.; Marzi, M.J.; Puppo, M.; Flinois, A.; Spadaro, D.; Citi, S.; Emionite, L.; et al. LncRNA EPR controls epithelial proliferation by coordinating Cdkn1a transcription and mRNA decay response to TGF-β. *Nat. Commun.* **2019**, *10*, 1969. [CrossRef]

46. Zapparoli, E.; Briata, P.; Rossi, M.; Brondolo, L.; Bucci, G.; Gherzi, R. Comprehensive multi-omics analysis uncovers a group of TGF-β-regulated genes among lncRNA EPR direct transcriptional targets. *Nucleic Acids Res.* **2020**, gkaa628. [CrossRef]

47. Zou, Y.; Xu, S.; Xiao, Y.; Qiu, Q.; Shi, M.; Wang, J.; Liang, L.; Zhan, Z.; Yang, X.; Olsen, N.; et al. Long noncoding RNA LERFS negatively regulates rheumatoid synovial aggression and proliferation. *J. Clin. Investig.* **2018**, *128*, 4510–4524. [CrossRef]

48. Kumar, P.P.; Emechebe, U.; Smith, R.; Franklin, S.; Moore, B.; Yandell, M.; Lessnick, S.L.; Moon, A.M. Coordinated control of senescence by lncRNA and a novel T-box3 co-repressor complex. *eLife* **2014**, *3*, e02805. [CrossRef]

49. Kawasaki, Y.; Komiya, M.; Matsumura, K.; Negishi, L.; Suda, S.; Okuno, M.; Yokota, N.; Osada, T.; Nagashima, T.; Hiyoshi, M.; et al. MYU, a Target lncRNA for Wnt/c-Myc Signaling, Mediates Induction of CDK6 to Promote Cell Cycle Progression. *Cell Rep.* **2016**, *16*, 2554–2564. [CrossRef]

50. Huang, J.; Zhang, A.; Ho, T.T.; Zhang, Z.; Zhou, N.; Ding, X.; Zhang, X.; Xu, M.; Mo, Y.Y. Linc-RoR promotes c-Myc expression through hnRNP I and AUF1. *Nucleic Acids Res.* **2016**, *44*, 3059–3069. [CrossRef]

51. Xu, H.; Jiang, Y.; Xu, X.; Su, X.; Liu, Y.; Ma, Y.; Zhao, Y.; Shen, Z.; Huang, B.; Cao, X. Inducible degradation of lncRNA Sros1 promotes IFN-γ-mediated activation of innate immune responses by stabilizing Stat1 mRNA. *Nat. Immunol.* **2019**, *20*, 1621–1630. [CrossRef] [PubMed]

52. Tahmasebi, S.; Khoutorsky, A.; Mathews, M.B.; Sonenberg, N. Translation deregulation in human disease. *Nat. Rev. Mol. Cell Biol.* **2018**, *19*, 791–807. [CrossRef] [PubMed]

53. Gumireddy, K.; Li, A.; Yan, J.; Setoyama, T.; Johannes, G.J.; Orom, U.A.; Tchou, J.; Liu, Q.; Zhang, L.; Speicher, D.W.; et al. Identification of a long non-coding RNA-associated RNP complex regulating metastasis at the translational step. *EMBO J.* **2013**, *32*, 2672–2684. [CrossRef] [PubMed]

54. Zhang, A.; Zhou, N.; Huang, J.; Liu, Q.; Fukuda, K.; Ma, D.; Lu, Z.; Bai, C.; Watabe, K.; Mo, Y.Y. The human long non-coding RNA-RoR is a p53 repressor in response to DNA damage. *Cell Res.* **2013**, *23*, 340–350. [CrossRef]

55. Huang, J.; Zhou, N.; Watabe, K.; Lu, Z.; Wu, F.; Xu, M.; Mo, Y.Y. Long non-coding RNA UCA1 promotes breast tumor growth by suppression of p27 (Kip1). *Cell Death Dis.* **2014**, *5*, e1008. [CrossRef]

56. Gong, C.; Li, Z.; Ramanujan, K.; Clay, I.; Zhang, Y.; Lemire-Brachat, S.; Glass, D.J. A long non-coding RNA, LncMyoD, regulates skeletal muscle differentiation by blocking IMP2-mediated mRNA translation. *Dev. Cell* **2015**, *34*, 181–191. [CrossRef]

57. Martone, J.; Mariani, D.; Santini, T.; Setti, A.; Shamloo, S.; Colantoni, A.; Capparelli, F.; Paiardini, A.; Dimartino, D.; Morlando, M.; et al. SMaRT lncRNA controls translation of a G-quadruplex-containing mRNA antagonizing the DHX36 helicase. *EMBO Rep.* **2020**, *21*, e49942. [CrossRef] [PubMed]

58. Jang, S.; Shin, H.; Lee, J.; Kim, Y.; Bak, G.; Lee, Y. Regulation of BC200 RNA-mediated translation inhibition by hnRNP E1 and E2. *FEBS Lett.* **2017**, *591*, 393–405. [CrossRef]

59. Deribe, Y.L.; Pawson, T.; Dikic, I. Post-translational modifications in signal integration. *Nat. Struct. Mol. Biol.* **2010**, *17*, 666–672. [CrossRef]

60. Loughlin, F.E.; Wilce, J.A. TDP-43 and FUS-structural insights into RNA recognition and self-association. *Curr. Opin. Struct. Biol.* **2019**, *59*, 134–142. [CrossRef]

61. Lo Piccolo, L.; Mochizuki, H.; Nagai, Y. The lncRNA hsrω regulates arginine dimethylation of human FUS to cause its proteasomal degradation in Drosophila. *J. Cell Sci.* **2019**, *132*, jcs236836. [CrossRef] [PubMed]

62. Liu, C.; Peng, Z.; Li, P.; Fu, H.; Feng, J.; Zhang, Y.; Liu, T.; Liu, Y.; Liu, Q.; Liu, Q.; et al. lncRNA RMST Suppressed GBM Cell Mitophagy through Enhancing FUS SUMOylation. *Mol. Ther. Nucleic Acids* **2020**, *19*, 1198–1208. [CrossRef] [PubMed]

63. Lan, Y.; Xiao, X.; He, Z.; Luo, Y.; Wu, C.; Li, L.; Song, X. Long noncoding RNA OCC-1 suppresses cell growth through destabilizing HuR protein in colorectal cancer. *Nucleic Acids Res.* **2018**, *46*, 5809–5821. [CrossRef]

64. Grammatikakis, I.; Abdelmohsen, K.; Gorospe, M. Posttranslational control of HuR function. *Wiley Interdiscip. Rev. RNA* **2017**, *8*, e1372. [CrossRef] [PubMed]

65. Ha, M.; Kim, V.N. Regulation of microRNA biogenesis. *Nat. Rev. Mol. Cell Biol.* **2014**, *15*, 509–524. [CrossRef]

66. Krol, J.; Krol, I.; Alvarez, C.P.; Fiscella, M.; Hierlemann, A.; Roska, B.; Filipowicz, W. A network comprising short and long noncoding RNAs and RNA helicase controls mouse retina architecture. *Nat. Commun.* **2015**, *6*, 7305. [CrossRef]

67. Lawson, H.; Vuong, E.; Miller, R.M.; Kiontke, K.; Fitch, D.H.; Portman, D.S. The Makorin lep-2 and the lncRNA lep-5 regulate lin-28 to schedule sexual maturation of the C. elegans nervous system. *eLife* **2019**, *8*, e43660. [CrossRef]

68. Knott, G.J.; Bond, C.S.; Fox, A.H. The DBHS proteins SFPQ, NONO and PSPC1: A multipurpose molecular scaffold. *Nucleic Acids Res.* **2016**, *44*, 3989–4004. [CrossRef]

69. Jiang, L.; Shao, C.; Wu, Q.J.; Chen, G.; Zhou, J.; Yang, B.; Li, H.; Gou, L.T.; Zhang, Y.; Wang, Y.; et al. NEAT1 scaffolds RNA-binding proteins and the Microprocessor to globally enhance pri-miRNA processing. *Nat. Struct. Mol. Biol.* **2017**, *24*, 816–824. [CrossRef]

70. Zhang, L.; Yang, Z.; Huang, W.; Wu, J. H19 potentiates let-7 family expression through reducing PTBP1 binding to their precursors in cholestasis. *Cell Death Dis.* **2019**, *10*, 168. [CrossRef]

71. Nair, S.J.; Yang, L.; Meluzzi, D.; Oh, S.; Yang, F.; Friedman, M.J.; Wang, S.; Suter, T.; Alshareedah, I.; Gamliel, A.; et al. Phase separation of ligand-activated enhancers licenses cooperative chromosomal enhancer assembly. *Nat. Struct. Mol. Biol.* **2019**, *26*, 193–203. [CrossRef] [PubMed]

72. André, A.A.M.; Spruijt, E. Liquid-Liquid Phase Separation in Crowded Environments. *Int. J. Mol. Sci.* **2020**, *21*, 5908. [CrossRef]

Review

Endogenous Double-Stranded RNA

Shaymaa Sadeq, Surar Al-Hashimi, Carmen M. Cusack and Andreas Werner *

Biosciences Institute, Medical School, Newcastle University, Newcastle upon Tyne NE2 4HH, UK;
s.k.sadeq2@newcastle.ac.uk (S.S.); s.o.t.al-hashimi2@newcastle.ac.uk (S.A.-H.);
Carmenmariacusack@gmail.com (C.M.C.)
* Correspondence: andreas.werner@ncl.ac.uk

Abstract: The birth of long non-coding RNAs (lncRNAs) is closely associated with the presence and activation of repetitive elements in the genome. The transcription of endogenous retroviruses as well as long and short interspersed elements is not only essential for evolving lncRNAs but is also a significant source of double-stranded RNA (dsRNA). From an lncRNA-centric point of view, the latter is a minor source of bother in the context of the entire cell; however, dsRNA is an essential threat. A viral infection is associated with cytoplasmic dsRNA, and endogenous RNA hybrids only differ from viral dsRNA by the 5′ cap structure. Hence, a multi-layered defense network is in place to protect cells from viral infections but tolerates endogenous dsRNA structures. A first line of defense is established with compartmentalization; whereas endogenous dsRNA is found predominantly confined to the nucleus and the mitochondria, exogenous dsRNA reaches the cytoplasm. Here, various sensor proteins recognize features of dsRNA including the 5′ phosphate group of viral RNAs or hybrids with a particular length but not specific nucleotide sequences. The sensors trigger cellular stress pathways and innate immunity via interferon signaling but also induce apoptosis via caspase activation. Because of its central role in viral recognition and immune activation, dsRNA sensing is implicated in autoimmune diseases and used to treat cancer.

Keywords: double-stranded RNA (dsRNA); innate immunity; repetitive DNA elements (RE); antisense transcript

Citation: Sadeq, S.; Al-Hashimi, S.; Cusack, C.M.; Werner, A. Endogenous Double-Stranded RNA. *Non-coding RNA* **2021**, *7*, 15. https://doi.org/10.3390/ncrna7010015

Received: 31 January 2021
Accepted: 17 February 2021
Published: 19 February 2021

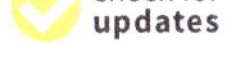

1. Introduction

If an endeavor has "Buckley's chance", no one in Melbourne would bet any money on it, as the odds to succeed are close to zero. The phrase "Buckley's chance" refers to William Buckley, an English convict who was deported to Australia. He escaped and lived with an Aboriginal tribe for more than 30 years. The chances of survival were, indeed, very slim for Buckley from the start; he was pursued and shot at when he escaped, and then he had to survive in the scorching Australian summer with little water and no food. Finally, he had to learn to communicate with the Aboriginal people and win their respect. In many ways, the unlikely survival story of William Buckley could stand as a metaphor for the development of spurious transcripts into "functional" long non-coding RNAs (lncRNAs) in a treacherous cellular environment.

The genome of complex organisms is riddled with repetitive sequences related to endogenous retroviruses (ERVs) and DNA transposons. They constitute a large part of the genome; in humans, 50–70% are repetitive or repeat-derived [1,2] and are largely responsible for the variation in genome size of complex organisms [3,4]. Despite the fact that the two classes of transposable elements (ERV and DNA transposons) can be grouped into superfamilies that are present in all taxa and then further into families and subfamilies, particular variants of transposable elements are species-specific.

The vast majority of transposons and retroviruses are inactivated through truncations and point mutations. In humans, only about 100 L1 retrotransposons (of about 500,000) are full-length, and less than 10 retained retro-transposition potential [5,6]. Hence, the repetitive, low-complexity part of the genome is often referred to as "junk DNA" [7].

Whether the vast graveyard of transposable elements actually represents "junk", functional elements or recyclable material constitutes an ongoing scientific debate [8,9]. Two important observations, however, are uncontested and particularly relevant in the context of long non-coding RNAs. First, the insertion of an ERV into the host genome affects transcriptional activity around the insertion site, thus creating the pressure to mitigate the overwhelmingly deleterious consequences of the interference [10]. Second, the remnants of transposable elements contain regulatory sequences such as weak promoters and enhancers or polyadenylation sites, and thus, a large proportion of the repetitive genome is being transcribed at a very low level [11,12]. In a sense, pervasive transcription may create opportunities to salvage genetic material in the form of long non-coding RNAs [13]. Accordingly, 75% of mature human lncRNA sequences contain an exon originating from transposable elements (TEs) [14,15]. Comparatively, the percentage of transcripts with TE material in 5′ and 3′ untranslated regions (UTRs) is substantially lower, with 8.44% in the 5′ UTR and 26.74% in the 3′ UTR [15]. The vast majority of the transcripts are quickly degraded because they lack protective modifications such as splicing, polyadenylation and capping that would also license them for export from the nucleus. Because of repetitive sequence content as well as bi-directional transcription, the spurious transcripts are prone to form both intra- and intermolecular double-stranded RNA (dsRNA) structures. Alternatively, the association with local cellular components such as chromatin remodeling complexes [16,17] may increase the stability and chances to escape degradation [18]. This brief review discusses the former outcome of pervasive transcription, the formation of endogenous dsRNA, which may trigger a cellular antiviral response; the focus will be on observations in humans and mice. It aims to draw a bigger picture rather than drilling into details.

2. Sources of Endogenous dsRNA

The detection and quantification of dsRNA requires specific tools such as specific antibodies or dsRNA-binding proteins [19,20]. After immune purification, RNA can be analyzed by high-throughput sequencing or conventional methods such as cloning or RT-PCR. An alternative strategy to investigate nuclear dsRNA uses adenosine-to-inosine (A-to-I) editing to identify double strand formation [21,22]. Single- or double-strand specific RNases in combination with RT-qPCR provide an additional tool to demonstrate RNA hybrids. Unfortunately, RNA purification prior to nuclease treatment introduces a positive or negative bias for dsRNA (depending on the specific methodology), making quantitative assays difficult to interpret [23].

There are three main sources of endogenous dsRNA: mitochondrial transcripts, repetitive nuclear sequences, including short and long interspersed elements (SINEs, LINEs), and endogenous retroviruses (ERVs) as well as natural sense–antisense transcript pairs.

2.1. Mitochondrial Transcripts

Human mitochondria have a circular genome of 16,566 bp, with a guanine-rich heavy strand and a guanine-poor light strand, depending on buoyant density. Both strands are equally transcribed, resulting in complimentary transcripts that may bind to each other, though the light strand undergoes rapid degradation. Complementarity encompasses the length of the entire mitochondrial genome, as shown by electron microscopic analysis [24,25]. The mitochondrial DNA encodes 13 genes, 12 of which are encoded by the heavy strand and one by the light strand [26]. Under physiological circumstances, the light strand is rapidly degraded by two enzymes, polynucleotide phosphorylase (PNPase) and the helicase HSuv3 [27]. PNPase is located in the inter-mitochondrial membrane space, thus being well-placed to play an important role in preventing the escape of dsRNA into the cytoplasm. Mitochondrial RNA is a potent stimulator of the innate immune system, especially in dendritic cells and Toll-like receptor (TLR)-expressing cells [28] via a protein kinase R (PKR)-modulated interferon response. Conversely, inhibition of HSuv3 resulted in an increase in dsRNA without triggering an interferon response, which suggests that the increased levels of dsRNA remained sequestered within the mitochondria [19]. These findings are underpinned by the knockout of PNPase or Suv3 that leads to an accumulation of dsRNA in the cytoplasm and an altered immune response [29]. Moreover, patients with bi-allelic PNPase variants showed increased levels of unprocessed mitochondrial transcripts and an enhanced expression of interferon-stimulated genes [30].

Mitochondrial dsRNA formation was also demonstrated using fCLIP-seq, an approach which entails formaldehyde cross-linking of PKR-bound dsRNA followed by high-throughput sequencing. Most of the dsRNA bound to PKR mapped to the mitochondrial genome. The mitochondrial origin of the RNA was corroborated by the lack of A-to-I edited nucleotides, as mitochondrial dsRNA is not subjected to adenosine deaminase acting on RNA (ADAR)-dependent editing [31]. Collectively, these findings established that mitochondria are an important source of dsRNA which may be released into the cytoplasm upon stress-mediated mitochondrial permeabilization [32].

2.2. Repetitive DNA Sequences

For dsRNA originating from nuclear DNA, A-to-I editing provides an accurate readout to assess genome-wide dsRNA formation [33]. In humans, 62.9% of all edited sites map to repeat regions, including SINEs, LINEs, endogenous retroviruses and DNA transposons, whereas protein coding transcripts are hardly edited at all (Figure 1). Overall, editing shows distinct species' variability and depends on the nature of the repetitive elements rather than the complexity of the organism [33,34].

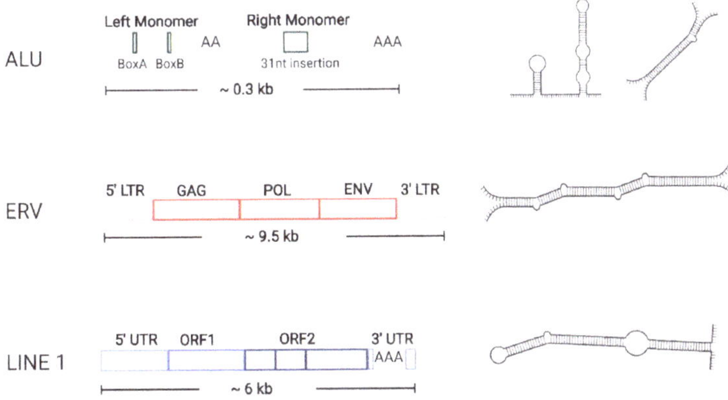

Figure 1. Schematic representation of repetitive elements in the human genome associated with double-stranded RNA (dsRNA) formation. LINE 1 and endogenous retroviruses (ERVs) give potential rise to long dsRNA structures formed from convergent transcripts or hairpin structures from read-through transcription of head-to-head/tail-to-tail arranged elements. Alu elements are much shorter and form hairpin structures as well as "open" dsRNA hybrids, though the intermolecular duplexes are rare. Alu elements are the predominant target for adenosine deaminase acting on RNA (ADAR)-mediated adenosine-to-inosine (A-to-I) editing. LTRs function as bi-directional promoters. ORF, open reading frame; GAG (group specific antigen), POL (reverse transcriptase), ENV (envelope protein), retroviral proteins; UTR, untranslated region; LTR, long terminal repeat. Figure created with Biorender.com.

SINEs: The most common sources of dsRNA in human cells are Alu repeats, the most abundant class of short interspersed nuclear elements [35] (Figure 1). Alu elements are approximately 300 nucleotides in length and contain two 7SL RNA genes including short A-rich stretches [36,37].

Alu repeats are commonly found in intergenic regions (autonomous) as well as in introns and UTRs of genes (mRNA-embedded elements) [38]. Autonomous Alu elements constitute a small portion of the repetitive genome and are highly induced by viral infection, heat shock and cycloheximide treatment [39]. Stress enhances the activity of the RNA polymerase lll (viral infection) or increases the chromatin accessibility of Alu elements (heat shock), which is reversed with recovery from stress [40]. As compared to autonomous Alus, embedded Alu elements represent a higher proportion of repeated sequences. Because of their enrichment in UTRs, embedded Alus play an important function in gene expression via the stabilization of mRNA, as well as its localization and translation [38,41].

The repetitive nature of Alu insertions allows the formation of predominantly intramolecular dsRNA, which is recognized by the nuclear isoform of ADAR [42,43]. In addition, PKR-fCLIP sequencing showed that more than 20% of dsRNAs associated with PKR derive from Alu repeats [31]. The Alu-dependent dsRNAs are not long enough to trigger efficient oligomerization and activation of melanoma differentiation-associated gene 5 (MDA5). In contrast, a mutated form of MDA5 that shows greater tolerance towards mismatches in the RNA hybrid has been linked to immune hypersensitivity and autoimmune disease (Aicardi–Goutières syndrome, [44]).

ERVs: Human endogenous retroviruses share a comparable structure with exogenous retroviruses, the protein coding genes gag, pro (protease), pol and env flanked by two terminal repeats (5′ and 3′ LTR) (Figure 1). ERVs comprise up to 8% of the human genome; however, most open reading frames (ORFs) are mutated [45]. Nevertheless, ERV-related transcripts can be detected in most human tissues [46], particularly when repressive DNA methylation is inhibited. In contrast to the mutated protein coding genes, ERV-related LTRs have retained their promoter activity and provide alternative transcriptional control elements for cellular genes or drive the production of non-coding cellular RNA [45,47].

LTR promoters are bi-directional and can lead to widespread dsRNA formation [48,49]; alternatively, two adjacent ERVs in opposite orientations could fold back and form a hairpin structure [31]. Although ERVs are not a very common source of dsRNA, the activation of LTR promoters and subsequent dsRNA formation still have significant clinical consequences. For example, transcription of ERVs can be triggered by DNA methyl transferase inhibitors such as Azacitdine and Decitabine through demethylation and activation of ERV promotors [50]. Induction of ERV expression results in activation of the mitochondrial antiviral signaling protein/interferon regulatory factors (MAVS-IRFs) pathway via MDA5 and, to lesser extent, retinoic acid-inducible gene I (RIG1). This "viral mimicry" is exploited for the treatment of many cancers such as melanoma and colorectal carcinoma by activating an innate immune response against cancer cells [51].

LINEs: Long interspersed nuclear elements (LINEs) are 6–7 kb in size and constitute up to 20% of the human genome. Full-length copies contain two open reading frames (ORF1 and ORF2) which encode proteins essential for retro-transposition [52] (Figure 1). ORF1 makes a 40-kDa RNA-binding protein (RBP 40) which plays an important role in activating the host innate immune system, while ORF2 encodes an endonuclease and the reverse transcriptase [53]. Transcription is driven by a promoter that harbors several transcription factor binding sites as well as a CpG island. Most LINEs are inactive because of truncations, mutations and rearranged copies; however, a small number of elements are functional [54].

The exact mechanisms by which LINEs form a double-strand configuration is unknown; some studies hypothesize that they form hairpin structures when two complementary LINEs are present in the same transcript. Alternatively, two LINEs on two different transcripts close to each other can hybridize [55]. This idea is supported by fCLIP sequencing data showing that the distance between two LINEs interacting with PKR is much shorter than the space between random copies [31]. Furthermore, LINE elements have the ability to fold back on their 5′ region, forming stable hairpin structures that are recognized by PKR [20].

LINEs associate with various dsRNA binding proteins, mostly PKR and MDA5, and their expression has been linked to the activation of an interferon 1 response [31]. Moreover, extensive editing of LINEs by ADAR has been shown using ADAR–CLIP sequencing [56,57]. Although LINEs only give rise to 3% of cellular dsRNA as compared to 67% from SINEs, they are linked to many human diseases [44].

Natural antisense transcripts: According to the gencode biotype definition, antisense transcripts are "transcripts that overlap the genomic span (i.e., exon or introns) of a protein-coding locus on the opposite strand". This definition excludes protein-coding antisense transcripts and read-through transcripts from tail-to-tail arranged gene pairs; if those are included, 40–70% of loci show bi-directional transcription [58,59]. Hence, if a sense/antisense transcript pair is co-expressed in the same cell, dsRNA structures are potentially formed (Figure 2). To what extend hybridization actually occurs is controversial and rather challenging to demonstrate experimentally.

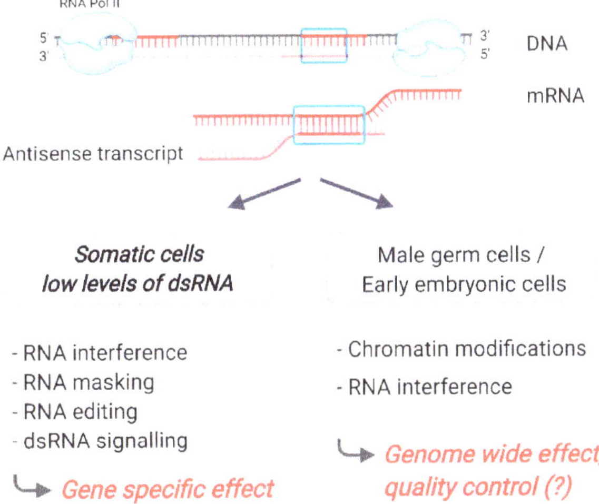

Figure 2. Double-stranded RNA (dsRNA) formation from sense–antisense transcripts. Natural antisense transcripts are processed and potentially reach the cytoplasm, where they interact with the sense transcript. In somatic cells, the level of sense–antisense hybrids is low, and there is no evidence of ADAR editing, for example, nor is dsRNA immune signaling triggered. Various mechanisms (RNA interference, RNA masking, RNA editing and dsRNA signaling) are potentially triggered by the dsRNA, depending on the cellular context. In male germ cells and during early embryogenesis, sense–antisense dsRNA formation may play a general, system-relevant role. Figure created with Biorender.com.

Before the dawn of the genomics era, natural antisense transcripts were studied in the context of parental imprinting. Early ground-breaking work demonstrated that the expression of the antisense transcript was associated with the silencing of the related sense transcript on the same allele. Experimental silencing of the antisense transcript (Airn, Kcnq1ot1, for example) abolished parental imprinting and led to bi-allelic expression of the entire cluster, not only of the complementary gene [60,61]. Similar observations were made with non-imprinted genes; a deletion in the genome of a patient with α-thalassemia placed the constitutively active LUC7L (Putative RNA-Binding Protein Luc7-Like) directly downstream of the HBA2 (Hemoglobin 2A) gene. The ectopic expression of LUC7L produced an antisense transcript complementary to HBA2, causing hypermethylation of the CpG-rich promoter and transcriptional silencing of the gene [62]. Likewise, the promotor of the tumor suppressor gene p15 (CDKN2B, Cyclin Dependent Kinase Inhibitor 2B) is hypermethylated and silenced in various tumors, associated with the expression of the antisense transcript p15-AS (CDKN2B-AS1) [63]. Silencing was found to be independent of Dicer, and the fact that the entire CDKN2B gene is imbedded in an intron of CDKN2B-AS1 argues against a role of dsRNA formation in an antisense transcript-mediated regulatory mechanism [63,64].

On the other hand, there is increasing evidence of dsRNA formation as the result of antisense transcription from both genomics studies and examples of specific sense–antisense transcript pairs. Early studies on the genome-wide expression of natural antisense transcripts followed a strategy where complementary full-length transcripts and expressed sequence tags in whole-data repository searches were identified [65,66]. The formation of dsRNA is inferred by the observation that natural antisense transcripts are significantly under-represented on the X chromosome of both humans and mice, whereas no such bias was found for sense–antisense pairs that lacked exonic complementarity [65,66]. Accordingly, dsRNA formation between processed transcripts represents a feature with a positive (accumulation on autosomes) or negative impact (reduction on X chromosomes) on evolutionary selection. The implications of dsRNA formation in the context of antisense transcription have been discussed including RNA masking, RNA editing, RNA interference as well as the stimulation of an innate immune response [67]. RNA masking is generally associated with a concordant expression of sense and antisense transcripts, often by interfering with the inhibitory action of miRNAs [68,69]. The latter three mechanisms (RNA editing, RNA interference and immune response) induce a discordant expression of the sense–antisense transcript pair ("antisense inhibits sense") (Figure 2).

There is a steadily increasing number of reports on specific sense–antisense pairs where dsRNA formation is implicated in a regulatory interaction between the two transcripts. In line with the proposed mechanisms, both concordant and discordant expression of the complementary transcripts have been observed [59]. An example of a stimulatory interaction described in detail is the interplay between the transcript for β secretase-1 (BACE1) and its natural antisense transcript (BACE1-AS) in the context of Alzheimer's disease pathophysiology. The antisense transcript protects BACE1 mRNA from miR-485-5p-induced degradation, and because of the increased β secretase, more β amyloid 1-42 was produced. In line with the mechanism, the levels of BACE1-AS were elevated in patients with Alzheimer's disease [70,71]. Other selected examples of antisense transcripts masking miRNA binding sites are listed in Piatek et al. [72]. Natural antisense transcripts can also stabilize the sense transcript by blocking the binding of RNA decay-promoting factors [73]. This mode of action is exemplified by the interaction between the tumor suppressor gene PDCD4 (Programmed Cell Death 4) and its antisense transcript PDCD4-AS1 in mammary epithelial cells. The antisense transcript blocks the binding of human antigen R (HuR), which, in turn, stabilizes the sense mRNA and leads to increased PDCD4 expression [73]. Accordingly, PDCD4-AS1 expression is decreased in breast cancer patients and is low in mammary epithelial cells.

The mechanisms that lead to the degradation of the sense transcript generate specific products that can be experimentally assessed at a large scale, i.e., A-to-I conversions for editing, short RNAs for RNA interference and sequencing of RNA bound to protein kinase R. However, only limited evidence supports that these mechanisms are involved in processing RNA hybrids between genic sense and antisense transcripts—at least in a specific experimental context [74,75]. There are a few examples where the involvement of Dicer or ADAR has been experimentally tested for specific bi-directionally transcribed loci including the gene pairs glutaminase (GLS)/GLS-AS or sodium/phosphate co-transporter and a read-through transcript from profilin 3 (Slc34a1/Pfn3) [74,76]. Low levels of GLS-AS and enhanced expression of GLS in patients with pancreatic cancer predict a poor clinical outcome. The underlying mechanism was investigated in PANC-1 cells (human pancreatic cancer cell line-1). Accordingly, dsRNA formation occurs in the nucleus and both ADAR and Dicer can process the hybrid, resulting in a decrease in GLS sense mRNA and encoded glutaminase. Enhanced levels of glutaminase are observed under nutrient stress and related to tumorigenesis [74]. With regard to the Slc34a1/Pfn3 locus, there is little evidence that the antisense transcript is involved in the physiological regulation of the Na-phosphate cotransporter. Depending on the model system, both RNA interference and transcriptional interference can be observed. The fact that both transcripts are lowly expressed in testis may indicate that the sense–antisense interaction is biologically relevant in male germ cells, where the vast majority of natural antisense transcripts are expressed [76].

Despite the ever-increasing number of mechanistically established sense–antisense interactions, there is still a huge gap between the number of characterized examples and the thousands of sense–antisense gene pairs. An interesting set of articles have recently revived the idea that natural antisense transcripts and the potential dsRNA formation feed into a common mechanism(s) that merits selection, as seen with the X-chromosome bias or—more generally—the weak evolutionary conservation of sense–antisense arrangements [77].

Work in a preprint by S Pillay investigated the role of natural antisense transcript expression during early zebrafish embryogenesis and divided the RNAs into two groups with negative and positive correlation with sense transcript abundance, respectively [78]. Positively correlated transcripts are predominantly associated with house-keeping genes, whereas the transcripts with discordant expression are maternally expressed and are complementary to developmental genes. Based on the finding that the discordantly regulated transcripts were enriched in the cytosol, the authors speculate that these natural antisense transcripts act in a similar way as miRNAs to silence ectopic expression of developmental genes [78]. Another study in our own lab focused on dsRNA formation in mouse testis and involved enrichment of dsRNA using the J2 antibody followed by deep sequencing. We found that dsRNA was predominately present in pachytene spermatocytes and that the dsRNA transcriptome in testis was fundamentally different from the one in somatic liver cells. In both cell types, dsRNA was derived from mitochondrial transcription, though in testis, mRNA-related signals were clearly more abundant than in liver. Moreover, we could establish an association between dsRNA, antisense genes and endogenous siRNAs (small interfering RNAs)—again, the link was weaker or insignificant in liver cells (Werner et al., under revision). Importantly, both investigations focused on native tissues and cells, developing male germ cells and early zebrafish embryos, respectively. Both systems display low levels of DNA methylation [79,80] and transcriptional activity that is distinct from "normal" somatic cells. Moreover, testis male germ cells are immune privileged and tolerate dsRNA without activating innate immunity [81]. It is intriguing to speculate that natural antisense transcripts and dsRNA formation play a role in mitigating the consequences of the genome-wide transcriptional changes. Findings in zebrafish and mouse testis also suggest that dsRNA may have a fundamentally different impact in somatic cells.

The different handling of dsRNA in germ cells versus somatic cells has been experimentally corroborated using transgenic mice expressing a construct with a long hairpin the 3′ UTR. In mouse oocytes, dsRNA was processed into siRNAs, whereas in somatic cells, a small fraction was A-to-I edited. An interferon (IFN) response was only observed after high-level expression of the hairpin construct in a transfected human cell line (HEK293) [82]. A germ cell-specific biological role of dsRNA and endo-siRNAs is also supported by low siRNA sequencing of both female and male germ cells [83,84].

3. Proteins Binding dsRNA

The structure of dsRNA adopts an A-form duplex with a narrow major groove (4-Å width) and wide minor groove (10–11-Å width). As a consequence, dsRNA-binding proteins are generally unable to form base pair-specific interactions and recognize the backbone rather than sequence motifs [85]. However, a few examples such as ADAR2 or STAUFEN recognize specific base pairs in the minor grove of the duplex [86,87]. Moreover, additional structures such as the Cap or RNA base modifications affect the binding of dsRNA-binding proteins and help the distinction between viral RNA and endogenous RNA hybrids. The dsRNA-binding protein families include RIG-I-like receptors (RLRs), PKR, ADAR, oligo adenylate synthetase (OAS), Dicer, Drosha and other helicases [20]. We focus here on the dsRNA-binding proteins that create a link between pathogenic dsRNA formation and the immune system (Figure 3). As part of the host defense against invading pathogens, these dsRNA sensors are also linked to a number of inflammatory and autoimmune diseases [88,89].

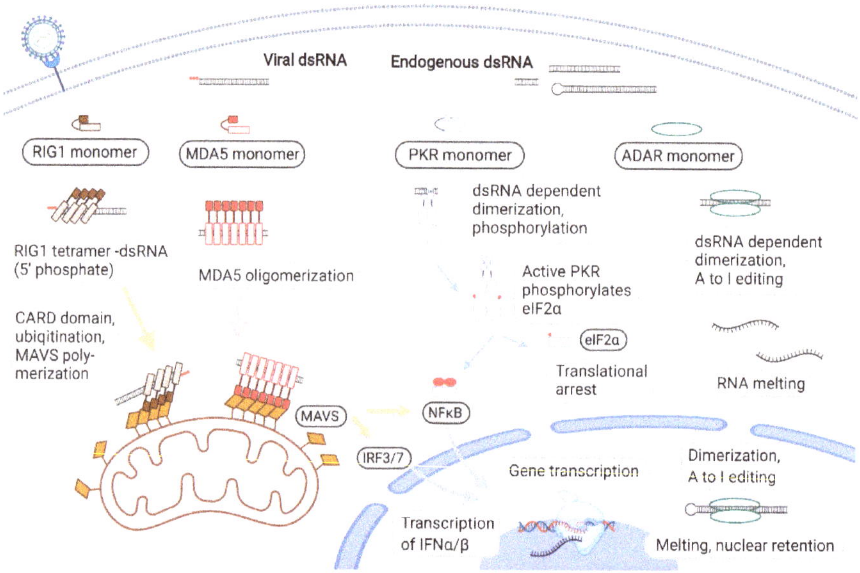

Figure 3. Double-stranded RNA (dsRNA) sensor proteins and activation of innate immunity. Viral dsRNA (including 5′ phosphorylation) or dsRNA from mitochondria and repetitive elements in the cytoplasm are recognized by dsRNA sensors retinoic acid-inducible gene I (RIG1), melanoma differentiation-associated gene 5 (MDA5), protein kinase R (PKR) and ADAR. RIG1 requires the 5′ phosphate group to initiate oligomerization, and MDA5 forms long dsRNA-dependent polymers. Both structures induce mitochondrial antiviral signaling (MAVS) polymerization and, eventually, caspase and interferon signaling. PKR binds short dsRNA molecules, dimerizes and becomes activated through autophosphorylation. Activated PKR dissociates from dsRNA, phosphorylates eukaryotic initiation factor 2α (eIF2α) (which, in turn, inhibits translation globally) and triggers an interferon response. ADAR is present in both the nucleus and cytoplasm and antagonizes dsRNA signaling by melting the RNA hybrid. Figure created with Biorender.com.

RIG-I-like receptors (RLRs): The protein family of retinoic acid-inducible gene-like receptors (RLRs), also called cytosolic RNA sensors, includes RIG1, MDA5 and laboratory of genetics and physiology 2 (LGP2). The latter lacks two caspase recruitment domains which are essential for downstream signaling. Consequently, LGP2 plays a regulatory role rather than an effector function in a dsRNA response [90]. The two main sensors that trigger a dsRNA inflammatory response are RIG1 and MDA5, which will be briefly introduced here [91].

RIG1 and MDA5 are members of the DExD/H box helicase family and contain five specific protein domains: from the N terminus, two caspase recruitment domains (CARDs), which participate in antiviral signaling, a DEAD-like helicase superfamily ATP-binding domain (DExDc), a helicase domain (HELICc) and a zinc-binding C-terminal domain [92]. In the non-signaling state, the two N-terminal domains are auto-repressed and unable to bind to mitochondrial antiviral signaling (MAVS) protein, a protein involved in the cellular innate antiviral defense. The auto-repression is abolished by the release of the N-terminal domains upon binding to dsRNA via helicase and the C-terminal domains [93].

RIG1 and MDA5 share the same signaling pathway but identify a discriminate group of dsRNA. Dimerization of RIG1 only takes a 300-base-pair duplex but requires a 5′ triphosphate group at the RNA end [94]. The triphosphate group is normally found in RNAs but is 7-methyl guanosine-capped in most eukaryotic mRNAs in the cytosol. Viral

RNA usually lacks this modification. Recognition of dsRNA by MDA5 does not depend on the triphosphate group but requires a longer stretch of dsRNA (500–1000 bp) for a process of nucleation and filament assembly to be activated [95].

RIG1 and MDA5 activation leads to oligomerization of CARD domains, which, in turn, produces a platform for the generation of MAVS filaments at the mitochondrial membrane [96]. This triggers two main cascades, one activating nuclear factor κB and the transcription of proinflammatory genes, the other leads to the phosphorylation of interferon regulatory factors 3/7 (IRF3/7) and the stimulation of interferon gene expression [97].

Protein kinase regulated by RNA: PKR, also referred to as eukaryotic translation initiation factor 2-alpha kinase 2, EIF2AK2, is activated by binding to dsRNA and its gene expression is induced by interferon [98]. PKR includes two N-terminal RNA-binding motifs (RI and RII) and a catalytic kinase domain at the C-terminus [99]. The dsRNA-binding domains can interact with adjacent minor grooves of dsRNA by binding to the phosphate and ribose backbone independent of the base sequence [100].

Activation of the enzymatic activity of PKR requires an RNA duplex of at least 33 bp. Activation efficiency increases up to 85 bp and decreases with longer duplexes or high concentrations of dsRNA because of a dilution effect that reduces the chances of PKR dimerization [101]. PKR recognizes all types of dsRNA, but the majority of PKR was bound to dsRNA of mitochondrial origin, followed by IRAlus (inverted- repeat Alu elements, 20%) [31].

Binding of PKR to dsRNA induces a conformational change which displaces the inhibitory dsRNA binding domain from the catalytic kinase domain [102]. Moreover, homodimerization results in auto-phosphorylation and activation of PKR. The activated kinase dissociates from dsRNA and phosphorylates eukaryotic translation initiation factor 2A (EIF2A) at serine 51 and triggers global translational shut-down [103]. Alternatively, PKR phosphorylation may lead to Fas-associated via death domain (FADD)/caspase 8-mediated activation of caspases 3/7 and, ultimately, apoptosis [104,105].

Adenosine deaminase acting on RNA (ADAR): Members of the ADAR protein family catalyze the conversion of A to I in dsRNA. In humans, there are three ADAR genes: ADAR1, ADAR2 and ADAR3, with ADAR1 being interferon-inducible [106,107]. All of the three ADAR proteins contain two or three dsRNA-binding domains and a C-terminal deaminase domain. Moreover, ADAR1 has one or two N-terminal Z-DNA-binding domains and ADAR3 contains an arginine-rich region [108].

Transcription of ADAR is driven by interferon inducible- and constitutively active promoters [109]. ADAR1 is ubiquitously expressed in human tissues and predominantly targets dsRNA formed by IRAlus in the 3′ UTR of the mRNAs. Around 97.7% of editing occurs in non-protein-coding regions [110,111]. ADAR2 expression is highest in the brain and is directly linked to site-specific base changes of neurotransmitter receptor transcripts with functional and phenotypic consequences [112,113]. Additional targets have been identified in the brain and other tissues, but the consequences of editing are less well established [114]. ADAR2 accounts for 25% of the editing in non-repetitive sites in protein-coding transcripts [111]. ADAR3 is exclusively expressed in the brain; the enzyme lacks catalytic activity and its main role appears to be the inhibition of ADAR2 by competition for dsRNA binding [115].

ADAR antagonizes apoptosis by counter-balancing the activation dsRNA sensors and the stimulation of inflammatory and apoptotic signaling [116]. In a negative feedback mechanism, interferon stimulates ADAR that binds to and melts dsRNA, thus competing with other dsRNA sensors [117]. Despite compartmentalization of dsRNA and the various other strategies to distinguish intrinsic dsRNA from viral insurgents, there are still various pathologies with an underlying inflammatory phenotype potentially linked to endogenous dsRNA. Two examples where dsRNA plays a role in disease development but also offers treatment avenues are cancer and autoimmune diseases.

4. Physiological and Pathophysiological Roles of dsRNA

Apart from stimulating an antiviral response, there is growing evidence to suggest that dsRNA contributes to physiological cell growth and function, depending on the length, abundance and location of dsRNA within the cell [118,119]. In this context, the activation of PKR and downstream interferon signaling as well as TLR3 activation by cytoplasmic long dsRNA are particularly relevant [118].

PKR is ubiquitously expressed in mitochondria as well as in the cytoplasm in its unphosphorylated inactive form; its physiological role extends beyond an antiviral response [31,120]. PKR activation is strictly regulated during mitosis, and its activity is essential for proper cell division. The disruption of the nuclear structure during mitosis means that IRAlus escape compartmentalization and activate PKR. As a consequence, eukaryotic initiation factor 2α (eIF2α) becomes phosphorylated, with subsequent suppression of the global translation [121]. Inhibition of PKR by RNA interference or expression of a transdominant-negative mutant alleviating translation suppression during M phase lead to the dysregulation of several mitotic factors (cyclins A and B and polo-like kinase 1). The reduced phosphorylation of histone 3 and stabilization of G2-specific cell cycle regulators cause a delay in the progression from G2 to M phase [121]. Activated PKR also induces phosphorylation of p53, a tumor-suppressor protein with a pivotal role in controlling cell cycle and apoptosis, which leads to a 25–35% increase in cells arrested in G1. On the other hand, a reduction in PKR expression by doxorubicin decreases p53 stability [122,123].

Wound-induced hair neogenesis (WIHN) is a rare example of adult organogenesis in which dsRNA plays a central role [124]. The activation of TLR3 by endogenous dsRNA contributes essentially to wound healing and hair regeneration. Full thickness wounds in mice result in the release of dsRNA from damaged skin that activates TLR3 and triggers downstream signaling via interleukin 6 and STAT3 (Signal transducer and activator of transcription 3), which promote hair neogenesis. Moreover, activated TLR3 induces intrinsic synthesis of retinoic acid (RA) that orchestrates skin appendages' growth and regeneration [125,126]. Injection of poly(I:C), a dsRNA analogue, into mouse wounded skin results in a significant increase in new hair formation, while TLR3-deficient mice failed to generate new hair upon skin wounding [124,126]. Furthermore, human skin biopsies taken after rejuvenation laser treatment display increased endogenous RA synthesis and enhanced gene expression signatures for dsRNA and RA [125].

Endogenous dsRNA and autoimmune diseases: Autoimmune diseases are pathologies where the immune system mistakenly attacks healthy cells. Around 50% of autoimmune diseases are of unknown etiology, while others are attributed to genetic pre-disposition or hormonal and environmental factors [127]. The contribution of dsRNA to autoimmune diseases was inferred by Schur and colleagues, who detected antibodies against dsRNA in the sera of 51% of patients with systemic lupus erythematosus (SLE) and 9% with rheumatoid arthritis as compared to 6% of normal people [128]. Elevated interferon levels and enhanced expression of IFN-stimulated genes in the blood of SLE patients have been shown more recently [129–131]. Furthermore, the presence of anti-MDA5 antibodies in dermatomyositis patients is considered as a prognostic marker associated with high death rate due to interstitial lung disease [132].

Myasthenia gravis is an autoimmune disease characterized by auto-antibodies against the acetyl choline receptor AChR. Injection of poly (I:C) in mice stimulates the expression of αAChR via TLR3 and PKR activation. Accordingly, the expressions of TLR3, PKR, IFR7, IRF5 and IFN-β are all upregulated in the thymus of patients with myasthenia gravis, indicating an important role of dsRNA signaling in the disease etiology [133]. PKR, MDA5 and RIG1 expression are all increased in psoriatic lesional skin, paralleled by high IFNα levels [134]. IFNα treatment for hepatitis C virus infection is well known to trigger autoimmune diseases such as psoriasis, antiphospholipid syndrome or sarcoidosis, highlighting the contribution of innate immunity to the pathogenesis of these diseases [135,136].

A-to-I RNA editing enhances transcriptome and protein diversity; conversely, editing in protein-coding regions generates auto-antigens and potentially causes or aggravates autoimmune diseases. Accordingly, increased editing was observed in SLE and rheumatoid arthritis [137,138]. On the other hand, there is a global reduction in A-to-I editing in psoriatic lesional skin and an accumulation of dsRNA feeding into an antiviral response, highlighting the fine balance between protective and detrimental consequences of dsRNA signaling.

dsRNA in cancer: Somatic mutations and escaping immune surveillance are essential steps in tumor initiation and progression. Recent studies have highlighted that RNA mutations constitute an additional cause for transition to malignant tumor, with RNA editing being a major cause for the underlying sequence changes. Adenosine-to-inosine changes in dsRNA by ADAR can give rise to transcriptomic alterations via point mutations, alternative splicing, altered RNA targeting and defects in microRNA synthesis [139]. Accordingly, many cancer types such as liver and breast cancer as well as some gastrointestinal malignancies express high levels of ADAR, which also promotes cancer growth and metastasis [140].

Although both ADAR1 and ADAR2 are linked to tumorigenesis, ADAR1 appears to play the major role due to its ubiquitous expression [139]. ADAR1 expression is stimulated by interferon as a negative feedback to control inflammation and cell survival, potentially also promoting tumor growth and invasiveness [141,142]. ADAR1 has been found to edit disease-relevant transcripts in a number of cancers [143]. For example, in prostate cancer, A-to-I editing in the androgen receptor transcript affects interaction of the receptor with androgens and androgen antagonists, which results in the reactivation of androgen signaling, tumor development and growth [144]. In hepatocellular carcinoma, increased levels of ADAR lead to editing of Antizyme Inhibitor 1 (AZIN1) and consequently enhanced nuclear import of the edited protein and stabilized interaction with its binding partner (Antizyme). The reduced inhibitory potential of the complex promotes tumor formation and is associated with aggressive behavior [145] (for a comprehensive review, see [143]).

dsRNA cancer therapies: There is a relation between autoimmune diseases and cancer—for example, long-standing autoimmune diseases may results in cancer transformation. Interestingly, the upregulation of ERV transcription is a common feature between these two pathologies [146,147]. The majority of ERVs are transcriptionally inactive, though 7% of their sequences can be reactivated by exogenous viruses or hypoxia [148]. Unlike in the autoimmune diseases discussed above, cancer cells mitigate the impact of moderate levels of ERV-related dsRNA formation and escape immune surveillance. However, drug-induced stimulation of ERV transcription can trigger a dsRNA-mediated immune response and make the cancerous cells visible to a variety of immune cells [149]. Hence, host dsRNA-binding proteins and the associated signaling cascades are widely used drug targets [150].

Transcription of ERVs is efficiently silenced through DNA hypermethylation in normal somatic cells [48]. Hypomethylating drugs such as azacytidine or decitabine induce transcription of ERVs and the formation of dsRNA, which, in turn, activates innate immune signaling. Both drugs are widely used to treat hematological cancers and have been investigated to treat other types of solid tissue tumors [151]. The consequences of bi-directional transcription of ERVs have been established in various cancer cell lines including epithelial ovarian cancers, colonic cancer cell lines and melanoma [48,152]. Accordingly, azacytidine causes an interferon response and increased expression of programmed death-ligand 1 (PD-L1), an important target in cancer immunotherapy [152].

A novel approach to treat various cancers combines ERV re-activation using histone deacetylase inhibitors (HDACi) in combination with immune checkpoint inhibitors targeting PD-1 or Cytotoxic T-Lymphocyte Associated Protein 4 (CTLA-4) [153]. Accordingly, ERV activation triggers a dsRNA-mediated interferon response that leads to increased expression of Major Histocompatibility Complex type I (MHC-I) on cancer cells; hence, the cell becomes "visible" to a T cell-mediated response [48]. Immune checkpoint inhibitors such as Atezolizumab and Avelumab or Ipilimumab (monoclonal antibodies against PD-1 or CTLA-4, respectively) used in combination dampen the inhibitory immune response and enhance anti-tumor activity [154].

The viral dsRNA analogues poly(I:C) and poly(A:U) are being used as adjuvants in anti-tumor therapy for their potential to stimulate an interferon response. There are two main mechanisms by which cancer cells are affected: first, by inducing cancer cell apoptosis through an IFN-β autocrine loop, and second, by IFN-β-mediated signaling. This leads to stimulation of the major players in anti-cancer immunity, including maturation and differentiation of dendritic cells, promotion of a T cell response and activation of natural killer cells [155]. Hence, immune-stimulatory adjuvants are key components of cancer vaccines together with tumor-specific antigens [156].

5. Conclusions

The pathways by which viral dsRNA activates innate immunity have been established for quite some time. In this context, the discovery of widespread dsRNA formation from endogenous sources such as repetitive elements or natural antisense transcripts raised questions of how the different stimulators of innate immunity are controlled. Compartmentalization and specialized dsRNA sensor proteins that integrate structural information and dsRNA abundance to elicit a physiologically sensible response have evolved as a protective strategy. Nonetheless, cellular dsRNA homeostasis is often challenged in disease and these observations have disclosed an interplay between repetitive genomic elements, long non-coding RNA and innate immune signaling that can jeopardize the well-being of cells, organs and the entire organism. A detailed understanding of dsRNA expression and processing can inform strategies to avoid ectopic dsRNA formation and inflammation through stress, drugs or malnutrition, for example. Alternatively, therapeutic stimulation of dsRNA expression shows great promise in directing an immune response against cancer cells.

Author Contributions: All authors have written parts of the manuscript and generated figures. All authors have read and agreed to the published version of the manuscript.

Funding: This work was partly funded by The Northern Counties Kidney Research Fund, Grant 18.011 to A.W., and the Iraqi Ministry of Higher Education (S.S. and S.A-H.).

Acknowledgments: We would like to thank Aikaterini Gatsiou for the critical discussion of the manuscript.

Conflicts of Interest: The authors declare no conflict of interest.

References

1. de Koning, A.P.; Gu, W.; Castoe, T.A.; Batzer, M.A.; Pollock, D.D. Repetitive elements may comprise over two-thirds of the human genome. *PLoS Genet.* **2011**, *7*, e1002384. [CrossRef] [PubMed]
2. Haubold, B.; Wiehe, T. How repetitive are genomes? *BMC Bioinform.* **2006**, *7*, 541. [CrossRef]
3. Canapa, A.; Barucca, M.; Biscotti, M.A.; Forconi, M.; Olmo, E. Transposons, Genome Size, and Evolutionary Insights in Animals. *Cytogenet. Genome Res.* **2015**, *147*, 217–239. [CrossRef] [PubMed]
4. Kidwell, M.G.; Lisch, D.R. Transposable elements and host genome evolution. *Trends Ecol. Evol.* **2000**, *15*, 95–99. [CrossRef]
5. Beck, C.R.; Collier, P.; Macfarlane, C.; Malig, M.; Kidd, J.M.; Eichler, E.E.; Badge, R.M.; Moran, J.V. LINE-1 retrotransposition activity in human genomes. *Cell* **2010**, *141*, 1159–1170. [CrossRef]
6. Sanchez-Luque, F.J.; Kempen, M.H.C.; Gerdes, P.; Vargas-Landin, D.B.; Richardson, S.R.; Troskie, R.L.; Jesuadian, J.S.; Cheetham, S.W.; Carreira, P.E.; Salvador-Palomeque, C.; et al. LINE-1 Evasion of Epigenetic Repression in Humans. *Mol. Cell* **2019**, *75*, 590–604.e12. [CrossRef] [PubMed]
7. Ohno, S. So much "junk" DNA in our genome. *Brookhaven Symp. Biol.* **1972**, *23*, 366–370.
8. Clark, M.B.; Amaral, P.P.; Schlesinger, F.J.; Dinger, M.E.; Taft, R.J.; Rinn, J.L.; Ponting, C.P.; Stadler, P.F.; Morris, K.V.; Morillon, A.; et al. The Reality of Pervasive Transcription. *PLoS Biol.* **2011**, *9*, e1000625. [CrossRef]
9. Lu, J.Y.; Shao, W.; Chang, L.; Yin, Y.; Li, T.; Zhang, H.; Hong, Y.; Percharde, M.; Guo, L.; Wu, Z.; et al. Genomic Repeats Categorize Genes with Distinct Functions for Orchestrated Regulation. *Cell Rep.* **2020**, *30*, 3296–3311.e5. [CrossRef]
10. Bourque, G.; Burns, K.H.; Gehring, M.; Gorbunova, V.; Seluanov, A.; Hammell, M.; Imbeault, M.; Izsvák, Z.; Levin, H.L.; Macfarlan, T.S.; et al. Ten things you should know about transposable elements. *Genome Biol.* **2018**, *19*, 199. [CrossRef]
11. Dinger, M.E.; Amaral, P.P.; Mercer, T.R.; Mattick, J.S. Pervasive transcription of the eukaryotic genome: Functional indices and conceptual implications. *Brief. Funct. Genom. Proteom.* **2009**, *8*, 407–423. [CrossRef] [PubMed]

12. Ponting, C.P.; Belgard, T.G. Transcribed dark matter: Meaning or myth? *Hum. Mol. Genet.* **2010**, *19*, R162–R168. [CrossRef] [PubMed]

13. Palazzo, A.F.; Koonin, E.V. Functional Long Non-coding RNAs Evolve from Junk Transcripts. *Cell* **2020**, *183*, 1151–1161. [CrossRef]

14. Ganesh, S.; Svoboda, P. Retrotransposon-associated long non-coding RNAs in mice and men. *Pflug. Arch* **2016**, *468*, 1049–1060. [CrossRef] [PubMed]

15. Kapusta, A.; Kronenberg, Z.; Lynch, V.J.; Zhuo, X.; Ramsay, L.; Bourque, G.; Yandell, M.; Feschotte, C. Transposable elements are major contributors to the origin, diversification, and regulation of vertebrate long noncoding RNAs. *PLoS Genet.* **2013**, *9*, e1003470. [CrossRef] [PubMed]

16. Achour, C.; Aguilo, F. Long non-coding RNA and Polycomb: An intricate partnership in cancer biology. *Front. Biosci. (Landmark Ed.)* **2018**, *23*, 2106–2132.

17. Rinn, J.L.; Kertesz, M.; Wang, J.K.; Squazzo, S.L.; Xu, X.; Brugmann, S.A.; Goodnough, L.H.; Helms, J.A.; Farnham, P.J.; Segal, E.; et al. Functional demarcation of active and silent chromatin domains in human HOX loci by noncoding RNAs. *Cell* **2007**, *129*, 1311–1323. [CrossRef]

18. Fabbri, M.; Girnita, L.; Varani, G.; Calin, G.A. Decrypting noncoding RNA interactions, structures, and functional networks. *Genome Res.* **2019**, *29*, 1377–1388. [CrossRef]

19. Dhir, A.; Dhir, S.; Borowski, L.S.; Jimenez, L.; Teitell, M.; Rotig, A.; Crow, Y.J.; Rice, G.I.; Duffy, D.; Tamby, C.; et al. Mitochondrial double-stranded RNA triggers antiviral signalling in humans. *Nature* **2018**, *560*, 238–242. [CrossRef] [PubMed]

20. Hur, S. Double-Stranded RNA Sensors and Modulators in Innate Immunity. *Annu. Rev. Immunol.* **2019**, *37*, 349–375. [CrossRef]

21. Barak, M.; Porath, H.T.; Finkelstein, G.; Knisbacher, B.A.; Buchumenski, I.; Roth, S.H.; Levanon, E.Y.; Eisenberg, E. Purifying selection of long dsRNA is the first line of defense against false activation of innate immunity. *Genome Biol.* **2020**, *21*, 26. [CrossRef] [PubMed]

22. Cohen-Fultheim, R.; Levanon, E.Y. Detection of A-to-I Hyper-edited RNA Sequences. In *RNA Editing: Methods and Protocols*; Picardi, E., Pesole, G., Eds.; Springer: New York, NY, USA, 2021; pp. 213–227. [CrossRef]

23. Molder, T.; Speek, M. [Letter to the Editor] Accelerated RNA-RNA hybridization by concentrated guanidinium thiocyanate solution in single-step RNA isolation. *BioTechniques* **2016**, *61*, 61–65. [CrossRef]

24. Aloni, Y.; Attardi, G. Symmetrical in vivo transcription of mitochondrial DNA in HeLa cells. *Proc. Natl. Acad. Sci. USA* **1971**, *68*, 1757–1761. [CrossRef]

25. Young, P.G.; Attardi, G. Characterization of double-stranded RNA from HeLa cell mitochondria. *Biochem. Biophys. Res. Commun.* **1975**, *65*, 1201–1207. [CrossRef]

26. Krishnan, K.J.; Turnbull, D.M. Mitochondrial DNA and genetic disease. *Essays Biochem.* **2010**, *47*, 139–151. [CrossRef] [PubMed]

27. Borowski, L.S.; Dziembowski, A.; Hejnowicz, M.S.; Stepien, P.P.; Szczesny, R.J. Human mitochondrial RNA decay mediated by PNPase-hSuv3 complex takes place in distinct foci. *Nucleic Acids Res.* **2013**, *41*, 1223–1240. [CrossRef] [PubMed]

28. Karikó, K.; Buckstein, M.; Ni, H.; Weissman, D. Suppression of RNA recognition by Toll-like receptors: The impact of nucleoside modification and the evolutionary origin of RNA. *Immunity* **2005**, *23*, 165–175. [CrossRef]

29. Pajak, A.; Laine, I.; Clemente, P.; El-Fissi, N.; Schober, F.A.; Maffezzini, C.; Calvo-Garrido, J.; Wibom, R.; Filograna, R.; Dhir, A.; et al. Defects of mitochondrial RNA turnover lead to the accumulation of double-stranded RNA in vivo. *PLoS Genet.* **2019**, *15*, e1008240. [CrossRef]

30. Rius, R.; Van Bergen, N.J.; Compton, A.G.; Riley, L.G.; Kava, M.P.; Balasubramaniam, S.; Amor, D.J.; Fanjul-Fernandez, M.; Cowley, M.J.; Fahey, M.C.; et al. Clinical Spectrum and Functional Consequences Associated with Bi-Allelic Pathogenic PNPT1 Variants. *J. Clin. Med.* **2019**, *8*, 2020. [CrossRef]

31. Kim, Y.; Park, J.; Kim, S.; Kim, M.; Kang, M.G.; Kwak, C.; Kang, M.; Kim, B.; Rhee, H.W.; Kim, V.N. PKR Senses Nuclear and Mitochondrial Signals by Interacting with Endogenous Double-Stranded RNAs. *Mol. Cell* **2018**, *71*, 1051–1063.e6. [CrossRef]

32. Arnaiz, E.; Miar, A.; Dias, A.G.; Prasad, N.; Schulze, U.; Waithe, D.; Rehwinkel, J.; Harris, A. Hypoxia regulates endogenous double-stranded RNA production via reduced mitochondrial DNA transcription. *bioRxiv* **2020**. [CrossRef]

33. Porath, H.T.; Knisbacher, B.A.; Eisenberg, E.; Levanon, E.Y. Massive A-to-I RNA editing is common across the Metazoa and correlates with dsRNA abundance. *Genome Biol.* **2017**, *18*, 185. [CrossRef]

34. Reich, D.P.; Bass, B.L. Mapping the dsRNA World. *Cold Spring Harb. Perspect. Biol.* **2019**, *11*. [CrossRef] [PubMed]

35. Quentin, Y. Origin of the Alu family: A family of Alu-like monomers gave birth to the left and the right arms of the Alu elements. *Nucleic Acids Res.* **1992**, *20*, 3397–3401. [CrossRef] [PubMed]

36. Deininger, P. Alu elements: Know the SINEs. *Genome Biol.* **2011**, *12*, 236. [CrossRef]

37. Ullu, E.; Tschudi, C. Alu sequences are processed 7SL RNA genes. *Nature* **1984**, *312*, 171–172. [CrossRef]

38. Batzer, M.A.; Deininger, P.L. Alu repeats and human genomic diversity. *Nat. Rev.* **2002**, *3*, 370–379. [CrossRef]

39. Li, T.-H.; Schmid, C.W. Differential stress induction of individual Alu loci: Implications for transcription and retrotransposition. *Gene* **2001**, *276*, 135–141. [CrossRef]

40. Berger, A.; Strub, K. Multiple Roles of Alu-Related Noncoding RNAs. *Prog. Mol. Subcell. Biol.* **2011**, *51*, 119–146. [CrossRef]

41. Caudron-Herger, M.; Pankert, T.; Seiler, J.; Németh, A.; Voit, R.; Grummt, I.; Rippe, K. Alu element-containing RNAs maintain nucleolar structure and function. *EMBO J.* **2015**, *34*, 2758–2774. [CrossRef] [PubMed]

42. Bazak, L.; Levanon, E.Y.; Eisenberg, E. Genome-wide analysis of Alu editability. *Nucleic Acids Res.* **2014**, *42*, 6876–6884. [CrossRef]

43. Kawahara, Y.; Nishikura, K. Extensive adenosine-to-inosine editing detected in Alu repeats of antisense RNAs reveals scarcity of sense-antisense duplex formation. *FEBS Lett.* **2006**, *580*, 2301–2305. [CrossRef]

44. Ahmad, S.; Mu, X.; Yang, F.; Greenwald, E.; Park, J.W.; Jacob, E.; Zhang, C.Z.; Hur, S. Breaching Self-Tolerance to Alu Duplex RNA Underlies MDA5-Mediated Inflammation. *Cell* **2018**, *172*, 797–810.e13. [CrossRef] [PubMed]

45. Zhang, M.; Liang, J.Q.; Zheng, S. Expressional activation and functional roles of human endogenous retroviruses in cancers. *Rev. Med Virol.* **2019**, *29*, e2025. [CrossRef]

46. Hurst, T.P.; Magiorkinis, G. Activation of the innate immune response by endogenous retroviruses. *J. Gen. Virol* **2015**, *96*, 1207–1218. [CrossRef] [PubMed]

47. Di Cristofano, A.; Strazzullo, M.; Longo, L.; La Mantia, G. Characterization and genomic mapping of the ZNF80 locus: Expression of this zinc-finger gene is driven by a solitary LTR of ERV9 endogenous retroviral family. *Nucleic Acids Res.* **1995**, *23*, 2823–2830. [CrossRef] [PubMed]

48. Chiappinelli, K.B.; Strissel, P.L.; Desrichard, A.; Li, H.; Henke, C.; Akman, B.; Hein, A.; Rote, N.S.; Cope, L.M.; Snyder, A.; et al. Inhibiting DNA Methylation Causes an Interferon Response in Cancer via dsRNA Including Endogenous Retroviruses. *Cell* **2015**, *162*, 974–986. [CrossRef] [PubMed]

49. Domansky, A.N.; Kopantzev, E.P.; Snezhkov, E.V.; Lebedev, Y.B.; Leib-Mosch, C.; Sverdlov, E.D. Solitary HERV-K LTRs possess bi-directional promoter activity and contain a negative regulatory element in the U5 region. *FEBS Lett.* **2000**, *472*, 191–195. [CrossRef]

50. Strick, R.; Strissel, P.L.; Baylin, S.B.; Chiappinelli, K.B. Unraveling the molecular pathways of DNA-methylation inhibitors: Human endogenous retroviruses induce the innate immune response in tumors. *Oncoimmunology* **2016**, *5*, e1122160. [CrossRef]

51. Roulois, D.; Loo Yau, H.; Singhania, R.; Wang, Y.; Danesh, A.; Shen, S.Y.; Han, H.; Liang, G.; Jones, P.A.; Pugh, T.J.; et al. DNA-Demethylating Agents Target Colorectal Cancer Cells by Inducing Viral Mimicry by Endogenous Transcripts. *Cell* **2015**, *162*, 961–973. [CrossRef]

52. Richardson, S.R.; Doucet, A.J.; Kopera, H.C.; Moldovan, J.B.; Garcia-Perez, J.L.; Moran, J.V. The Influence of LINE-1 and SINE Retrotransposons on Mammalian Genomes. *Microbiol. Spectr.* **2015**, *3*, 1165–1208. [CrossRef]

53. Martin, S.L. The ORF1 protein encoded by LINE-1: Structure and function during L1 retrotransposition. *J. Biomed. Biotechnol.* **2006**, *2006*, 45621. [CrossRef] [PubMed]

54. Faulkner, G.J.; Billon, V. L1 retrotransposition in the soma: A field jumping ahead. *Mob. DNA* **2018**, *9*, 22. [CrossRef] [PubMed]

55. Kim, S.; Ku, Y.; Ku, J.; Kim, Y. Evidence of Aberrant Immune Response by Endogenous Double-Stranded RNAs: Attack from Within. *Bioessays* **2019**, *41*, e1900023. [CrossRef] [PubMed]

56. Bahn, J.H.; Ahn, J.; Lin, X.; Zhang, Q.; Lee, J.-H.; Civelek, M.; Xiao, X. Genomic analysis of ADAR1 binding and its involvement in multiple RNA processing pathways. *Nat. Commun.* **2015**, *6*, 6355. [CrossRef] [PubMed]

57. Orecchini, E.; Doria, M.; Antonioni, A.; Galardi, S.; Ciafrè, S.A.; Frassinelli, L.; Mancone, C.; Montaldo, C.; Tripodi, M.; Michienzi, A. ADAR1 restricts LINE-1 retrotransposition. *Nucleic Acids Res.* **2017**, *45*, 155–168. [CrossRef]

58. Balbin, O.A.; Malik, R.; Dhanasekaran, S.M.; Prensner, J.R.; Cao, X.; Wu, Y.M.; Robinson, D.; Wang, R.; Chen, G.; Beer, D.G.; et al. The landscape of antisense gene expression in human cancers. *Genome Res.* **2015**, *25*, 1068–1079. [CrossRef]

59. Katayama, S.; Tomaru, Y.; Kasukawa, T.; Waki, K.; Nakanishi, M.; Nakamura, M.; Nishida, H.; Yap, C.C.; Suzuki, M.; Kawai, J.; et al. Antisense transcription in the mammalian transcriptome. *Science* **2005**, *309*, 1564–1566.

60. Sleutels, F.; Zwart, R.; Barlow, D.P. The non-coding Air RNA is required for silencing autosomal imprinted genes. *Nature* **2002**, *415*, 810–813. [CrossRef]

61. Thakur, N.; Tiwari, V.K.; Thomassin, H.; Pandey, R.R.; Kanduri, M.; Gondor, A.; Grange, T.; Ohlsson, R.; Kanduri, C. An antisense RNA regulates the bidirectional silencing property of the Kcnq1 imprinting control region. *Mol. Cell. Biol.* **2004**, *24*, 7855–7862. [CrossRef] [PubMed]

62. Tufarelli, C.; Frischauf, A.M.; Hardison, R.; Flint, J.; Higgs, D.R. Characterization of a widely expressed gene (LUC7-LIKE; LUC7L) defining the centromeric boundary of the human alpha-globin domain. *Genomics* **2001**, *71*, 307–314. [CrossRef] [PubMed]

63. Wong, I.H.; Lo, Y.M.; Yeo, W.; Lau, W.Y.; Johnson, P.J. Frequent p15 promoter methylation in tumor and peripheral blood from hepatocellular carcinoma patients. *Clin. Cancer Res. Off. J. Am. Assoc. Cancer Res.* **2000**, *6*, 3516–3521.

64. Yu, W.; Gius, D.; Onyango, P.; Muldoon-Jacobs, K.; Karp, J.; Feinberg, A.P.; Cui, H. Epigenetic silencing of tumour suppressor gene p15 by its antisense RNA. *Nature* **2008**, *451*, 202–206. [CrossRef] [PubMed]

65. Chen, J.; Sun, M.; Kent, W.J.; Huang, X.; Xie, H.; Wang, W.; Zhou, G.; Shi, R.Z.; Rowley, J.D. Over 20% of human transcripts might form sense-antisense pairs. *Nucleic Acids Res.* **2004**, *32*, 4812–4820. [CrossRef]

66. Kiyosawa, H.; Yamanaka, I.; Osato, N.; Kondo, S.; Hayashizaki, Y. Antisense transcripts with FANTOM2 clone set and their implications for gene regulation. *Genome Res.* **2003**, *13*, 1324–1334. [CrossRef]

67. Wight, M.; Werner, A. The functions of natural antisense transcripts. *Essays Biochem.* **2013**, *54*, 91–101. [CrossRef]

68. Faghihi, M.A.; Wahlestedt, C. Regulatory roles of natural antisense transcripts. *Nat. Rev. Mol. Cell Biol.* **2009**, *10*, 637–643. [CrossRef] [PubMed]

69. Zinad, H.S.; Natasya, I.; Werner, A. Natural Antisense Transcripts at the Interface between Host Genome and Mobile Genetic Elements. *Front. Microbiol.* **2017**, *8*, 2292. [CrossRef]

70. Faghihi, M.A.; Modarresi, F.; Khalil, A.M.; Wood, D.E.; Sahagan, B.G.; Morgan, T.E.; Finch, C.E.; St Laurent, G., 3rd; Kenny, P.J.; Wahlestedt, C. Expression of a noncoding RNA is elevated in Alzheimer's disease and drives rapid feed-forward regulation of beta-secretase. *Nat. Med.* **2008**, *14*, 723–730. [CrossRef]

71. Faghihi, M.A.; Zhang, M.; Huang, J.; Modarresi, F.; Van der Brug, M.P.; Nalls, M.A.; Cookson, M.R.; St-Laurent, G., 3rd; Wahlestedt, C. Evidence for natural antisense transcript-mediated inhibition of microRNA function. *Genome Biol.* **2010**, *11*, R56. [CrossRef]

72. Piatek, M.J.; Henderson, V.; Zynad, H.S.; Werner, A. Natural antisense transcription from a comparative perspective. *Genomics* **2016**. [CrossRef]

73. Jadaliha, M.; Gholamalamdari, O.; Tang, W.; Zhang, Y.; Petracovici, A.; Hao, Q.; Tariq, A.; Kim, T.G.; Holton, S.E.; Singh, D.K.; et al. A natural antisense lncRNA controls breast cancer progression by promoting tumor suppressor gene mRNA stability. *PLoS Genet.* **2018**, *14*, e1007802. [CrossRef]

74. Deng, S.J.; Chen, H.Y.; Zeng, Z.; Deng, S.; Zhu, S.; Ye, Z.; He, C.; Liu, M.L.; Huang, K.; Zhong, J.X.; et al. Nutrient Stress-Dysregulated Antisense lncRNA GLS-AS Impairs GLS-Mediated Metabolism and Represses Pancreatic Cancer Progression. *Cancer Res.* **2019**, *79*, 1398–1412. [CrossRef] [PubMed]

75. Werner, A.; Cockell, S.; Falconer, J.; Carlile, M.; Alnumeir, S.; Robinson, J. Contribution of natural antisense transcription to an endogenous siRNA signature in human cells. *BMC Genom.* **2014**, *15*, 19. [CrossRef]

76. Piatek, M.J.; Henderson, V.; Fearn, A.; Chaudhry, B.; Werner, A. Ectopically expressed Slc34a2a sense-antisense transcripts cause a cerebellar phenotype in zebrafish embryos depending on RNA complementarity and Dicer. *PLoS ONE* **2017**, *12*, e0178219. [CrossRef]

77. Dahary, D.; Elroy-Stein, O.; Sorek, R. Naturally occurring antisense: Transcriptional leakage or real overlap? *Genome Res.* **2005**, *15*, 364–368. [CrossRef]

78. Pillay, S.; Takahashi, H.; Carninci, P.; Kanhere, A. Antisense ncRNAs during early vertebrate development are divided in groups with distinct features. *bioRxiv* **2020**. [CrossRef]

79. Goll, M.G.; Halpern, M.E. DNA methylation in zebrafish. *Prog. Mol. Biol. Transl. Sci.* **2011**, *101*, 193–218. [CrossRef] [PubMed]

80. Morgan, H.D.; Santos, F.; Green, K.; Dean, W.; Reik, W. Epigenetic reprogramming in mammals. *Hum. Mol. Genet.* **2005**, *14*, R47–R58. [CrossRef]

81. Li, N.; Wang, T.; Han, D. Structural, cellular and molecular aspects of immune privilege in the testis. *Front. Immunol.* **2012**, *3*, 152. [CrossRef] [PubMed]

82. Nejepinska, J.; Malik, R.; Filkowski, J.; Flemr, M.; Filipowicz, W.; Svoboda, P. dsRNA expression in the mouse elicits RNAi in oocytes and low adenosine deamination in somatic cells. *Nucleic Acids Res.* **2012**, *40*, 399–413. [CrossRef] [PubMed]

83. Song, R.; Hennig, G.W.; Wu, Q.; Jose, C.; Zheng, H.; Yan, W. Male germ cells express abundant endogenous siRNAs. *Proc. Natl. Acad. Sci. USA* **2011**, *108*, 13159–13164. [CrossRef]

84. Watanabe, T.; Totoki, Y.; Toyoda, A.; Kaneda, M.; Kuramochi-Miyagawa, S.; Obata, Y.; Chiba, H.; Kohara, Y.; Kono, T.; Nakano, T.; et al. Endogenous siRNAs from naturally formed dsRNAs regulate transcripts in mouse oocytes. *Nature* **2008**, *453*, 539–543. [CrossRef] [PubMed]

85. Masliah, G.; Barraud, P.; Allain, F.H. RNA recognition by double-stranded RNA binding domains: A matter of shape and sequence. *Cell. Mol. Life Sci.* **2013**, *70*, 1875–1895. [CrossRef] [PubMed]

86. Stefl, R.; Oberstrass, F.C.; Hood, J.L.; Jourdan, M.; Zimmermann, M.; Skrisovska, L.; Maris, C.; Peng, L.; Hofr, C.; Emeson, R.B.; et al. The solution structure of the ADAR2 dsRBM-RNA complex reveals a sequence-specific readout of the minor groove. *Cell* **2010**, *143*, 225–237. [CrossRef]

87. Yadav, D.K.; Zigáčková, D.; Zlobina, M.; Klumpler, T.; Beaumont, C.; Kubíčková, M.; Vaňáčová, Š.; Lukavsky, P.J. Staufen1 reads out structure and sequence features in ARF1 dsRNA for target recognition. *Nucleic Acids Res.* **2020**, *48*, 2091–2106. [CrossRef]

88. Kawasaki, T.; Kawai, T.; Akira, S. Recognition of nucleic acids by pattern-recognition receptors and its relevance in autoimmunity. *Immunol. Rev.* **2011**, *243*, 61–73. [CrossRef] [PubMed]

89. Matz, K.M.; Guzman, R.M.; Goodman, A.G. The Role of Nucleic Acid Sensing in Controlling Microbial and Autoimmune Disorders. *Int. Rev. Cell Mol. Biol.* **2019**, *345*, 35–136. [CrossRef]

90. Zhu, Z.; Zhang, X.; Wang, G.; Zheng, H. The laboratory of genetics and physiology 2: Emerging insights into the controversial functions of this RIG-I-like receptor. *BioMed Res. Int.* **2014**, *2014*, 960190. [CrossRef]

91. Cui, J.; Chen, Y.; Wang, H.Y.; Wang, R.F. Mechanisms and pathways of innate immune activation and regulation in health and cancer. *Hum. Vaccines Immunother.* **2014**, *10*, 3270–3285. [CrossRef]

92. Sohn, J.; Hur, S. Filament assemblies in foreign nucleic acid sensors. *Curr. Opin. Struct. Biol.* **2016**, *37*, 134–144. [CrossRef]

93. Kowalinski, E.; Lunardi, T.; McCarthy, A.A.; Louber, J.; Brunel, J.; Grigorov, B.; Gerlier, D.; Cusack, S. Structural basis for the activation of innate immune pattern-recognition receptor RIG-I by viral RNA. *Cell* **2011**, *147*, 423–435. [CrossRef]

94. Schlee, M.; Roth, A.; Hornung, V.; Hagmann, C.A.; Wimmenauer, V.; Barchet, W.; Coch, C.; Janke, M.; Mihailovic, A.; Wardle, G.; et al. Recognition of 5′ triphosphate by RIG-I helicase requires short blunt double-stranded RNA as contained in panhandle of negative-strand virus. *Immunity* **2009**, *31*, 25–34. [CrossRef]

95. Peisley, A.; Jo, M.H.; Lin, C.; Wu, B.; Orme-Johnson, M.; Walz, T.; Hohng, S.; Hur, S. Kinetic mechanism for viral dsRNA length discrimination by MDA5 filaments. *Proc. Natl. Acad. Sci. USA* **2012**, *109*, E3340–E3349. [CrossRef] [PubMed]

96. Reikine, S.; Nguyen, J.B.; Modis, Y. Pattern Recognition and Signaling Mechanisms of RIG-I and MDA5. *Front. Immunol.* **2014**, *5*, 342. [CrossRef] [PubMed]

97. Ren, Z.; Ding, T.; Zuo, Z.; Xu, Z.; Deng, J.; Wei, Z. Regulation of MAVS Expression and Signaling Function in the Antiviral Innate Immune Response. *Front. Immunol.* **2020**, *11*, 1030. [CrossRef] [PubMed]

98. Chukwurah, E.; Farabaugh, K.T.; Guan, B.J.; Ramakrishnan, P.; Hatzoglou, M. A tale of two proteins: PACT and PKR and their roles in inflammation. *FEBS J.* **2021**. [CrossRef]

99. Toth, A.M.; Zhang, P.; Das, S.; George, C.X.; Samuel, C.E. Interferon action and the double-stranded RNA-dependent enzymes ADAR1 adenosine deaminase and PKR protein kinase. *Prog. Nucleic Acid Res. Mol. Biol.* **2006**, *81*, 369–434. [CrossRef]

100. Bevilacqua, P.C.; Cech, T.R. Minor-groove recognition of double-stranded RNA by the double-stranded RNA-binding domain from the RNA-activated protein kinase PKR. *Biochemistry* **1996**, *35*, 9983–9994. [CrossRef] [PubMed]

101. Manche, L.; Green, S.R.; Schmedt, C.; Mathews, M.B. Interactions between double-stranded RNA regulators and the protein kinase DAI. *Mol. Cell. Biol.* **1992**, *12*, 5238–5248. [CrossRef] [PubMed]

102. Cole, J.L. Activation of PKR: An open and shut case? *Trends Biochem. Sci.* **2007**, *32*, 57–62. [CrossRef] [PubMed]

103. García, M.A.; Gil, J.; Ventoso, I.; Guerra, S.; Domingo, E.; Rivas, C.; Esteban, M. Impact of protein kinase PKR in cell biology: From antiviral to antiproliferative action. *Microbiol. Mol. Biol. Rev.* **2006**, *70*, 1032–1060. [CrossRef] [PubMed]

104. Gal-Ben-Ari, S.; Barrera, I.; Ehrlich, M.; Rosenblum, K. PKR: A Kinase to Remember. *Front. Mol. Neurosci.* **2018**, *11*, 480. [CrossRef]

105. von Roretz, C.; Gallouzi, I.E. Protein kinase RNA/FADD/caspase-8 pathway mediates the proapoptotic activity of the RNA-binding protein human antigen R (HuR). *J. Biol. Chem.* **2010**, *285*, 16806–16813. [CrossRef]

106. Patterson, J.B.; Samuel, C.E. Expression and regulation by interferon of a double-stranded-RNA-specific adenosine deaminase from human cells: Evidence for two forms of the deaminase. *Mol. Cell. Biol.* **1995**, *15*, 5376–5388. [CrossRef]

107. Walkley, C.R.; Li, J.B. Rewriting the transcriptome: Adenosine-to-inosine RNA editing by ADARs. *Genome Biol.* **2017**, *18*, 205. [CrossRef] [PubMed]

108. Mannion, N.; Arieti, F.; Gallo, A.; Keegan, L.P.; O'Connell, M.A. New Insights into the Biological Role of Mammalian ADARs; the RNA Editing Proteins. *Biomolecules* **2015**, *5*, 2338–2362. [CrossRef]

109. George, C.X.; Samuel, C.E. Human RNA-specific adenosine deaminase ADAR1 transcripts possess alternative exon 1 structures that initiate from different promoters, one constitutively active and the other interferon inducible. *Proc. Natl. Acad. Sci. USA* **1999**, *96*, 4621–4626. [CrossRef]

110. Hundley, H.A.; Bass, B.L. ADAR editing in double-stranded UTRs and other noncoding RNA sequences. *Trends Biochem. Sci.* **2010**, *35*, 377–383. [CrossRef]

111. Tan, M.H.; Li, Q.; Shanmugam, R.; Piskol, R.; Kohler, J.; Young, A.N.; Liu, K.I.; Zhang, R.; Ramaswami, G.; Ariyoshi, K.; et al. Dynamic landscape and regulation of RNA editing in mammals. *Nature* **2017**, *550*, 249–254. [CrossRef]

112. Hood, J.L.; Emeson, R.B. Editing of neurotransmitter receptor and ion channel RNAs in the nervous system. *Curr. Top. Microbiol. Immunol.* **2012**, *353*, 61–90. [CrossRef] [PubMed]

113. Higuchi, M.; Maas, S.; Single, F.N.; Hartner, J.; Rozov, A.; Burnashev, N.; Feldmeyer, D.; Sprengel, R.; Seeburg, P.H. Point mutation in an AMPA receptor gene rescues lethality in mice deficient in the RNA-editing enzyme ADAR2. *Nature* **2000**, *406*, 78–81. [CrossRef] [PubMed]

114. Stulić, M.; Jantsch, M.F. Spatio-temporal profiling of Filamin A RNA-editing reveals ADAR preferences and high editing levels outside neuronal tissues. *RNA Biol.* **2013**, *10*, 1611–1617. [CrossRef] [PubMed]

115. Oakes, E.; Anderson, A.; Cohen-Gadol, A.; Hundley, H.A. Adenosine Deaminase That Acts on RNA 3 (ADAR3) Binding to Glutamate Receptor Subunit B Pre-mRNA Inhibits RNA Editing in Glioblastoma. *J. Biol. Chem.* **2017**, *292*, 4326–4335. [CrossRef]

116. Pfaller, C.K.; Li, Z.; George, C.X.; Samuel, C.E. Protein kinase PKR and RNA adenosine deaminase ADAR1: New roles for old players as modulators of the interferon response. *Curr. Opin. Immunol.* **2011**, *23*, 573–582. [CrossRef]

117. Yang, S.; Deng, P.; Zhu, Z.; Zhu, J.; Wang, G.; Zhang, L.; Chen, A.F.; Wang, T.; Sarkar, S.N.; Billiar, T.R.; et al. Adenosine deaminase acting on RNA 1 limits RIG-I RNA detection and suppresses IFN production responding to viral and endogenous RNAs. *J. Immunol.* **2014**, *193*, 3436–3445. [CrossRef]

118. Wang, Q.; Carmichael, G.G. Effects of length and location on the cellular response to double-stranded RNA. *Microbiol. Mol. Biol. Rev.* **2004**, *68*, 432–452. [CrossRef] [PubMed]

119. Zhang, Z.; Carmichael, G.G. The fate of dsRNA in the nucleus: A p54(nrb)-containing complex mediates the nuclear retention of promiscuously A-to-I edited RNAs. *Cell* **2001**, *106*, 465–475. [CrossRef]

120. Jeffrey, I.W.; Kadereit, S.; Meurs, E.F.; Metzger, T.; Bachmann, M.; Schwemmle, M.; Hovanessian, A.G.; Clemens, M.J. Nuclear localization of the interferon-inducible protein kinase PKR in human cells and transfected mouse cells. *Exp. Cell Res.* **1995**, *218*, 17–27. [CrossRef]

121. Kim, Y.; Lee, J.H.; Park, J.E.; Cho, J.; Yi, H.; Kim, V.N. PKR is activated by cellular dsRNAs during mitosis and acts as a mitotic regulator. *Genes Dev.* **2014**, *28*, 1310–1322. [CrossRef]

122. Bennett, R.L.; Pan, Y.; Christian, J.; Hui, T.; May, W.S., Jr. The RAX/PACT-PKR stress response pathway promotes p53 sumoylation and activation, leading to G_1 arrest. *Cell Cycle* **2012**, *11*, 407–417. [CrossRef]

123. Cuddihy, A.R.; Wong, A.H.; Tam, N.W.; Li, S.; Koromilas, A.E. The double-stranded RNA activated protein kinase PKR physically associates with the tumor suppressor p53 protein and phosphorylates human p53 on serine 392 in vitro. *Oncogene* **1999**, *18*, 2690–2702. [CrossRef] [PubMed]

124. Zhu, A.S.; Li, A.; Ratliff, T.S.; Melsom, M.; Garza, L.A. After Skin Wounding, Noncoding dsRNA Coordinates Prostaglandins and Wnts to Promote Regeneration. *J. Investig. Dermatol.* **2017**, *137*, 1562–1568. [CrossRef]

125. Kim, D.; Chen, R.; Sheu, M.; Kim, N.; Kim, S.; Islam, N.; Wier, E.M.; Wang, G.; Li, A.; Park, A.; et al. Noncoding dsRNA induces retinoic acid synthesis to stimulate hair follicle regeneration via TLR3. *Nat. Commun.* **2019**, *10*, 2811. [CrossRef]
126. Nelson, A.M.; Reddy, S.K.; Ratliff, T.S.; Hossain, M.Z.; Katseff, A.S.; Zhu, A.S.; Chang, E.; Resnik, S.R.; Page, C.; Kim, D.; et al. dsRNA Released by Tissue Damage Activates TLR3 to Drive Skin Regeneration. *Cell Stem Cell* **2015**, *17*, 139–151. [CrossRef] [PubMed]
127. Stojanovich, L.; Marisavljevich, D. Stress as a trigger of autoimmune disease. *Autoimmun. Rev.* **2008**, *7*, 209–213. [CrossRef] [PubMed]
128. Schur, P.H.; Stollar, B.D.; Steinberg, A.D.; Talal, N. Incidence of antibodies to double-stranded RNA in systemic lupus erythematosus and related diseases. *Arthritis Rheum.* **1971**, *14*, 342–347. [CrossRef]
129. Baechler, E.C.; Batliwalla, F.M.; Karypis, G.; Gaffney, P.M.; Ortmann, W.A.; Espe, K.J.; Shark, K.B.; Grande, W.J.; Hughes, K.M.; Kapur, V.; et al. Interferon-inducible gene expression signature in peripheral blood cells of patients with severe lupus. *Proc. Natl. Acad. Sci. USA* **2003**, *100*, 2610–2615. [CrossRef]
130. Bennett, L.; Palucka, A.K.; Arce, E.; Cantrell, V.; Borvak, J.; Banchereau, J.; Pascual, V. Interferon and granulopoiesis signatures in systemic lupus erythematosus blood. *J. Exp. Med.* **2003**, *197*, 711–723. [CrossRef]
131. Crow, M.K.; Kirou, K.A.; Wohlgemuth, J. Microarray analysis of interferon-regulated genes in SLE. *Autoimmunity* **2003**, *36*, 481–490. [CrossRef] [PubMed]
132. Fenando, A.; Firn, K.; Louissa, S.; Hussain, A. Case of anti-MDA-5 positive dermatomyositis with rapidly progressive interstitial lung disease. *BMJ Case Rep.* **2020**, *13*, e235493. [CrossRef]
133. Cufi, P.; Dragin, N.; Weiss, J.M.; Martinez-Martinez, P.; De Baets, M.H.; Roussin, R.; Fadel, E.; Berrih-Aknin, S.; Le Panse, R. Implication of double-stranded RNA signaling in the etiology of autoimmune myasthenia gravis. *Ann. Neurol.* **2013**, *73*, 281–293. [CrossRef]
134. Rácz, E.; Prens, E.P.; Kant, M.; Florencia, E.; Jaspers, N.G.; Laman, J.D.; de Ridder, D.; van der Fits, L. Narrowband ultraviolet B inhibits innate cytosolic double-stranded RNA receptors in psoriatic skin and keratinocytes. *Br. J. Dermatol.* **2011**, *164*, 838–847. [CrossRef]
135. Afshar, M.; Martinez, A.D.; Gallo, R.L.; Hata, T.R. Induction and exacerbation of psoriasis with Interferon-alpha therapy for hepatitis C: A review and analysis of 36 cases. *J. Eur. Acad. Dermatol. Venereol.* **2013**, *27*, 771–778. [CrossRef] [PubMed]
136. Shinohara, M.M.; Davis, C.; Olerud, J. Concurrent antiphospholipid syndrome and cutaneous [corrected] sarcoidosis due to interferon alfa and ribavirin treatment for hepatitis C. *J. Drugs Dermatol.* **2009**, *8*, 870–872.
137. Roth, S.H.; Danan-Gotthold, M.; Ben-Izhak, M.; Rechavi, G.; Cohen, C.J.; Louzoun, Y.; Levanon, E.Y. Increased RNA Editing May Provide a Source for Autoantigens in Systemic Lupus Erythematosus. *Cell Rep.* **2018**, *23*, 50–57. [CrossRef]
138. Vlachogiannis, N.I.; Gatsiou, A.; Silvestris, D.A.; Stamatelopoulos, K.; Tektonidou, M.G.; Gallo, A.; Sfikakis, P.P.; Stellos, K. Increased adenosine-to-inosine RNA editing in rheumatoid arthritis. *J. Autoimmun.* **2020**, *106*, 102329. [CrossRef]
139. Paz-Yaacov, N.; Bazak, L.; Buchumenski, I.; Porath, H.T.; Danan-Gotthold, M.; Knisbacher, B.A.; Eisenberg, E.; Levanon, E.Y. Elevated RNA Editing Activity Is a Major Contributor to Transcriptomic Diversity in Tumors. *Cell Rep.* **2015**, *13*, 267–276. [CrossRef]
140. Peng, X.; Xu, X.; Wang, Y.; Hawke, D.H.; Yu, S.; Han, L.; Zhou, Z.; Mojumdar, K.; Jeong, K.J.; Labrie, M.; et al. A-to-I RNA Editing Contributes to Proteomic Diversity in Cancer. *Cancer Cell* **2018**, *33*, 817–828.e7. [CrossRef]
141. Han, L.; Diao, L.; Yu, S.; Xu, X.; Li, J.; Zhang, R.; Yang, Y.; Werner, H.M.J.; Eterovic, A.K.; Yuan, Y.; et al. The Genomic Landscape and Clinical Relevance of A-to-I RNA Editing in Human Cancers. *Cancer Cell* **2015**, *28*, 515–528. [CrossRef]
142. Sakurai, M.; Shiromoto, Y.; Ota, H.; Song, C.; Kossenkov, A.V.; Wickramasinghe, J.; Showe, L.C.; Skordalakes, E.; Tang, H.Y.; Speicher, D.W.; et al. ADAR1 controls apoptosis of stressed cells by inhibiting Staufen1-mediated mRNA decay. *Nat. Struct. Mol. Biol.* **2017**, *24*, 534–543. [CrossRef]
143. Xu, L.D.; Öhman, M. ADAR1 Editing and its Role in Cancer. *Genes* **2018**, *10*, 12. [CrossRef] [PubMed]
144. Martinez, H.D.; Jasavala, R.J.; Hinkson, I.; Fitzgerald, L.D.; Trimmer, J.S.; Kung, H.J.; Wright, M.E. RNA editing of androgen receptor gene transcripts in prostate cancer cells. *J. Biol. Chem.* **2008**, *283*, 29938–29949. [CrossRef]
145. Chen, L.; Li, Y.; Lin, C.H.; Chan, T.H.; Chow, R.K.; Song, Y.; Liu, M.; Yuan, Y.F.; Fu, L.; Kong, K.L.; et al. Recoding RNA editing of AZIN1 predisposes to hepatocellular carcinoma. *Nat. Med.* **2013**, *19*, 209–216. [CrossRef]
146. Valencia, J.C.; Egbukichi, N.; Erwin-Cohen, R.A. Autoimmunity and Cancer, the Paradox Comorbidities Challenging Therapy in the Context of Preexisting Autoimmunity. *J. Interferon Cytokine Res.* **2019**, *39*, 72–84. [CrossRef]
147. Staege, M.S.; Emmer, A. Editorial: Endogenous Viral Elements-Links Between Autoimmunity and Cancer? *Front. Microbiol.* **2018**, *9*, 3171. [CrossRef]
148. Brütting, C.; Emmer, A.; Kornhuber, M.E.; Staege, M.S. Cooccurrences of Putative Endogenous Retrovirus-Associated Diseases. *BioMed Res. Int.* **2017**, *2017*, 7973165. [CrossRef] [PubMed]
149. Chen, M.; Hu, S.; Li, Y.; Jiang, T.T.; Jin, H.; Feng, L. Targeting nuclear acid-mediated immunity in cancer immune checkpoint inhibitor therapies. *Signal Transduct. Target. Ther.* **2020**, *5*, 270. [CrossRef]
150. Jin, B.; Cheng, L.F.; Wu, K.; Yu, X.H.; Yeo, A.E. Application of dsRNA in cancer immunotherapy: Current status and future trends. *Anti-Cancer Agents Med. Chem.* **2014**, *14*, 241–255. [CrossRef]

151. Tsai, H.C.; Li, H.; Van Neste, L.; Cai, Y.; Robert, C.; Rassool, F.V.; Shin, J.J.; Harbom, K.M.; Beaty, R.; Pappou, E.; et al. Transient low doses of DNA-demethylating agents exert durable antitumor effects on hematological and epithelial tumor cells. *Cancer Cell* **2012**, *21*, 430–446. [CrossRef]
152. Li, H.; Chiappinelli, K.B.; Guzzetta, A.A.; Easwaran, H.; Yen, R.W.; Vatapalli, R.; Topper, M.J.; Luo, J.; Connolly, R.M.; Azad, N.S.; et al. Immune regulation by low doses of the DNA methyltransferase inhibitor 5-azacitidine in common human epithelial cancers. *Oncotarget* **2014**, *5*, 587–598. [CrossRef]
153. Banik, D.; Moufarrij, S.; Villagra, A. Immunoepigenetics Combination Therapies: An Overview of the Role of HDACs in Cancer Immunotherapy. *Int. J. Mol. Sci.* **2019**, *20*, 2241. [CrossRef]
154. Covre, A.; Coral, S.; Nicolay, H.; Parisi, G.; Fazio, C.; Colizzi, F.; Fratta, E.; Di Giacomo, A.M.; Sigalotti, L.; Natali, P.G.; et al. Antitumor activity of epigenetic immunomodulation combined with CTLA-4 blockade in syngeneic mouse models. *Oncoimmunology* **2015**, *4*, e1019978. [CrossRef]
155. Ammi, R.; De Waele, J.; Willemen, Y.; Van Brussel, I.; Schrijvers, D.M.; Lion, E.; Smits, E.L. Poly(I:C) as cancer vaccine adjuvant: Knocking on the door of medical breakthroughs. *Pharmacol. Ther.* **2015**, *146*, 120–131. [CrossRef]
156. Cheever, M.A. Twelve immunotherapy drugs that could cure cancers. *Immunol. Rev.* **2008**, *222*, 357–368. [CrossRef]

non-coding **RNA**

Review

Epigenetic Regulation of Alternative Splicing: How LncRNAs Tailor the Message

Giuseppina Pisignano [1],* and Michael Ladomery [2],*

1   Department of Biology and Biochemistry, University of Bath, Claverton Down, Bath BA2 7AY, UK
2   Faculty of Health and Applied Sciences, University of the West of England, Coldharbour Lane, Frenchay, Bristol BS16 1QY, UK
*   Correspondence: gp529@bath.ac.uk (G.P.); michael.ladomery@uwe.ac.uk (M.L.)

Abstract: Alternative splicing is a highly fine-tuned regulated process and one of the main drivers of proteomic diversity across eukaryotes. The vast majority of human multi-exon genes is alternatively spliced in a cell type- and tissue-specific manner, and defects in alternative splicing can dramatically alter RNA and protein functions and lead to disease. The eukaryotic genome is also intensively transcribed into long and short non-coding RNAs which account for up to 90% of the entire transcriptome. Over the years, lncRNAs have received considerable attention as important players in the regulation of cellular processes including alternative splicing. In this review, we focus on recent discoveries that show how lncRNAs contribute significantly to the regulation of alternative splicing and explore how they are able to shape the expression of a diverse set of splice isoforms through several mechanisms. With the increasing number of lncRNAs being discovered and characterized, the contribution of lncRNAs to the regulation of alternative splicing is likely to grow significantly.

Keywords: long non-coding RNAs; alternative splicing; splicing factors; post-transcriptional regulation

check for **updates**

Citation: Pisignano, G.; Ladomery, M. Epigenetic Regulation of Alternative Splicing: How LncRNAs Tailor the Message. *Non-coding RNA* **2021**, 7, 21. https://doi.org/10.3390/ncrna7010021

Received: 21 January 2021
Accepted: 22 February 2021
Published: 11 March 2021

Publisher's Note: MDPI stays neutral with regard to jurisdictional claims in published maps and institutional affiliations.

1. Introduction

In the late 1970s, researchers were interested in gaining a better understanding of the mechanisms of adenoviral gene expression when they noticed something unusual, a long adenoviral transcript hybridized to the viral genome forming a three-stranded mRNA:DNA hybrid structure with an intervening DNA sequence that did not match the mature mRNA [1]. It then became apparent that these intervening sequences were in fact introns and that the primary transcript is made by a succession of exonic and intronic sequences. In what is now thought to be a mainly co-transcriptional process, introns are 'spliced' out from the mRNA precursor (pre-mRNA) and the exons joined together through two transesterification reactions catalysed by a complex molecular machinery consisting of five small nuclear ribonucleoproteins (snRNPs), called the spliceosome [2]. Over the past decades it has become clear that pre-mRNA splicing is a widespread phenomenon across eukaryotes and that a single gene can generate multiple transcripts often encoding different proteins by a process known as alternative splicing (AS) [3]. Many types of AS are possible, including "cassette exons", "alternative 5′ and 3′ splice sites", "alternative first exons" (through different promoters), "alternative last exons" (through different polyadenylation sites), "mutually exclusive exons" and "retained introns" [4]. Fundamental to the process of AS is the definition of the precise location of 5′ (donor) and 3′ (acceptor) splice sites and the assembly of the spliceosome complex. The first relies on splicing factors (SFs), a category of RNA-binding proteins (RBPs) expressed in a tissue and stage-specific way that recognize regulatory elements within exons and introns. Importantly, SF activity is in turn modified by splicing factor kinases and phosphatases activated through cell signaling mechanisms.

AS greatly enhances proteome diversity and represents an essential aspect of gene expression in development, normal physiology and disease across eukaryotes [5], from single-celled yeast to humans [6]. The advent of high-throughput sequencing technologies

69

revealed that ~92–94% of human multi-exon genes are alternatively spliced [7], increasing the interest in understanding the mechanisms underpinning its regulation. With the discovery and growing importance of non-coding RNAs, the nature of AS regulation has become more complex. Both short (<200 nt) and long (>200 nt) non-coding RNAs can contribute to the regulation of AS in many different ways; either indirectly by regulating the activity of splice factors; or directly, by interacting with pre-mRNAs. Long non-coding RNAs (lncRNAs) are particularly well suited to these roles due to their demonstrated capacity to act as regulatory molecules that modulate gene expression at every level. Either alone, or in association with partner proteins, these long RNA polymerase II transcripts have been shown to take part in a wide range of developmental processes and disease in complex organisms [8–10]. Here, we review the current knowledge of the multiple mechanisms through which lncRNAs contribute to the regulation of AS (Figure 1 and Table 1).

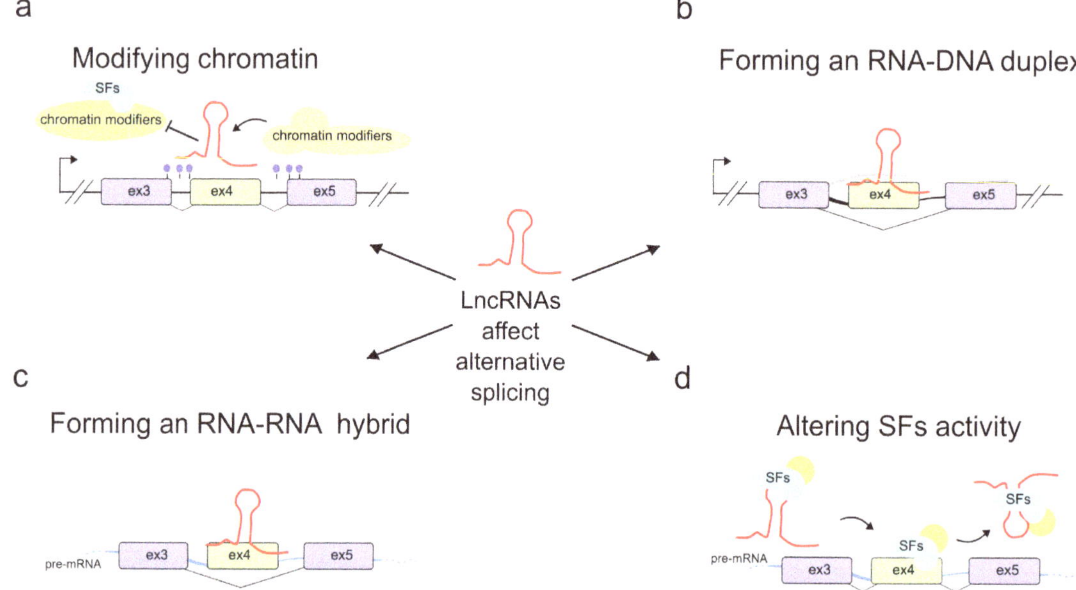

Figure 1. Regulation of pre-mRNA splicing by lncRNAs. LncRNAs (red) are able to control pre-mRNA splicing by (**a**) modifying chromatin accessibility through recruiting or impeding access to chromatin modifying complexes at the transcribed genomic locus. In some cases, this might result in more drastic long-range structural changes; (**b**) interacting with the transcribed genomic locus through an RNA-DNA hybrid; (**c**) hybridizing with the pre-mRNA molecule (light blue); (**d**) promoting SF recruitment or by sequestering SFs into specific subnuclear compartments, thereby interfering with SF activities.

2. LncRNAs Regulate Alternative Splicing through Chromatin Modification

The eukaryotic genome is tightly packaged into chromatin fibers consisting of DNA wrapped around nucleosomes made of histone proteins. Post-translational modifications (PTMs), such as methylation, acetylation, phosphorylation, and ubiquitination, occur on the histone tails that are functionally linked to the epigenetic regulation of gene expression. By defining the accessibility to chromatin, histone modifications demarcate amenable or silenced chromatin domains which ultimately reflect the activity of gene transcription. An intimate relationship exists between lncRNAs and chromatin conformation [11,12]. LncRNAs regulate chromatin modifications by recruiting or directly interacting with histone-modifying complexes or enzymes at specific chromosomal loci; these in turn modulate gene transcription [13–18]. Histone modification signatures can also influence AS through a chromatin-reading protein which acts as an adaptor linker between the RNA polymerase II (RNAPII) and the pre-mRNA splicing machinery [19]. Several studies have also demonstrated that the local chromatin context influences the RNAPII elongation rate which in turn affects AS [20–22].

A possible lncRNA-mediated crosstalk between histone modifications and the pre-mRNA splicing machinery has also been proposed [23]. Cell type–specific splicing of the gene encoding the fibroblast growth factor receptor 2 (*FGFR2*) is now known to rely on the methylation state of the *FGFR2* locus. In mesenchymal stem cells, *FGFR2* is enriched in di- (me2) and tri-methylated (me3) histone H3K36, which inhibits the inclusion of the alternatively spliced exon IIIb. *FGFR2* is, in contrast, devoid of H3K36 methylation in epithelial cells. The cell-specific switch in splicing is made possible by an evolutionarily-conserved nuclear antisense lncRNA (*asFGFR2*), transcribed within the human *FGFR2* locus and exclusively expressed in epithelial cells. By recruiting Polycomb-group proteins and the histone lysine-specific demethylase 2a (KDM2a) to the locus, *asFGFR2* ensures the deposition of H3K27me3 and a decrease in H3K36me2/3. This impairs both the binding of the chromatin-binding protein MRG15 for H3K36me2/3 [24] and the recruitment, via protein-protein interactions, of the negative splicing regulator PTBP1 to exon IIIb [19]. Through this combined action, the chromatin-splicing adaptor complex MRG15–PTBP1 can no longer inhibit the inclusion of exon IIIb favoring the epithelial-specific AS of *FGFR2* [23] (Figure 2a).

Chromatin structure is itself likely to play an important role in modulating the effects of transcription on AS [25]. In particular, the tri-dimensional chromatin organizer CCCTC-binding factor (CTCF) has been shown to bind target DNA sites located within an alternative exon creating a roadblock to transcriptional elongation that favors exon inclusion into mature mRNA [26]. Several lncRNAs appear to control important aspects of chromatin organization including chromatin looping, either remaining tethered to the site of transcription or moving over distant loci [27,28]. Interestingly, lncRNAs can efficiently remove structural roadblocks in chromatin by CTCF eviction [29,30]. A fascinating lncRNA-mediated mechanism modulates the diversity of transcripts at the complex Protocadherin (*Pcdh*) α gene cluster [31]. Each *Pcdh*α gene of the cluster functions as a 'variable' first exon (out of 13) that is individually spliced to a downstream constant region to form distinct transcripts, differentially expressed in individual neurons and important for neuronal self-identity. The stochastic expression of 13 alternate exons is driven by their own promoter, each of which is equally likely to be activated by a long-range DNA loop interaction between a selected *Pcdh*α promoter and a downstream enhancer, called "hypersensitivity site 5-1" (HS5-1) [32–35]. The *Pcdh*α gene choice involves the selective activation of a specific antisense lncRNA located at the promoter of the first exon of each *Pcdh*α alternate gene. By promoting DNA demethylation, the antisense transcript recruits CTCF at sites proximal to the relative promoter and favors the promoter-enhancer interaction which ultimately triggers the sense transcription of the corresponding selected first exon [31] (Figure 2b). Further studies will be required to understand if other clustered genes share a similar mechanism. It will also be of interest to determine how frequently mechanisms involving lncRNAs, among the thousands transcribed, mediate chromatin structure changes that result in AS.

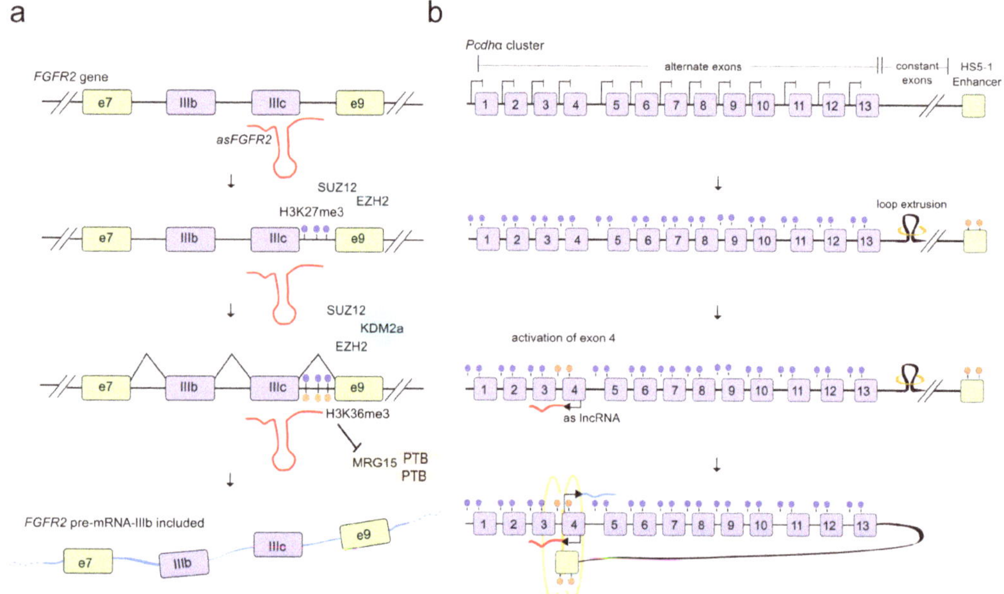

Figure 2. LncRNAs regulate alternative splicing through chromatin modification. (**a**) In epithelial cells, the antisense lncRNA *asFGFR2* (red), recruits the Polycomb-group proteins EZH2 and SUZ12 to the *FGFR2* gene locus and allows H3K27me3 deposition (blue lollipop) and a decrease in methylation of H3K36me3 (orange lollipop) by the recruitment of the H3K36 demethylase KDM2a. As a result, the chromatin-splicing adaptor complex MRG15–PTB1 can no longer bind to exon IIIb, which is then included in the *FGFR2* transcript (light blue). (**b**) The activation of a specific antisense lncRNA (as lncRNA; red) at the *Pcdhα* promoter of one (out of 13) alternate first exon promotes proximal DNA demethylation (orange lollipop) and CTCF (turquoise) recruitment and favors the interaction between the selected promoter and a distant HS5-1 enhancer by a long-range DNA loop. This ultimately triggers sense transcription (light blue) of the corresponding selected *Pcdhα* first-exon which is individually spliced to a downstream constant region to form a distinct transcript.

3. LncRNAs Regulate Pre-mRNA Splicing through RNA-DNA Interactions

LncRNAs can tether DNA forming an RNA-dsDNA triplex by targeting specific DNA sequences and inserting themselves as a third strand into the major groove of the DNA duplex [30,36]. These are known as R-loops; three-stranded nucleic acid structures, composed of RNA–DNA hybrids, frequently formed during transcription. Aberrant R-loops are generally associated with DNA damage, transcription elongation defects, hyper-recombination and genome instability [37].

Recent lines of evidence indicate a potential role for R-loops in alternative pre-mRNA splicing. A class of lncRNAs, the so-called circular RNAs (circRNAs), have recently been characterized [38–40]. These abundant, conserved transcripts originate from a non-canonical AS process (back-splicing) leading to the formation of head-to-tail splice junctions, joined together to form circular transcripts. Recent studies suggest that they are clearly involved in multiple aspects of normal physiology, development and disease [41]. Since most circRNAs are derived from the middle exons of protein-coding genes [42], their biogenesis can itself affect splicing of their precursor transcripts and lead to altered gene expression outcomes [43]. For example, in *Arabidopsis thaliana*, the circular RNA derived from exon 6 of the *SEPALLATA3* (*SEP3*) gene increases the abundance of the cognate exon-skipped alternative splicing variant (SEP3.3 isoform) which in turn drives floral homeotic phenotypes [44]. This is made possible because *SEP3* exon 6 circRNA tethers to its cognate DNA locus through an R-loop promoting transcriptional pausing, which coincides with

SF recruitment and AS [45–47] (Figure 3). Whether or not other lncRNAs are involved in similar processes in plants or other organisms remains to be investigated.

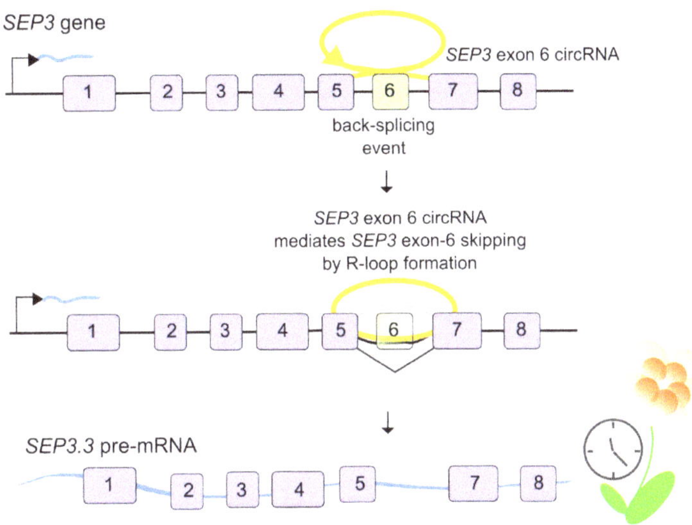

Figure 3. LncRNAs regulate pre-mRNA splicing through an RNA-DNA interaction. In *Arabidopsis thaliana*, when the *SEP3* gene is transcribed, exon 6 can be back-spliced into a circular RNA (*SEP3* exon 6 circRNA, yellow) which interacts directly with its parental genomic locus. By forming RNA–DNA hybrids (R-loops), *SEP3* exon 6 circRNA favors exon-6 skipping of its linear cognate and promotes the *SEP3.3* mRNA (light blue) isoform accumulation which in turn affects flowering time.

4. LncRNAs Regulate Pre-mRNA Splicing through RNA-RNA Interactions

Over the past decades, antisense transcripts have been characterized as being widespread throughout the genomes of the vast majority of organisms [48–50]. It is estimated that more than 30% of annotated human transcripts have at least one cognate antisense transcript [50]. Although generally low in abundance and over 10-fold less expressed than their counterpart sense transcripts [50], antisense RNAs have been widely implicated at almost all stages of gene expression, from transcription and translation to RNA degradation [51]. A considerable proportion of genes that express multiple spliced isoforms has been associated with antisense transcription, suggesting that antisense-mediated processes could be a common mechanism to regulate AS [52]. Therapeutic strategies based on antisense-mediated exon skipping and aimed at changing the levels of alternatively spliced isoforms or at disrupting open reading frames have been also developed [20]. For example, an antisense oligoribonucleotide (AON) approach efficiently restores the open reading frame of the *DMD* gene and generates functional dystrophin by inducing exon skipping [53].

Identified in multiple eukaryotes, Natural Antisense Transcripts (NATs) are a class of long non-coding RNA molecules, transcribed from both coding and non-coding genes on the opposite strand of protein-coding ones [54]. Regardless of their genomic origin, NATs can hybridize with pre-mRNAs and form RNA-RNA duplexes. In some cases, a double function is also possible, and NATs can encode for proteins on one hand, while at the same time working as non-coding molecules modulating the splicing of a neighbouring gene's transcript [55]. At the oncogene *NMYC* locus, for example, the cis-antisense gene *NCYM* located at the first *NMYC* intron has recently been shown to encode a protein that regulates the genesis and progression of human neuroblastomas that is associated with unfavorable prognosis [56]. However, previous studies have classified the corresponding transcript as a NAT able to modulate, via sense/antisense RNA-RNA duplexes, the processing of *NMYC* pre-mRNA resulting in a population of *NMYC* mRNA splice isoforms that retain the first intron [57] (Figure 4a).

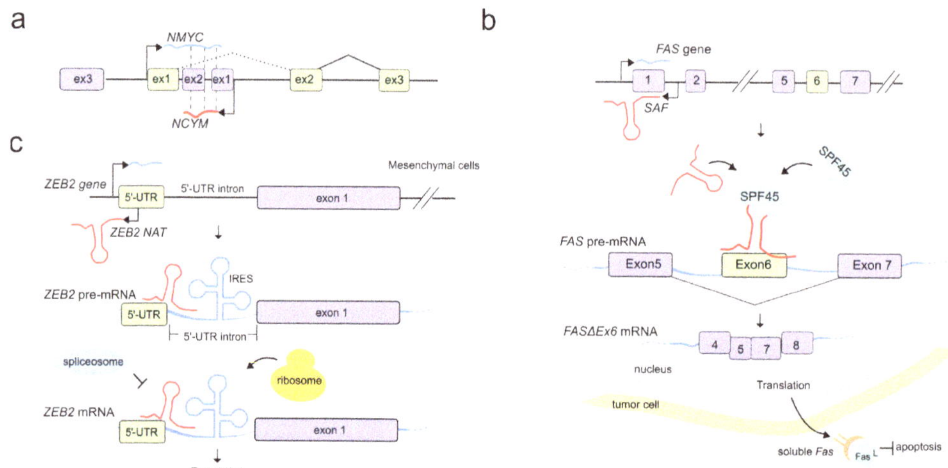

Figure 4. LncRNAs regulate pre-mRNA splicing through an RNA-RNA interaction. (**a**) NAT (red) at the *NCYM* gene modulates splicing of the *NMYC* mRNA (light blue) forming a sense-antisense RNA-RNA duplex which results in an intron-retained *NMYC* mRNA isoform population. (**b**) In tumor cells the natural antisense *SAF* (red) is transcribed from the first intron of *FAS* gene and interacts with both *FAS* pre-mRNA (light blue) at 5–6 and 6–7 exon junctions and the human SFP45 to facilitate the AS and exclusion of exon 6. The accumulation of the exon 6-skipped alternatively spliced variant of *FAS* pre-mRNA (*FASΔEx6* mRNA) leads to the production of a soluble Fas (sFas) protein that binds FasL and makes tumor cells resistant to FasL-induced apoptosis. (**c**) After EMT, Snail1 transcription factor induces the co-transcription of *ZEB2* NAT (red) in mesenchymal cells. *ZEB2* NAT hybridises with a region of the *ZEB2* pre-mRNA (light blue) encompassing the 5' splice site of a 3 kb-long 5'-UTR intron. This RNA-RNA duplex prevents both the binding of the spliceosome and the subsequent removal of the 5'-UTR intron. The resulting mRNA contains the full isoform of the 5'-UTR, including an internal ribosome entry site (IRES) proximal to the *ZEB2* AUG, which favors translation. In absence of *ZEB2* NAT (epithelial cells) instead the removal of the 5'-UTR intron results in an mRNA containing a sequence that inhibits scanning by the ribosomes and therefore prevents translation of ZEB2 protein (not shown).

Overlapping antisense transcription has been shown to modulate AS at the thyroid hormone receptor alpha (*THRA*) locus [58]. This locus encodes two overlapping mRNAs, *α1* and *α2* corresponding to TR-α1 and its splice variant TR-α2, which differ at the 3'-end because of the presence of a third overlapping mRNA, *NR1D1* (also known as *Rev-erbAα*). The latter is transcribed in the opposite direction at the 3'-end of *α2*, but not *α1* mRNA. It has been suggested that the relative abundance of the *NR1D1* RNA prevents the splicing of *α2*, likely through RNA-RNA base pairing, thereby favoring the formation of the non-overlapping *α1*. Consistent with this hypothesis, other studies noted a positive correlation between the *α2/α1* isoform ratio and the level of *NR1D1* mRNA in cells [58,59]. Therefore, relatively modest changes in splice site selection of *α1* and *α2* caused by naturally occurring antisense RNAs might cause major changes in cellular thyroid hormone-responsiveness with a broader physiological impact.

NATs that drive AS during programmed cell death (apoptosis) have also been reported. The FAS gene encodes for a receptor protein which usually binds its Fas ligand (FasL) and triggers the apoptotic process. At the *FAS* locus, the lncRNA SAF is transcribed in reverse orientation and from the opposite strand of the first intron of FAS. In tumor cells, *SAF* transcription promotes the formation of the exon 6-skipped spliced variant of *FAS* pre-mRNA (*FASΔEx6*) by interacting with both the *FAS* pre mRNA, predominantly at exon 5/6 and exon 6/7 junctions, and the human splicing factor 45 (SPF45). The resulting splicing variant lacks the transmembrane domain which gives more solubility to the isoform (sFas) and protects tumor cells against FasL-induced apoptosis [60] (Figure 4b).

Reverse transcription can affect pre-mRNA splicing by masking specific splice sites and preventing their processing. A remarkable example of how NATs can affect the splicing and in turn increase mRNA translation efficiency is the human *ZEB2* gene (zinc-finger E-box-binding homeobox 2). Boosting the translation of ZEB2 repressor is one of the ways by which E-cadherin repression is initiated by the transcriptional factor Snail1 during epithelial-mesenchymal transition (EMT). Normally, the *ZEB2* 5′-UTR contains a structural intronic motif that works as an internal ribosome entry site (IRES) which is spliced out to hinder ZEB2 translation. However, once EMT is triggered, Snail1 induces the transcription of a *ZEB2* NAT which is transcribed from the opposite strand of the *ZEB2* locus, covering the 5′ splice site of the *ZEB2* 5′-UTR. *ZEB2* NAT prevents the recognition of the splice sites by the spliceosome by RNA-RNA duplex interaction with *ZEB2* mRNA and promotes the subsequent inclusion of the intron present in the *ZEB2* 5′ UTR, thereby promoting ZEB2 translation [61] (Figure 4c).

Masking canonical splicing sites has also been linked with the most common form of dementia, Alzheimer's disease (AD). Sortilin-related receptor 1 (SORL1) expression is generally reduced in brain tissues from individuals with AD [62] suggesting a potential role in AD pathogenesis [63,64]. The importance of this receptor is underlined by the recent demonstration that SORL1 downregulation promotes amyloid precursor protein (APP) secretion and subsequently an increase of neurotoxic β-amyloid peptide (Aβ) [65,66]. A 300 nt antisense non-coding RNA transcribed by RNA polymerase III, called *51A*, maps to the intron 1 of the *SORL1* gene and, by pairing with the *SORL1* pre-mRNA, drives a splicing shift of SORL1 from the canonical full-length protein variant A to an alternatively spliced shorter protein form (variant B). This process results in the decreased synthesis of SORL1 variant A and is associated with impairing processing of APP, leading to increase of Aβ formation [67].

Antisense transcripts that cause a shift in isoform balance occur also at the *GPR51* locus, hosting the antisense lncRNA *17A* on its intron 3. LncRNA *17A* expression is induced by inflammatory molecules and leads to the production of the GADAD R2 protein isoform devoid of transduction activity and the concomitant down-regulation of the canonical full-length GABAB R2 variant, which impairs GABAB signaling. The change in the ratio of the two isoforms was found to be linked to AD. Increased levels of *17 A* expression have been found in patient brains, suggesting a role of this lncRNA in *GPR51* splicing regulation to preserve cerebral function [68].

Alternative isoform expression can also be controlled by antisense transcription via transcription attenuation (transcription RNAPII pausing and/or premature termination). A recent study shows that during specific differentiation stages in mouse embryonic stem cells (mESCs), the expression of two novel antisense enhancer-associated RNAs, *Zmynd8as* and *Brd1as*, is associated with shorter overlapping sense transcript isoforms with alternative termination sites [69], a phenomenon similarly found affecting the length of sense mRNAs of genes in a single operon in some bacteria [70]. Whereas the mechanism through which isoform specificity is achieved via enhancer-associated antisense RNAs has not been totally elucidated, this example enhances the corollary of antisense-mediated splicing mechanisms. A similar transcription attenuation mechanism mediating splicing is likely to occur at other genomic loci occupied by overlapping coding and non-coding genes [52].

5. LncRNAs Regulate Pre-mRNA Splicing by Modulating the Activity of Splicing Factors

As well as modifying AS by altering the chromatin landscape, through transcription, or through direct nucleic acid interactions, lncRNAs also interact in a dynamic network with many SFs and their pre-mRNA target sequences to modulate transcriptome reprogramming in eukaryotes.

LncRNAs that are notoriously associated with pre-mRNA splicing are the nuclear *MALAT1/NEAT2* and *NEAT1*, both known to regulate the localization and phosphorylation status of SFs, and differentially expressed in a wide range of tissues in human and mouse. They are localized to specific subnuclear domains mainly in the nuclear speckle periphery, also known as paraspeckles (*NEAT1*); while *MALAT1/NEAT2* is part of the polyadenylated component of nuclear speckles [71].

MALAT1/NEAT2 regulates splicing by modulating the activity of the conserved family of serine/arginine (SR) splicing factors by modifying their localization and phosphorylation [72] through shuttling between speckles and the sites of transcription, where splicing occurs [73]. In human cells, *MALAT1/NEAT2* knockdown enhances the phosphorylated pool of SR proteins, displaying a more homogeneous nuclear distribution resulting in the mislocalization of speckle components and altered patterns of AS of pre-mRNAs [74–76]. *MALAT1/NEAT2* binds to the SRSF1 splice factor through its RRM domain [77,78]. A correct phosphorylation/dephosphorylation cycle of SR proteins is fundamental to ensure the proper nucleocytoplasmic transport of mRNA–protein complexes (mRNPs). When SRSF1 is phosphorylated, it accumulates in nuclear speckles; while its dephosphorylation favors the interaction with mRNAs, transport and accumulation in the cytoplasm [79,80]. Although the exact mechanisms through which *MALAT1/NEAT2*-interacting with SRSF1 modulates the phosphorylated/dephosphorylated ratio of SR proteins remains unclear, it might occur through interaction with PP1/2A phosphatases or with the SRPK1 splice factor kinase [81–83] or alternatively, by the direct interaction with *MALAT1/NEAT2* [73] (Figure 5a). Beyond AS, controlled levels of phosphorylated SR proteins are also likely to regulate other SR-dependent post-transcriptional regulatory events such as RNA export, nonsense mediated decay, and translation [77,81]. Interestingly, additional studies have also shown that *MALAT1/NEAT2* can hybridize with many nascent pre-mRNAs at active gene loci and participate in pre-mRNA splicing of such actively transcribed genes by recruiting SFs to the pre-mRNAs [84]. Furthermore, according to the psoralen analysis of RNA interactions and structures (PARIS) [85] and to the more recent developed RIC-seq application [86], multiple interaction sites exist between *MALAT1* and the spliceosomal RNA, U1snRNA, raising the possibility that *MALAT1/NEAT2* influences RNA processing through the recruitment or modification of other proteins localized to these sites.

MALAT1/NEAT2 is abundantly expressed and widely associated with a variety of cancers. In hepatocellular carcinoma, *MALAT1/NEAT2* acts as a proto-oncogene through Wnt pathway activation and transcriptional induction of SRSF1. The latter leads to the over accumulation of its active form in the cell nucleus and the modulation of SRSF1 splicing targets, including the anti-apoptotic AS isoforms of S6K1 [87]. In colorectal cancer, instead, *MALAT1/NEAT2* triggers tumor growth and metastasis by binding to the splicing factor SFPQ causing the subsequent disruption of the splicing regulator complex SFPQ-PTBP2 and the release of the oncogene PTBP2 [88].

During adipocyte differentiation, the 4 kb lncRNA *NEAT1* tethers the SR protein SRp40 (now known as SRSF5) and retains it in paranuclear bodies to fine-tune the relative abundance of mRNA isoforms of the major transcription factor driving adipogenesis, *PPARγ*. It has been observed that the *NEAT1*-SRp40 association enhances SRp40 phosphorylation by CLK1 splicing factor kinase activity [89]. Conversely, depletion of *NEAT1* upon drug or siRNAs treatment, causes a decrease of both *PPARγ* isoforms (*PPARγ1* and especially *PPARγ2*) and SRp40 phosphorylation impairment, respectively. Furthermore, while SRp40 depletion resulted in deregulation of both *PPARγ* isoforms and, predominantly of *PPARγ2* mRNA levels, its overexpression increased exclusively *PPARγ2*. Therefore, fluxes in *NEAT1* levels during adipogenesis seem to modulate AS events likely by controlling the availability of phosphorylated SRp40 thereby affecting *PPARγ* splicing [90] (Figure 5b).

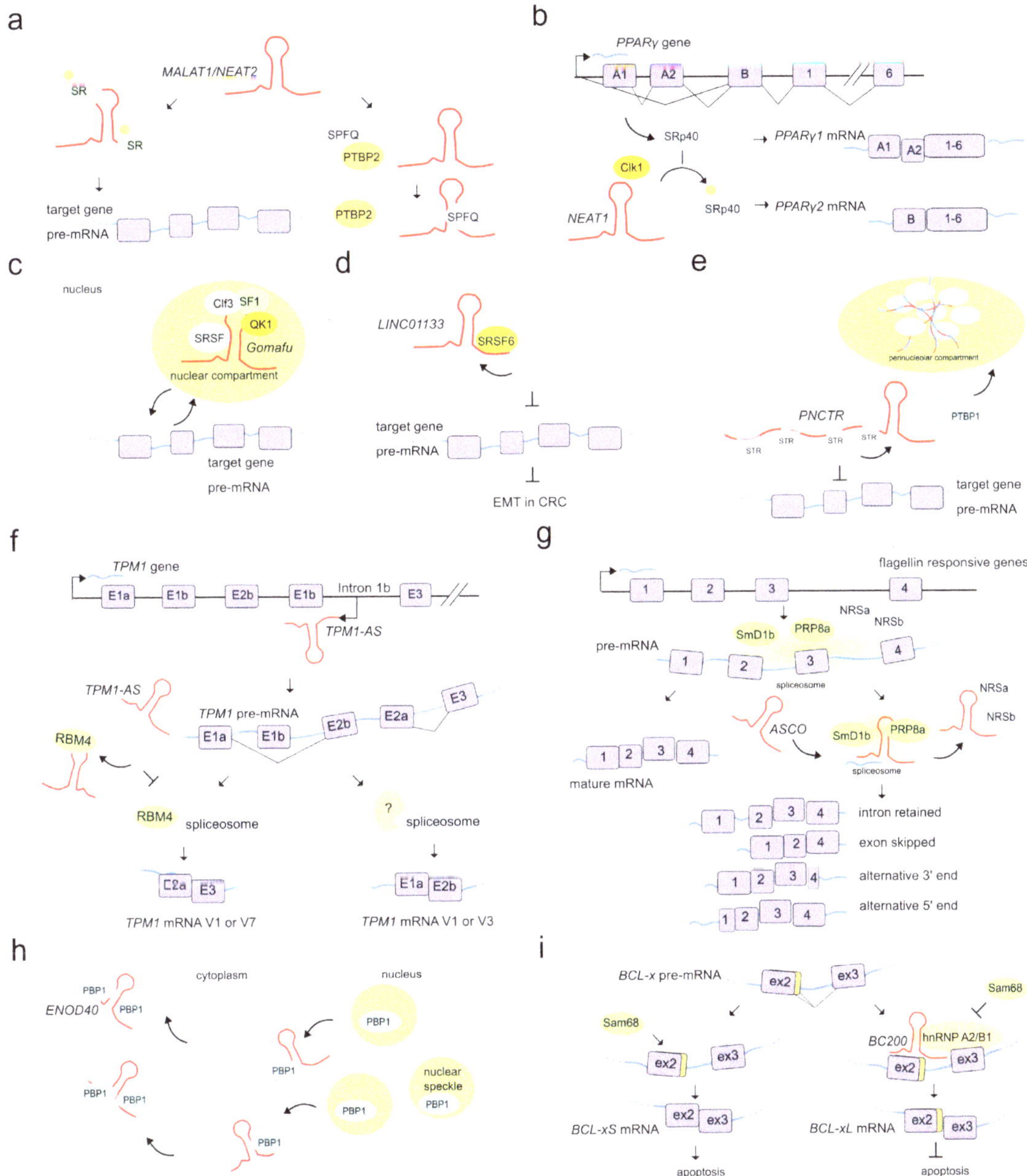

Figure 5. LncRNAs regulate pre-mRNA splicing by recruiting or sequestering splicing factors into subnuclear compartments.
(**a**) Left, *MALAT1/ NEAT2* (red) is responsible of phosphorylated/dephosphorylated SFs shuttle from nuclear speckles
to target mRNAs and cytoplasm. Right, *MALAT1/ NEAT2* in colon cancer. The binding of SFPQ with *MALAT1/ NEAT2*
causes the disruption of the splicing regulator complex SFPQ-PTBP2 and the release of PTBP2. (**b**) During adipogenesis,
the lncRNA *NEAT1* (red) interacts with the CLK1 splicing factor kinase (orange) and regulates *PPARγ* gene splicing by
modulating SRp40 (light blue, also known as SRSF5) phosphorylation status (light orange). When Srp40 is phosphorylated,
the *PPARγ* pre-mRNA is mainly processed into the *PPARγ2* mRNA, whereas when dephosphoryled, Srp40 promotes the
accumulation of the PPARγ1 isoform. (**c**) *Gomafu* (red) sequesters multiple splicing factors (e.g., QKI, SRSF1, SF1, Clf3) in

nuclear compartments and after specific stimuli/conditions it releases them in the nucleus to then direct the alternative splicing of pre-mRNA target genes (light blue) such as the schizophrenia-associated genes. (**d**) The lncRNA *LINC01133* (red), by sequestering the splicing factor SRSF6, impairs the alternative splicing events on target pre-mRNA genes which ultimately lead to the inhibition of EMT and metastasis in colorectal cancer (CRC). (**e**) *PNCTR* (red), contains hundreds of short tandem repeats (STR) to bind and sequester a substantial fraction of PTBP1 in the perinucleolar compartment. (**f**) Sense and antisense *TPM1* gene cotrascription results in both *TPM1* pre-mRNA (light blue) and lncRNA *TPM1-AS* (red). The latter is then able to sequester RBM4 protein, forcing the splicing of *TPM1* pre-mRNA (likely in cooperation with other protein partners) toward RBM4-deprived specific isoforms (V1 or V3). (**g**) LncRNA *ASCO* (red) associates with the two core components of the spliceosome SmD1b and PRP8a (green) and concomitantly sequesters NSRa and b proteins (light blue). By this mechanism *ASCO* enhances transcriptome diversity in response to flagellin, resulting in a variety pool of isoforms. (**h**) *ENOD40* is recognized by MtRBP1 (here RBP1 for simplicity) and is responsible of its nucleocytoplasmic trafficking and accumulation into cytoplasmic granules, likely modulating RBP1-dependent splicing. (**i**) Left, *BCL-x* pre-mRNA interacts with Sam68 that promotes pre-mRNA splicing in the apoptotic isoform *BCL-xS*. Right, the presence of *BC200* lncRNA and the recruitment of the hnRNP A2/B1 splicing factor interferes with the association of Sam68 and promote *BCL-x* splicing into the anti-apoptotic *BCL-xL*.

Another lncRNA abundantly localized to nuclear bodies is the lncRNA *Gomafu/RNCR2/MIAT* which is expressed in a distinct set of neurons in the mouse retina [91,92] and implicated in retinal cell specification [93,94] brain development [95] and post-mitotic neuronal function [92,96]. *Gomafu* was found to interact directly with the splicing factors QKI and SRSF1 and its dysregulation leads to aberrant AS patterns that resemble those observed in schizophrenia-associated genes (*DISC1* and *ERBB4*) [97]. In addition, *Gomafu* harbors a conserved tandem sequence of UACUAAC motifs that binds the splicing factor SF1, an early stage player of spliceosome assembly [98]. Furthermore, the splicing factor Clf3 was found to interact specifically with *Gomafu* in RNA–protein complexes containing the splicing factors SF1 and localize in specific nuclear bodies named CS bodies in the neuroblastoma cell line Neuro2A [99,100]. It has been proposed that *Gomafu* regulates splicing efficiency by changing the local concentration of SFs by sequestering them to separate regions of the nucleus [98] (Figure 5c).

An additional example of how lncRNAs may hijack SFs to fine-tune AS is the lncRNA *LINC01133*. This lncRNA binds the AS factor SRSF6, which induces EMT in colorectal cancer. By sequestering SRSF6 from other mRNA substrates, *LINC01133* modulates SRSF6 activity and reshapes the population of AS isoforms of SRSF6 mRNA targets which finally leads to the inhibition of EMT and metastasis [101] (Figure 5d). Similarly, the lncRNA *PNCTR*, over-expressed in a variety of cancer cells, contains hundreds of short tandem repeats to bind and sequester a consistent fraction of PTBP1 in the perinucleolar compartment [102]. This prevents PTBP1 from influencing splicing and therefore PTBP1-dependent pro-apoptotic events [103–105] (Figure 5e).

LncRNAs that act as sponge molecules can extensively rewire post-transcriptional gene regulatory networks by uncoupling the protein–RNA interaction landscape in a cell-type-specific manner. A recent study showed that the loss of 39 lncRNAs causes many thousands of skipped exons and retained intron splicing events affecting a total of 759 human genes at the post-transcriptional level. Interestingly, the alternatively spliced events were found associated with RBPs binding in proximal intron–exon junctions in a cell-type-specific manner [106]. Similarly, the natural antisense *TPM1-AS*, reverse-transcribed from the fourth intronic region of the tropomyosin I gene (*TPM1*), regulates *TPM1* alternative splicing through interaction with RNA-binding motif protein 4 (RBM4). The interaction prevents the binding of RBM4 to *TPM1* pre-mRNA and inhibits *TPM1* exon 2a inclusion (Figure 5f) [107]. Plant lncRNAs are also able to modulate AS by hijacking RBPs from their targets. In *A. thaliana*, an important number of intron retention events and a differential $5'$ or $3'$-end have been observed in a subset of genes in the plant-specific AS regulators (NSRa and NSRb) mutant compared to wild type plants [108]. In vitro experiments suggested that the lncRNA *ASCO* competes with other mRNA-target for its binding to these

NSR regulators [109]. More recently, researchers analyzed the genome-wide effect of the knock-down and overexpression of *ASCO* and found a large number of deregulated and differentially spliced genes related to flagellin responses and biotic stress [110]. During this splicing process, *ASCO* interacts with multiple SFs including the highly conserved core spliceosome component PRP8a and another spliceosome component, SmD1b (Figure 5g). The NSR's closest homolog in the model legume *Medicago truncatula*, MtRBP1/MtNSR1, has been characterized as a protein partner of the highly conserved and structured lncRNA *ENOD40*, which participates in root symbiotic nodule development [111]. *ENOD40* appears to re-localize MtRBP1 from nuclear speckles into cytoplasmic granules during nodulation thereby modulating MtRBP1-dependent splicing events [112] (Figure 5h).

SF-associated lncRNAs might also influence a specific splicing outcome depending on a given cellular context. For example, the prostate-specific lncRNA *PCGEM1* can mutually bind the splicing factors hnRNP A1 (silencer) and U2AF65 (enhancer) with opposite effects. While its interaction with hnRNP A1 suppresses the expression of androgen receptor (AR) splice variants such as AR3 by exon skipping, the interaction of *PCGEM1* with U2AF65 promotes AR3 by exonization and favors castration resistance [113]. In the brain, the cytoplasmic 200 long non-coding RNA *BC200* (*BCYRN1*) prevents apoptosis by modulating AS of a member of the Bcl-2 family proteins, the *BCL-x* gene [114]. AS of *BCL-x* leads to opposite effects on apoptosis when processed in either *BCL-xL* (anti-apoptotic) or *BCL-xS* (pro-apoptotic) [115]. Whereas *BC200* overexpression promotes *BCL-xL*, its depletion induces *BCL-xS* formation. A 17-nucleotide complementary sequence to *BCL-x* pre-mRNA in *BC200* appears to facilitate its binding to the pre-mRNA and promotes the recruitment of the hnRNP A2/B1 splicing factor. HnRNP A2/B1 binding interferes with association of *BCL-x* pre-mRNA with the *BCL-xS*-promoting factor Sam68 [116], leading to a blockade of Bcl-xS expression and anti-apoptotic conditions [117] (Figure 5i). Another example of a cellular context that causes isoform switching through lncRNAs is that of fibroblast growth factor receptors. FGF-2-sensitive cells arise following *lnc-Spry1* depletion. This lncRNA acts as an early mediator of TGF-β signaling-induced EMT and regulates the expression of TGF-β-regulated gene targets. However, *lnc-Spry1* has also been found to interact with the U2AF65 pyrimidine-tract binding splicing factor suggesting a dual role in affecting both transcriptional and post-transcriptional gene regulation in epithelial cells promoting a mesenchymal-like phenotype [118]. Recently, a link between stress-induced lncRNAs and AS has also been shown. The lncRNA *LASTR*, elevated in hypoxic breast cancer, is upregulated through the stress-induced JNK/c-JUN pathway. It interacts with SART3, a U4/U6 snRNP recycling factor, and promotes splicing efficiency. Depletion of *LASTR* leads to increased intron retention, with the resulting downregulation of essential genes to the detriment of cancer cells [119].

Ribosomal and RNA splicing complexes components, including YBX1, PCBP1, PCBP2, RPS6 and RPL7, have been shown to bind *LINC-HELLP*, a lncRNA implicated in the pregnancy-specific *HELLP* syndrome, through a splicing-mediated mechanism that is largely unknown. *HELLP* patient mutations within *LINC-HELLP*, alter the binding with these proteins depending to their location and negatively affect trophoblast differentiation. While mutations occurring from the 5′-end up to the middle of the *LINC-HELLP* are likely to cause loss of partner protein interactions, those at the far 3′-end increase their binding [120]. Among a cohort of breast cancer-associated and oestrogen-regulated lncRNAs, *DSCAM-AS1* has been recently found to be associated with tumor progression and tamoxifen resistance [121]. Researchers found over 2085 splicing events regulated by *DSCAM-AS1*, including alternative polyadenylation sites, 3′ UTR shortening and exon skipping events. *DSCAM-AS1* affects target gene expression and causes changes in the AS by interacting with hnRNPL which appears to mediate the exon skipping and 3′ UTR usage by a mechanism not yet fully elucidated [121].

Canonical splicing of the linear pre-mRNA can compete for SFs with circularization of exons in circRNAs by mechanisms that are tissue specific and conserved in animals [122]. In flies and humans, the second exon of the SF muscleblind (Mbl (fly)/MBNL1 (human)) is circularized in *circMbl*. The introns flanking this circRNA as well as the circRNA itself contain highly conserved Mbl/MBNL1 binding sites, which are strongly and specifically bound by Mbl. Modulation of Mbl levels regulates the splicing of its own pre-mRNA into *circMbl*, and this in turn relies on Mbl binding sites [123] (Figure 6a). A circRNA proposed to act as an angiogenesis regulator by sponging SFs, is *circSMARCA5*. *CircSMARCA5* interacts with SRSF1 and promotes the switching from pro- to anti-angiogenic splicing isoforms of VEGF-A in glioblastoma multiforme, representing an opportunity to develop a novel anti-angiogenic cancer therapy [124]. Interestingly, circRNAs have been also found associated with the splicing factor QKI during human EMT [123], and correlate with exon skipping throughout the genome in human endothelial cells [125].

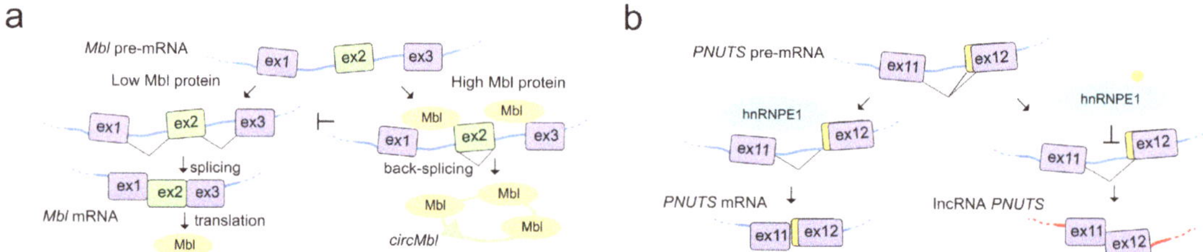

Figure 6. LncRNAs regulate pre-mRNA splicing by competing for splicing factors during their own splicing. (**a**) Left, In the presence of low amounts of *Mbl* (orange), the *Mbl* transcript is canonically spliced into a translatable mRNA encoding the Mbl protein. Right, when Mbl levels are high, Mbl binds to the pre-mRNA at the intronic regions flanking exon 2 and causes the exon2 back-splicing into *circMbl* (green), thereby preventing linear splicing and translation of the Mbl protein. *CircMbl* can also sequester Mbl protein, lowering its free cellular concentration, thereby providing a feedback mechanism to regulate Mbl levels. (**b**) The *PNUTS* gene can encode either the *PNUTS* mRNA or the lncRNA *PNUTS* depending on the usage of the 3′ alternative splice site located at the 5′-end of exon 12 which leads to the change of the ORF and the generation of a premature stop codon. Left, upon the binding of hnRNP E1 to a BAT consensus element located in the alternative splicing site that mask and prevents its usage, *PNUTS* pre-mRNA is spliced into *PNUTS* mRNA then translated into the PNUTS protein. Right, loss of hnRNP E1 binding to the alternative splice site uncovers the consensus element and allows its usage by the spliceosome machinery to achieve the splicing to yield the lncRNA *PNUTS* transcript.

LncRNAs can also interact with SFs to regulate their own splicing as is the case with the lncRNA *PNUTS*, also known as a competitive endogenous RNA (ce-RNA). The *PNUTS* gene can express a regular *PNUTS* mRNA encoding for the protein phosphatase 1 binding protein, but also to an alternatively spliced non-coding isoform called lncRNA-*PNUTS* with a distinct biological function. While *PNUTS* mRNA is ubiquitously expressed, the lncRNA-*PNUTS* one is more tumor-relevant and generally serves as a competitive sponge for miR-205 during EMT. The splicing decision to produce either mRNA or lncRNA relies on the binding of hnRNP E1 to a structural element located in exon 12 of *PNUTS* pre-RNA. Once released from this structural element, hnRNP E1 translocates from the nucleus to cytoplasm, allowing the AS and generation of the non-coding isoform of *PNUTS* to take place [126] (Figure 6b).

6. Concluding Remarks and Future Perspectives

Growing evidence suggests that lncRNAs control the regulation of AS in response to several physiological stimuli or during disease processes through changes in chromatin conformation, or by interfering with the overlapping antisense genes, genomic loci or SF activity. LncRNA antisense transcription pausing and elongation, as well as the capability of sponging RBPs, can also result in altered mRNA splice isoform expression patterns. The recent discovery of the circRNAs has also shown how a special class of lncRNAs can wholly integrate with the splicing process itself, affecting the splicing outcome of their linear cognates.

Some aspects of lncRNA-mediated AS regulation remain mostly unexplored. For instance most lncRNA sequences are not conserved across species, suggesting that most of their functionality might relies on their RNA structure. The role played by lncRNA secondary structure in determining their ability to regulate AS remains poorly investigated. Moreover, mRNA methylation is known to impact on AS by affecting the accessibility of hnRNPs to pre-mRNAs. Specifically, N6-methyladenosine (m6A) can serve as a switch to regulate gene expression and RNA maturation [127]. The existence of an interplay between RNA methylation and long non-coding RNA also raises the question of whether or not lncRNAs play a role in recruiting or reading mRNA methylation during AS processes. Furthermore, m6A modifications that occur on lncRNAs and circRNAs might change their function in AS regulation by providing a binding site for the m6A reader proteins or by modulating their structure–all of these questions remain unanswered.

Over the past years, our understanding of the mechanisms through which lncRNAs affect gene expression has been limited by their intrinsic properties (mainly length and low expression) and the lack of powerful experimental assays. With the increasing prevalence of splicing events and the discovery of over a hundred thousand lncRNAs, it is likely that the involvement of lncRNAs in regulating AS is far greater than the currently known. Further research is needed to gain a deeper understanding of how lncRNAs contribute to the regulation of AS in development and disease.

Table 1. List of lncRNAs involved in splicing regulation.

LncRNA Name	Splicing Target	Splicing Mechanism	Regulatory Effect or Associated Disease	Ref
		LncRNAs regulating AS by chromatin modifications		
asFGFR2	*FGFR2*	Recruiting Polycomb complexes and KDM2a to modify histone methylation and favor exon IIIb inclusion	Epithelial development	[23]
Antisense transcripts at each *Pcdhα* first exon	*Pcdhα*	First exon selection by histone modifications and distant DNA loop	Neuronal self-identity	[31]
		LncRNAs regulate AS through DNA-RNA interactions		
SEP3 exon6 circRNA (plant)	*SEP3*	Exon skipping through R-loop formation at exon 6	Flowering time	[44]
		LncRNAs regulate AS through RNA-RNA interactions		
NCYM NAT	*NMYC*	Intron I retention via antisense-sense RNA-RNA duplex	Cancer	[57]
NR1D1	*THRA*	Favoring α1 isoform by forming antisense-sense RNA-RNA duplex with the α2 mRNA	Thyroid hormone-responsiveness	[58,59]
SAF	*FAS*	Exon 6 skipping by forming RNA-RNA duplex with the target pre-mRNA and recruiting SPF45	Cancer Apoptosis	[60]
ZEB2 NAT	*ZEB2*	Preventing splicing of the IRES-containing intron through RNA-RNA interaction with the mRNA	EMT	[61]
51A	*SORL1*	Splicing shift from A to variant B by antisense-sense RNA-RNA duplex with an intronic sequence of the pre-mRNA	Alzheimer	[67]
17A	*GPR51*	Splicing shift from full-length to shorter GABAB R2 variant by antisense-sense RNA-RNA duplex	Alzheimer	[68]
		LncRNAs regulate AS by modulating the activity of splicing factors		
MALAT1/NEAT2		Modulation of SR localization and phosphorylation Uncoupling PTBP2 from SFPQ-PTBP2	Cancer	[73,88]
NEAT1	*PPARγ*	By interacting with CLK1 kinase to modulate SRp40 phosphorylation status	Adipocyte differentiation	[71,89,90]

Table 1. *Cont.*

LncRNA Name	Splicing Target	Splicing Mechanism	Regulatory Effect or Associated Disease	Ref
Gomafu/RNCR2/ MIAT		Interaction with QKI and SRSF1 Association with SF1 Localization of SF1 and Clf3 in CS bodies	Schizophrenia Retinal cell and brain development Post-mitotic neuronal function	[97–100]
LINC01133		Interaction and titration of SRSF6 splicing factor from target genes	EMT	[101]
PNCTR		Hijacking PTBP1 in the perinucleolar compartment	Cell survival	[102]
TPM1-AS	*TPM1*	Splicing shift to V1 or V3 isoforms by sequestering RBM4	Cancer	[107]
ASCO (plant)		Association with SmD1b and PRP8a and hijacking NSRa/b from the spliceosome	Lateral root formation	[109,110]
ENOD40 (plant)		Control nucleocytoplasmic of MtRBP1	Symbiotic nodule development	[111,112]
PCGEM1		Mutual bond with either hnRNP A1 or U2AF65 to promote or suppress specific AR splice variants	Castration resistance	[113]
BC200	*BCL-x*	Interaction with pre-mRNA and recruitment of the hnRNP A2/B1 which prevent Sam68 association	Apoptosis	[115,117]
Lnc-Spry1		Interaction with U2AF65	EMT	[118]
LASTR		Promoting splicing efficiency by interacting with SART3	Stress-induced JNK/c-JUN pathway	[119]
LINC-HELLP		Interaction with ribosomal and splicing complex components (eg: YBX1, PCBP1, PCBP2, RPS6 and RPL7)	*HELLP* syndrome	[120]
DSCAM-AS1		Exon skipping and 3′ UTR usage by interaction with hnRNPL	Tumor progression and anti-estrogen resistance	[121]
CircMbl	*Mbl*	Competing with the linear cognate by sequestering Mbl protein	Neuron Development	[123]
CircSMARCA5		Interaction with SRSF1 and promotion of the anti-angiogenic splicing isoforms of VEGF-A	Angiogenesis	[124]
PNUTS	*PNUTS*	Self-splicing regulation modulating the activity of hnRNP E1	EMT	[126]

Funding: This research received no external funding.

Conflicts of Interest: The authors declare no conflict of interest.

Abbreviations

Alternative splicing (AS); splicing factor (SF); RNA binding proteins (RBPs); long non-coding RNAs (lncRNAs); RNA polymerase II (RNAPII); CCCTC-binding factor (CTCF); Natural Antisense Transcripts (NATs); epithelial-mesenchymal transition (EMT); Alzheimer's disease (AD); amyloid precursor protein (APP); circular RNAs (circRNAs); N6-methyladenosine (m6A).

References

1. Berget, S.M.; Moore, C.; Sharp, P.A. Spliced segments at the 5′ terminus of adenovirus 2 late mRNA. *Proc. Natl. Acad. Sci. USA* **1977**, *74*, 3171–3175. [CrossRef]
2. Shi, Y. Mechanistic insights into precursor messenger RNA splicing by the spliceosome. *Nat. Rev. Mol. Cell Biol.* **2017**, *18*, 655–670. [CrossRef]
3. Rosenfeld, M.G.; Amara, S.G.; Roos, B.A.; Ong, E.S.; Evans, R.M. Altered expression of the calcitonin gene associated with RNA polymorphism. *Nature* **1981**, *290*, 63–65. [CrossRef] [PubMed]
4. Baralle, F.E.; Giudice, J. Alternative splicing as a regulator of development and tissue identity. *Nat. Rev. Mol. Cell Biol.* **2017**, *18*, 437–451. [CrossRef]
5. Ward, A.J.; Cooper, T.A. The pathobiology of splicing. *J. Pathol.* **2010**, *220*, 152–163. [CrossRef] [PubMed]

6.   Kim, E.; Magen, A.; Ast, G. Different levels of alternative splicing among eukaryotes. *Nucleic Acids Res.* **2007**, *35*, 125–131. [CrossRef] [PubMed]

7.   Wang, E.T.; Sandberg, R.; Luo, S.; Khrebtukova, I.; Zhang, L.; Mayr, C.; Kingsmore, S.F.; Schroth, G.P.; Burge, C.B. Alternative isoform regulation in human tissue transcriptomes. *Nature* **2008**, *456*, 470–476. [CrossRef] [PubMed]

8.   Mattick, J.S. RNA regulation: A new genetics? *Nat. Rev. Genet.* **2004**, *5*, 316–323. [CrossRef] [PubMed]

9.   Statello, L.; Guo, C.J.; Chen, L.L.; Huarte, M. Gene regulation by long non-coding RNAs and its biological functions. *Nat. Rev. Mol. Cell Biol.* **2021**, *22*, 96–118. [CrossRef] [PubMed]

10.   Mercer, T.R.; Dinger, M.E.; Mattick, J.S. Long non-coding RNAs: Insights into functions. *Nat. Rev. Genet.* **2009**, *10*, 155–159. [CrossRef]

11.   David Wang, X.Q.; Crutchley, J.L.; Dostie, J. Shaping the Genome with Non-coding RNAs. *Curr. Genom.* **2011**, 307–321. [CrossRef]

12.   Flynn, R.A.; Chang, H.Y. Active chromatin and noncoding RNAs: An intimate relationship. *Curr. Opin. Genet. Dev.* **2012**, *22*, 172–178. [CrossRef]

13.   Tsai, M.C.; Manor, O.; Wan, Y.; Mosammaparast, N.; Wang, J.K.; Lan, F.; Shi, Y.; Segal, E.; Chang, H.Y. Long noncoding RNA as modular scaffold of histone modification complexes. *Science* **2010**, *329*, 689–693. [CrossRef] [PubMed]

14.   Wapinski, O.; Chang, H.Y. Long noncoding RNAs and human disease. *Trends Cell Biol.* **2011**, *21*, 354–361. [CrossRef] [PubMed]

15.   Rinn, J.L.; Kertesz, M.; Wang, J.K.; Squazzo, S.L.; Xu, X.; Brugmann, S.A.; Goodnough, L.H.; Helms, J.A.; Farnham, P.J.; Segal, E.; et al. Functional Demarcation of Active and Silent Chromatin Domains in Human HOX Loci by Noncoding RNAs. *Cell* **2007**, *129*, 1311–1323. [CrossRef]

16.   Goff, L.A.; Rinn, J.L. Linking RNA biology to lncRNAs. *Genome Res.* **2015**, *25*, 1456–1465. [CrossRef] [PubMed]

17.   Ponting, C.P.; Oliver, P.L.; Reik, W. Evolution and Functions of Long Noncoding RNAs. *Cell* **2009**, *136*, 629–641. [CrossRef]

18.   Marchese, F.P.; Raimondi, I.; Huarte, M. The multidimensional mechanisms of long noncoding RNA function. *Genome Biol.* **2017**, *18*, 206. [CrossRef]

19.   Luco, R.F.; Pan, Q.; Tominaga, K.; Blencowe, B.J.; Pereira-Smith, O.M.; Misteli, T. Regulation of alternative splicing by histone modifications. *Science* **2010**, *327*, 996–1000. [CrossRef]

20.   Schor, I.E.; Rascovan, N.; Pelisch, F.; Alió, M.; Kornblihtt, A.R. Neuronal cell depolarization induces intragenic chromatin modifications affecting NCAM alternative splicing. *Proc. Natl. Acad. Sci. USA* **2009**, *106*, 4325–4330. [CrossRef]

21.   Batsché, E.; Yaniv, M.; Muchardt, C. The human SWI/SNF subunit Brm is a regulator of alternative splicing. *Nat. Struct Mol. Biol.* **2006**. [CrossRef] [PubMed]

22.   Luco, R.F.; Allo, M.; Schor, I.E.; Kornblihtt, A.R.; Misteli, T. Epigenetics in alternative pre-mRNA splicing. *Cell* **2011**, *144*, 16–26. [CrossRef]

23.   Gonzalez, I.; Munita, R.; Agirre, E.; Dittmer, T.A.; Gysling, K.; Misteli, T.; Luco, R.F. A lncRNA regulates alternative splicing via establishment of a splicing-specific chromatin signature. *Nat. Struct. Mol. Biol.* **2015**, *22*, 370–376. [CrossRef] [PubMed]

24.   Zhang, P.; Du, J.; Sun, B.; Dong, X.; Xu, G.; Zhou, J.; Huang, Q.; Liu, Q.; Hao, Q.; Ding, J. Structure of human MRG15 chromo domain and its binding to Lys36-methylated histone H3. *Nucleic Acids Res.* **2006**, *34*, 6621–6628. [CrossRef] [PubMed]

25.   Kornblihtt, A.R. Chromatin, transcript elongation and alternative splicing. *Nat. Struct. Mol. Biol.* **2006**, *13*, 5–7. [CrossRef]

26.   Shukla, S.; Kavak, E.; Gregory, M.; Imashimizu, M.; Shutinoski, B.; Kashlev, M.; Oberdoerffer, P.; Sandberg, R.; Oberdoerffer, S. CTCF-promoted RNA polymerase II pausing links DNA methylation to splicing. *Nature* **2011**, *479*, 74–79. [CrossRef]

27.   Amaral, P.P.; Leonardi, T.; Han, N.; Viré, E.; Gascoigne, D.K.; Arias-Carrasco, R.; Büscher, M.; Pandolfini, L.; Zhang, A.; Pluchino, S.; et al. Genomic positional conservation identifies topological anchor point RNAs linked to developmental loci. *Genome Biol.* **2018**, *19*, 32. [CrossRef]

28.   Pisignano, G.; Pavlaki, I.; Murrell, A. Being in a loop: How long non-coding RNAs organise genome architecture. *Essays Biochem.* **2019**, *63*, 177–186. [CrossRef]

29.   Lefevre, P.; Witham, J.; Lacroix, C.E.; Cockerill, P.N.; Bonifer, C. The LPS-Induced Transcriptional Upregulation of the Chicken Lysozyme Locus Involves CTCF Eviction and Noncoding RNA Transcription. *Mol. Cell.* **2008**, *32*, 129–139. [CrossRef]

30.   Blank-Giwojna, A.; Postepska-Igielska, A.; Grummt, I. lncRNA KHPS1 Activates a Poised Enhancer by Triplex-Dependent Recruitment of Epigenomic Regulators. *Cell Rep.* **2019**, *26*, 2904–2915. [CrossRef] [PubMed]

31.   Canzio, D.; Nwakeze, C.L.; Horta, A.; Rajkumar, S.M.; Coffey, E.L.; Duffy, E.E.; Duffié, R.; Monahan, K.; O'Keeffe, S.; Simon, M.D.; et al. Antisense lncRNA Transcription Mediates DNA Demethylation to Drive Stochastic Protocadherin α Promoter Choice. *Cell* **2019**, *177*, 639–653. [CrossRef]

32.   Guo, Y.; Monahan, K.; Wu, H.; Gertz, J.; Varley, K.E.; Li, W.; Myers, R.M.; Maniatis, T.; Wu, Q. CTCF/cohesin-mediated DNA looping is required for protocadherin α promoter choice. *Proc. Natl. Acad. Sci. USA* **2012**, *109*, 21081–21086. [CrossRef] [PubMed]

33.   Kehayova, P.; Monahan, K.; Chen, W.; Maniatis, T. Regulatory elements required for the activation and repression of the protocadherin-á gene cluster. *Proc. Natl. Acad. Sci. USA* **2011**, *108*, 17195–17200. [CrossRef] [PubMed]

34.   Monahan, K.; Rudnick, N.D.; Kehayova, P.D.; Pauli, F.; Newberry, K.M.; Myers, R.M.; Maniatis, T. Role of CCCTC binding factor (CTCF) and cohesin in the generation of single-cell diversity of Protocadherin-α gene expression. *Proc. Natl. Acad. Sci. USA* **2012**, *109*, 9125–9130. [CrossRef] [PubMed]

35.   Ribich, S.; Tasic, B.; Maniatis, T. Identification of long-range regulatory elements in the protocadherin-α gene cluster. *Proc. Natl. Acad. Sci. USA* **2006**, *103*, 19719–19724. [CrossRef] [PubMed]

36. Li, Y.; Syed, J.; Sugiyama, H. RNA-DNA Triplex Formation by Long Noncoding RNAs. *Cell Chem. Biol.* **2016**, *23*, 1325–1333. [CrossRef] [PubMed]

37. Niehrs, C.; Luke, B. Regulatory R-loops as facilitators of gene expression and genome stability. *Nat. Rev. Mol. Cell Biol.* **2020**, *21*, 167–178. [CrossRef]

38. Memczak, S.; Jens, M.; Elefsinioti, A.; Torti, F.; Krueger, J.; Rybak, A.; Maier, L.; Mackowiak, S.D.; Gregersen, L.H.; Munschauer, M.; et al. Circular RNAs are a large class of animal RNAs with regulatory potency. *Nature* **2013**, *495*, 333–338. [CrossRef] [PubMed]

39. Jeck, W.R.; Sorrentino, J.A.; Wang, K.; Slevin, M.K.; Burd, C.E.; Liu, J.; Marzluff, W.F.; Sharpless, N.E. Circular RNAs are abundant, conserved, and associated with ALU repeats. *RNA* **2013**, *19*, 141–157. [CrossRef] [PubMed]

40. Barrett, S.P.; Wang, P.L.; Salzman, J. Circular RNA biogenesis can proceed through an exon-containing lariat precursor. *Elife* **2015**, *4*, e07540. [CrossRef]

41. Lee, E.C.S.; Elhassan, S.A.M.; Lim, G.P.L.; Kok, W.H.; Tan, S.W.; Leong, E.N.; Tan, S.H.; Chan, E.W.L.; Bhattamisra, S.K.; Rajendran, R.; et al. The roles of circular RNAs in human development and diseases. *Biomed. Pharmacother.* **2019**, *111*, 198–208. [CrossRef]

42. Zhang, X.O.; Wang HBin Zhang, Y.; Lu, X.; Chen, L.L.; Yang, L. Complementary sequence-mediated exon circularization. *Cell* **2014**, *159*, 134–147. [CrossRef]

43. Li, X.; Yang, L.; Chen, L.L. The Biogenesis, Functions, and Challenges of Circular RNAs. *Mol. Cell* **2018**, *71*, 428–442. [CrossRef] [PubMed]

44. Conn, V.M.; Hugouvieux, V.; Nayak, A.; Conos, S.A.; Capovilla, G.; Cildir, G.; Jourdain, A.; Tergaonkar, V.; Schmid, M.; Zubieta, C.; et al. A circRNA from SEPALLATA3 regulates splicing of its cognate mRNA through R-loop formation. *Nat. Plants* **2017**, *3*, 17053. [CrossRef]

45. Alexander, R.D.; Innocente, S.A.; Barrass, J.D.; Beggs, J.D. Splicing-Dependent RNA polymerase pausing in yeast. *Mol. Cell.* **2010**, *40*, 582–593. [CrossRef] [PubMed]

46. El Hage, A.; Webb, S.; Kerr, A.; Tollervey, D. Genome-Wide Distribution of RNA-DNA Hybrids Identifies RNase H Targets in tRNA Genes, Retrotransposons and Mitochondria. *PLoS Genet.* **2014**, *10*, e1004716. [CrossRef] [PubMed]

47. Wongsurawat, T.; Jenjaroenpun, P.; Kwoh, C.K.; Kuznetsov, V. Quantitative model of R-loop forming structures reveals a novel level of RNA-DNA interactome complexity. *Nucleic Acids Res.* **2012**. [CrossRef] [PubMed]

48. Sugino, A.; Hirose, S.; Okazaki, R. RNA-linked nascent DNA fragments in Escherichia coli. *Proc. Natl. Acad. Sci. USA* **1972**. [CrossRef] [PubMed]

49. Greider, C.W.; Blackburn, E.H. A telomeric sequence in the RNA of Tetrahymena telomerase required for telomere repeat synthesis. *Nature* **1989**. [CrossRef]

50. Williams, J.S.; Kunkel, T.A. Ribonucleotides in DNA: Origins, repair and consequences. *DNA Repair (Amst.)* **2014**. [CrossRef] [PubMed]

51. Pelechano, V.; Steinmetz, L.M. Gene regulation by antisense transcription. *Nat. Rev. Genet.* **2013**. [CrossRef] [PubMed]

52. Morrissy, A.S.; Griffith, M.; Marra, M.A. Extensive relationship between antisense transcription and alternative splicing in the human genome. *Genome Res.* **2011**. [CrossRef] [PubMed]

53. Aartsma-Rus, A.; Van Ommen, G.J.B. Antisense-mediated exon skipping: A versatile tool with therapeutic and research applications. *RNA* **2007**. [CrossRef] [PubMed]

54. Khorkova, O.; Myers, A.J.; Hsiao, J.; Wahlestedt, C. Natural antisense transcripts. *Hum. Mol. Genet.* **2014**. [CrossRef]

55. Bardou, F.; Merchan, F.; Ariel, F.; Crespi, M. Dual RNAs in plants. *Biochimie* **2011**. [CrossRef] [PubMed]

56. Suenaga, Y.; Islam, S.M.; Alagu, J.; Kaneko, Y.; Kato, M.; Tanaka, Y.; Kawana, H.; Hossain, S.; Matsumoto, D.; Yamamoto, M.; et al. NCYM, a Cis-Antisense Gene of MYCN, Encodes a De Novo Evolved Protein That Inhibits GSK3β Resulting in the Stabilization of MYCN in Human Neuroblastomas. *PLoS Genet.* **2014**, *10*. [CrossRef]

57. Krystal, G.W.; Armstrong, B.C.; Battey, J.F. N-myc mRNA forms an RNA-RNA duplex with endogenous antisense transcripts. *Mol. Cell Biol.* **1990**. [CrossRef] [PubMed]

58. Munroe, S.H.; Lazar, M.A. Inhibition of c-erbA mRNA splicing by a naturally occurring antisense RNA. *J. Biol Chem.* **1991**, *266*, 22083–22086. [CrossRef]

59. Chassande, O.; Fraichard, A.; Gauthier, K.; Flamant, F.; Legrand, C.; Savatier, P.; Laudet, V.; Samarut, J. Identification of Transcripts Initiated from an Internal Promoter in the c-erbAα Locus That Encode Inhibitors of Retinoic Acid Receptor-α and Triiodothyronine Receptor Activities. *Mol. Endocrinol.* **1997**, *11*, 1278–1290. [CrossRef]

60. Villamizar, O.; Chambers, C.B.; Riberdy, J.M.; Persons, D.A.; Wilber, A. Long noncoding RNA Saf and splicing factor 45 increase soluble Fas and resistance to apoptosis. *Oncotarget* **2016**. [CrossRef]

61. Beltran, M.; Puig, I.; Peña, C.; García, J.M.; Alvarez, A.B.; Peña, R.; Bonilla, F.; de Herreros, A.G. A natural antisense transcript regulates Zeb2/Sip1 gene expression during Snail1-induced epithelial-mesenchymal transition. *Genes Dev.* **2008**, *22*, 756–769. [CrossRef] [PubMed]

62. Ma, Q.L.; Galasko, D.R.; Ringman, J.M.; Vinters, H.V.; Edland, S.D.; Pomakian, J.; Ubeda, O.J.; Rosario, E.R.; Teter, B.; Frautschy, S.A.; et al. Reduction of SorLA/LR11, a sorting protein limiting β-amyloid production, in alzheimer disease cerebrospinal fluid. *Arch. Neurol.* **2009**, *66*, 448–457. [CrossRef] [PubMed]

63. Reitz, C.; Cheng, R.; Rogaeva, E.; Lee, J.H.; Tokuhiro, S.; Zou, F.; Bettens, K.; Sleegers, K.; Tan, E.K.; Kimura, R.; et al. Meta-analysis of the association between variants in SORL1 and Alzheimer disease. *Arch. Neurol.* **2011**, *68*, 99–106. [CrossRef] [PubMed]

64. Rogaeva, E.; Meng, Y.; Lee, J.H.; Gu, Y.; Kawarai, T.; Zou, F.; Katayama, T.; Baldwin, C.T.; Cheng, R.; Hasegawa, H.; et al. The neuronal sortilin-related receptor SORL1 is genetically associated with Alzheimer disease. *Nat. Genet.* **2007**, *39*, 168–177. [CrossRef] [PubMed]

65. Andersen, O.M.; Reiche, J.; Schmidt, V.; Gotthardt, M.; Spoelgen, R.; Behlke, J.; von Arnim, C.A.; Breiderhoff, T.; Jansen, P.; Wu, X.; et al. Neuronal sorting protein-related receptor sorLA/LR11 regulates processing of the amyloid precursor protein. *Proc. Natl. Acad. Sci. USA* **2005**, *102*, 13461–13466. [CrossRef]

66. Small, S.A.; Kent, K.; Pierce, A.; Leung, C.; Kang, M.S.; Okada, H.; Honig, L.; Vonsattel, J.P.; Kim, T.W. Model-guided microarray implicates the retromer complex in Alzheimer's disease. *Ann. Neurol.* **2005**, *58*, 909–919. [CrossRef]

67. Ciarlo, E.; Massone, S.; Penna, I.; Nizzari, M.; Gigoni, A.; Dieci, G.; Russo, C.; Florio, T.; Cancedda, R.; Pagano, A. An intronic ncRNA-dependent regulation of SORL1 expression affecting Aβ formation is upregulated in post-mortem Alzheimer's disease brain samples. *DMM Dis Model. Mech.* **2013**, *6*, 424–433. [CrossRef]

68. Massone, S.; Vassallo, I.; Fiorino, G.; Castelnuovo, M.; Barbieri, F.; Borghi, R.; Tabaton, M.; Robello, M.; Gatta, E.; Russo, C.; et al. 17A, a novel non-coding RNA, regulates GABA B alternative splicing and signaling in response to inflammatory stimuli and in Alzheimer disease. *Neurobiol Dis.* **2011**, *41*, 308–317. [CrossRef] [PubMed]

69. Onodera, C.S.; Underwood, J.G.; Katzman, S.; Jacobs, F.; Greenberg, D.; Salama, S.R.; Haussler, D. Gene isoform specificity through enhancer-associated antisense transcription. *PLoS ONE* **2012**, *7*, e43511. [CrossRef]

70. Stork, M.; Di Lorenzo, M.; Welch, T.J.; Crosa, J.H. Transcription termination within the iron transport-biosynthesis operon of Vibrio anguillarum requires an antisense RNA. *J. Bacteriol.* **2007**. [CrossRef]

71. Hutchinson, J.N.; Ensminger, A.W.; Clemson, C.M.; Lynch, C.R.; Lawrence, J.B.; Chess, A. A screen for nuclear transcripts identifies two linked noncoding RNAs associated with SC35 splicing domains. *BMC Genomics.* **2007**, *8*, 39. [CrossRef]

72. Bernard, D.; Prasanth, K.V.; Tripathi, V.; Colasse, S.; Nakamura, T.; Xuan, Z.; Zhang, M.Q.; Sedel, F.; Jourdren, L.; Coulpier, F.; et al. A long nuclear-retained non-coding RNA regulates synaptogenesis by modulating gene expression. *EMBO J.* **2010**, *29*, 3082–3093. [CrossRef] [PubMed]

73. Tripathi, V.; Ellis, J.D.; Shen, Z.; Song, D.Y.; Pan, Q.; Watt, A.T.; Freier, S.M.; Bennett, C.F.; Sharma, A.; Bubulya, P.A.; et al. The nuclear-retained noncoding RNA MALAT1 regulates alternative splicing by modulating SR splicing factor phosphorylation. *Mol. Cell.* **2010**, *39*, 925–938. [CrossRef] [PubMed]

74. Patton, J.G.; Porro, E.B.; Galceran, J.; Tempst, P.; Nadal-Ginard, B. Cloning and characterization of PSF, a novel pre-mRNA splicing factor. *Genes Dev.* **1993**. [CrossRef]

75. Gozani, O.; Patton, J.G.; Reed, R. A novel set of spliceosome-associated proteins and the essential splicing factor PSF bind stably to pre-mRNA prior to catalytic step II of the splicing reaction. *EMBO J.* **1994**. [CrossRef]

76. Misteli, T.; Cáceres, J.F.; Clement, J.Q.; Krainer, A.R.; Wilkinson, M.F.; Spector, D.L. Serine phosphorylation of SR proteins is required for their recruitment to sites of transcription in vivo. *J. Cell Biol.* **1998**. [CrossRef] [PubMed]

77. Long, J.C.; Caceres, J.F. The SR protein family of splicing factors: Master regulators of gene expression. *Biochem. J.* **2009**. [CrossRef]

78. Cao, W.; Jamison, S.F.; Garcia-Blanco, M.A. Both phosphorylation and dephosphorylation of ASF/SF2 are required for pre-mRNA splicing in vitro. *RNA* **1997**, *3*, 1456–1467. [PubMed]

79. Santord, J.R.; Ellis, J.D.; Cazalla, D.; Cáceres, J.F. Reversible phosphorylation differentially affects nuclear and cytoplasmic functions of splicing factor 2/alternative splicing factor. *Proc. Natl. Acad. Sci. USA* **2005**, *102*, 15042–15047. [CrossRef]

80. Huang, Y.; Yario, T.A.; Steitz, J.A. A molecular link between SR protein dephosphorylation and mRNA export. *Proc. Natl. Acad. Sci. USA* **2004**. [CrossRef]

81. Stamm, S. Regulation of alternative splicing by reversible protein phosphorylation. *J. Biol Chem.* **2008**. [CrossRef]

82. Shi, Y.; Manley, J.L. A Complex Signaling Pathway Regulates SRp38 Phosphorylation and Pre-mRNA Splicing in Response to Heat Shock. *Mol. Cell* **2007**. [CrossRef]

83. Zhong, X.Y.; Ding, J.H.; Adams, J.A.; Ghosh, G.; Fu, X.D. Regulation of SR protein phosphorylation and alternative splicing by modulating kinetic interactions of SRPK1 with molecular chaperones. *Genes Dev.* **2009**. [CrossRef]

84. Engreitz, J.M.; Sirokman, K.; McDonel, P.; Shishkin, A.A.; Surka, C.; Russell, P.; Grossman, S.R.; Chow, A.Y.; Guttman, M.; Lander, E.S. RNA-RNA interactions enable specific targeting of noncoding RNAs to nascent pre-mRNAs and chromatin sites. *Cell.* **2014**, *159*, 188–199. [CrossRef]

85. Lu, Z.; Zhang, Q.C.; Lee, B.; Flynn, R.A.; Smith, M.A.; Robinson, J.T.; Davidovich, C.; Gooding, A.R.; Goodrich, K.J.; Mattick, J.S.; et al. RNA Duplex Map in Living Cells Reveals Higher-Order Transcriptome Structure. *Cell.* **2016**, *165*, 1267–1279. [CrossRef] [PubMed]

86. Cai, Z.; Cao, C.; Ji, L.; Ye, R.; Wang, D.; Xia, C.; Wang, S.; Du, Z.; Hu, N.; Yu, X.; et al. RIC-seq for global in situ profiling of RNA–RNA spatial interactions. *Nature* **2020**, *582*, 432–437. [CrossRef] [PubMed]

87. Malakar, P.; Shilo, A.; Mogilevsky, A.; Stein, I.; Pikarsky, E.; Nevo, Y.; Benyamini, H.; Elgavish, S.; Zong, X.; Prasanth, K.V.; et al. Long noncoding RNA MALAT1 promotes hepatocellular carcinoma development by SRSF1 upregulation and mTOR activation. *Cancer Res.* **2017**, *77*, 155–1167. [CrossRef]

88. Ji, Q.; Zhang, L.; Liu, X.; Zhou, L.; Wang, W.; Han, Z.; Sui, H.; Tang, Y.; Wang, Y.; Liu, N.; et al. Long non-coding RNA MALAT1 promotes tumour growth and metastasis in colorectal cancer through binding to SFPQ and releasing oncogene PTBP2 from SFPQ/PTBP2 complex. *Br. J. Cancer.* **2014**, *111*, 736–748. [CrossRef]

89. Jiang, K.; Patel, N.A.; Watson, J.E.; Apostolatos, H.; Kleiman, E.; Hanson, O.; Hagiwara, M.; Cooper, D.R. Akt2 regulation of Cdc2-like kinases (Clk/Sty), serine/arginine-rich (SR) protein phosphorylation, and insulin-induced alternative splicing of PKCβJII messenger ribonucleic acid. *Endocrinology* **2009**, *150*, 2087–2097. [CrossRef]

90. Cooper, D.R.; Carter, G.; Li, P.; Patel, R.; Watson, J.E.; Patel, N.A. Long non-coding RNA NEAT1 associates with SRp40 to temporally regulate PPARγ2 splicing during adipogenesis in 3T3-L1 cells. *Genes* **2014**, *5*, 1050–1063. [CrossRef] [PubMed]

91. Blackshaw, S.; Harpavat, S.; Trimarchi, J.; Cai, L.; Huang, H.; Kuo, W.P.; Weber, G.; Lee, K.; Fraioli, R.E.; Cho, S.H.; et al. Genomic analysis of mouse retinal development. *PLoS Biol.* **2004**, *2*. [CrossRef] [PubMed]

92. Sone, M.; Hayashi, T.; Tarui, H.; Agata, K.; Takeichi, M.; Nakagawa, S. The mRNA-like noncoding RNA Gomafu constitutes a novel nuclear domain in a subset of neurons. *J. Cell Sci.* **2007**, *120*, 2498–2506. [CrossRef]

93. Rapicavoli, N.A.; Blackshaw, S. New meaning in the message: Noncoding RNAs and their role in retinal development. *Dev. Dyn.* **2009**. [CrossRef]

94. Rapicavoli, N.A.; Poth, E.M.; Blackshaw SRapicavoli, N.A.; Poth, E.M.; Blackshaw, S. The long noncoding RNA RNCR2 directs mouse retinal cell specification. *BMC Dev. Biol.* **2010**, *10*, 49. [CrossRef] [PubMed]

95. Mercer, T.R.; Qureshi, I.A.; Gokhan, S.; Dinger, M.E.; Li, G.; Mattick, J.S.; Mehler, M.F. Long noncoding RNAs in neuronal-glial fate specification and oligodendrocyte lineage maturation. *BMC Neurosci.* **2010**. [CrossRef] [PubMed]

96. Mercer, T.R.; Dinger, M.E.; Sunkin, S.M.; Mehler, M.F.; Mattick, J.S. Specific expression of long noncoding RNAs in the mouse brain. *Proc. Natl. Acad. Sci. USA* **2008**. [CrossRef]

97. Barry, G.; Briggs, J.A.; Vanichkina, D.P.; Poth, E.M.; Beveridge, N.J.; Ratnu, V.S.; Nayler, S.P.; Nones, K.; Hu, J.; Bredy, T.W.; et al. The long non-coding RNA Gomafu is acutely regulated in response to neuronal activation and involved in schizophrenia-associated alternative splicing. *Mol. Psychiatry* **2014**, *19*, 486–494. [CrossRef]

98. Tsuiji, H.; Yoshimoto, R.; Hasegawa, Y.; Furuno, M.; Yoshida, M.; Nakagawa, S. Competition between a noncoding exon and introns: Gomafu contains tandem UACUAAC repeats and associates with splicing factor-1. *Genes Cells* **2011**, *16*, 479–490. [CrossRef] [PubMed]

99. Ladd, A.N. CUG-BP, Elav-like family (CELF)-mediated alternative splicing regulation in the brain during health and disease. *Mol. Cell Neurosci.* **2013**, *56*, 456–464. [CrossRef]

100. Ishizuka, A.; Hasegawa, Y.; Ishida, K.; Yanaka, K.; Nakagawa, S. Formation of nuclear bodies by the lncRNA Gomafu-associating proteins Celf3 and SF1. *Genes Cells.* **2014**, *19*, 704–721. [CrossRef]

101. Kong, J.; Sun, W.; Li, C.; Wan, L.; Wang, S.; Wu, Y.; Xu, E.; Zhang, H.; Lai, M. Long non-coding RNA LINC01133 inhibits epithelial–mesenchymal transition and metastasis in colorectal cancer by interacting with SRSF6. *Cancer Lett.* **2016**, *380*, 476–484. [CrossRef]

102. Yap, K.; Mukhina, S.; Zhang, G.; Tan, J.S.C.; Ong, H.S.; Makeyev, E.V. A Short Tandem Repeat-Enriched RNA Assembles a Nuclear Compartment to Control Alternative Splicing and Promote Cell Survival. *Mol. Cell* **2018**, *72*, 525–540. [CrossRef]

103. Bushell, M.; Stoneley, M.; Kong, Y.W.; Hamilton, T.L.; Spriggs, K.A.; Dobbyn, H.C.; Qin, X.; Sarnow, P.; Willis, A.E. Polypyrimidine Tract Binding Protein Regulates IRES-Mediated Gene Expression during Apoptosis. *Mol. Cell* **2006**, *23*, 401–412. [CrossRef]

104. Bielli, P.; Bordi, M.; Di Biasio, V.; Sette, C. Regulation of BCL-X splicing reveals a role for the polypyrimidine tract binding protein (PTBP1/hnRNP I) in alternative 5′ splice site selection. *Nucleic Acids Res.* **2014**, *42*, 12070–12081. [CrossRef] [PubMed]

105. Zhang, J.; Bahi, N.; Llovera, M.; Comella, J.X.; Sanchis, D. Polypyrimidine tract binding proteins (PTB) regulate the expression of apoptotic genes and susceptibility to caspase-dependent apoptosis in differentiating cardiomyocytes. *Cell Death Differ.* **2009**, *16*, 1460–1468. [CrossRef] [PubMed]

106. Porto, F.W.; Daulatabad, S.V.; Janga, S.C. Long non-coding RNA expression levels modulate cell-type-specific splicing patterns by altering their interaction landscape with RNA-binding proteins. *Genes* **2019**, *10*, 593. [CrossRef]

107. Huang, G.W.; Zhang, Y.L.; Liao LDi Li, E.M.; Xu, L.Y. Natural antisense transcript TPM1-AS regulates the alternative splicing of tropomyosin I through an interaction with RNA-binding motif protein 4. *Int. J. Biochem. Cell Biol.* **2017**, *90*, 59–67. [CrossRef] [PubMed]

108. Bardou, F.; Ariel, F.; Simpson, C.G.; Romero-Barrios, N.; Laporte, P.; Balzergue, S.; Brown, J.W.; Crespi, M. Long Noncoding RNA Modulates Alternative Splicing Regulators in Arabidopsis. *Dev. Cell* **2014**, *30*, 166–176. [CrossRef]

109. Tran, V.D.T.; Souiai, O.; Romero-Barrios, N.; Crespi, M.; Gautheret, D. Detection of generic differential RNA processing events from RNA-seq data. *RNA Biol.* **2016**, *13*, 59–67. [CrossRef]

110. Rigo, R.; Bazin, J.; Romero-Barrios, N.; Moison, M.; Lucero, L.; Christ, A.; Benhamed, M.; Blein, T.; Huguet, S.; Charon, C.; et al. The Arabidopsis lncRNA ASCO modulates the transcriptome through interaction with splicing factors. *EMBO Rep.* **2020**, *21*, e48977. [CrossRef]

111. Crespi, M.D.; Jurkevitch, E.; Poiret, M.; d'Aubenton-Carafa, Y.; Petrovics, G.; Kondorosi, E.; Kondorosi, A. Enod40, a gene expressed during nodule organogenesis, codes for a non-translatable RNA involved in plant growth. *EMBO J.* **1994**, *13*, 5099–5112. [CrossRef]

112. Campalans, A.; Kondorosi, A.; Crespi, M. Enod40, a short open reading frame-containing mRNA, induces cytoplasmic localization of a nuclear RNA binding protein in Medicago truncatula. *Plant. Cell* **2004**, *16*, 1047–1059. [CrossRef] [PubMed]

113. Zhang, Z.; Zhou, N.; Huang, J.; Ho, T.T.; Zhu, Z.; Qiu, Z.; Zhou, X.; Bai, C.; Wu, F.; Xu, M.; et al. Regulation of androgen receptor splice variant AR3 by PCGEM1. *Oncotarget* **2016**, *7*, 15481–15491. [CrossRef] [PubMed]
114. Hetz, C. BCL-2 protein family. Essential regulators of cell death. Preface. *Adv. Exp. Med. Biol.* **2010**.
115. Minnt, A.J.; Boise, L.H.; Thompson, C.B. Bcl-XS Antagonizes the protective effects of Bcl-xL. *J. Biol. Chem.* **1996**, *271*, 6306–6312. [CrossRef] [PubMed]
116. Paronetto, M.P.; Achsel, T.; Massiello, A.; Chalfant, C.E.; Sette, C. The RNA-binding protein Sam68 modulates the alternative splicing of Bcl-x. *J. Cell Biol.* **2007**, *176*, 929–939. [CrossRef]
117. Singh, R.; Gupta, S.C.; Peng, W.X.; Zhou, N.; Pochampally, R.; Atfi, A.; Watabe, K.; Lu, Z.; Mo, Y.Y. Regulation of alternative splicing of Bcl-x by BC200 contributes to breast cancer pathogenesis. *Cell Death Dis.* **2016**, *7*, e2262. [CrossRef]
118. Rodríguez-Mateo, C.; Torres, B.; Gutiérrez, G.; Pintor-Toro, J.A. Downregulation of Lnc-Spry1 mediates TGF-β-induced epithelial-mesenchymal transition by transcriptional and posttranscriptional regulatory mechanisms. *Cell Death Differ.* **2017**, *24*, 785–797. [CrossRef]
119. De Troyer, L.; Zhao, P.; Pastor, T.; Baietti, M.F.; Barra, J.; Vendramin, R.; Dok, R.; Lechat, B.; Najm, P.; Van Haver, D.; et al. Stress-induced lncRNA LASTR fosters cancer cell fitness by regulating the activity of the U4/U6 recycling factor SART3. *Nucleic Acids Res.* **2020**, *48*, 2502–2517. [CrossRef]
120. van Dijk, M.; Visser, A.; Buabeng, K.M.L.; Poutsma, A.; van der Schors, R.C.; Oudejans, C.B.M. Mutations within the LINC-HELLP non-coding RNA differentially bind ribosomal and RNA splicing complexes and negatively affect trophoblast differentiation. *Hum. Mol. Genet.* **2015**, *24*, 5475–5485. [CrossRef] [PubMed]
121. Elhasnaoui, J.; Miano, V.; Ferrero, G.; Doria, E.; Leon, A.E.; Fabricio, A.S.C.; Annaratone, L.; Castellano, I.; Sapino, A.; De Bortoli, M. DSCAM-AS1-driven proliferation of breast cancer cells involves regulation of alternative exon splicing and 3′-end usage. *Cancers* **2020**, *12*, 1453. [CrossRef] [PubMed]
122. Chen, L.L. The biogenesis and emerging roles of circular RNAs. *Nat. Rev. Mol. Cell. Biol.* **2016**, *17*, 205–211. [CrossRef]
123. Xie, L.; Mao, M.; Xiong, K.; Jiang, B. Circular RNAs: A novel player in development and disease of the central nervous system. *Front. Cell Neurosci.* **2017**, *11*, 354. [CrossRef] [PubMed]
124. Barbagallo, D.; Caponnetto, A.; Cirnigliaro, M.; Brex, D.; Barbagallo, C.; D'Angeli, F.; Morrone, A.; Caltabiano, R.; Barbagallo, G.M.; Ragusa, M.; et al. CircSMARCA5 inhibits migration of glioblastoma multiforme cells by regulating a molecular axis involving splicing factors SRSF1/SRSF3/PTB. *Int. J. Mol. Sci.* **2018**, *19*, 480. [CrossRef]
125. Kelly, S.; Greenman, C.; Cook, P.R.; Papantonis, A. Exon Skipping Is Correlated with Exon Circularization. *J. Mol. Biol.* **2015**, *427*, 2414–2417. [CrossRef]
126. Grelet, S.; Link, L.A.; Howley, B.; Obellianne, C.; Palanisamy, V.; Gangaraju, V.K.; Diehl, J.A.; Howe, P.H. A regulated PNUTS mRNA to lncRNA splice switch mediates EMT and tumour progression. *Nat. Cell Biol.* **2017**, *19*, 1105–1115. [CrossRef] [PubMed]
127. Liu, N.; Dai, Q.; Zheng, G.; He, C.; Parisien, M.; Pan, T. N6 -methyladenosine-dependent RNA structural switches regulate RNA-protein interactions. *Nature* **2015**, *518*, 560–564. [CrossRef] [PubMed]

non-coding *RNA*

MDPI

Review

The Role of LncRNAs in Translation

Didem Karakas [1] and Bulent Ozpolat [2,*]

1   Department of Molecular Biology and Genetics, Faculty of Arts and Sciences, Istinye University, Istanbul 34010, Turkey; didemmkarakass@gmail.com
2   Department of Experimental Therapeutics, The University of Texas MD Anderson Cancer Center, Houston, TX 77030, USA
*   Correspondence: bozpolat@mdanderson.org; Tel.: +1-713-792-0362

Abstract: Long non-coding RNAs (lncRNAs), a group of non-protein coding RNAs with lengths of more than 200 nucleotides, exert their effects by binding to DNA, mRNA, microRNA, and proteins and regulate gene expression at the transcriptional, post-transcriptional, translational, and post-translational levels. Depending on cellular location, lncRNAs are involved in a wide range of cellular functions, including chromatin modification, transcriptional activation, transcriptional interference, scaffolding and regulation of translational machinery. This review highlights recent studies on lncRNAs in the regulation of protein translation by modulating the translational factors (i.e, eIF4E, eIF4G, eIF4A, 4E-BP1, eEF5A) and signaling pathways involved in this process as wells as their potential roles as tumor suppressors or tumor promoters.

Keywords: non-coding RNAs; long non-coding RNAs; ncRNAs; translation; cancer

Citation: Karakas, D.; Ozpolat, B. The Role of LncRNAs in Translation. *Non-coding RNA* **2021**, 7, 16. https://doi.org/10.3390/ ncrna7010016

Received: 4 January 2021
Accepted: 18 February 2021
Published: 20 February 2021

Publisher's Note: MDPI stays neutral with regard to jurisdictional claims in published maps and institutional affiliations.

1. Introduction

The majority of the mammalian genome consists of non-coding RNAs (ncRNAs), including long ncRNAs (lncRNAs), transfer RNAs (tRNAs), ribosomal RNAs (rRNAs), and small ncRNAs such as microRNAs (miRNAs), small nuclear RNAs (snRNA) and circular RNAs (circRNAs), while only a small portion (~1.5%) of it is comprised of protein-coding mRNAs [1].

lncRNA transcripts, which are a group of ncRNAs longer than 200 nucleotides, account for the majority (98%) of the ncRNAs. Currently, about 30,000 different lncRNA transcripts are belived to exist in the human genome [2]. Since most lncRNAs are transcribed by RNA polymerase II (RNAP II), they share some similarities with mRNAs, such as poly-adenylation and the presence of 5′-cap structure. Just like mRNAs, lncRNAs form secondary structures, undergo post-transcriptional processing (i.e., 5′-cap structure, polyadenylation) and splicing [3], present in the nucleus, cytosol, and mitochondria [4], and can have tissue-specific expression patterns.

lncRNAs have been shown to play a pivotal role in a wide range of cellular processes such as gene expression, translation regulation, splicing, chromosomal organization and X chromosome silencing [5–7]. Besides, specific lncRNAs are known to be dysregulated in various diseases, such as cancer, neurological diseases, and diabetes [8]. Considering their extensive roles in both health and disease, a better understanding of the functions of lncRNAs in the regulation of cellular events is needed.

In this review, we aim to discuss the role of lncRNAs in the regulation of protein translation by controlling translational factors and signaling pathways. Furthermore, because translational regulation is often dysregulated in cancer cells, we also briefly summarize the role of lncRNAs in tumorigenesis and cancer progression as tumor promoters or tumor suppressors.

Non-coding RNA **2021**, 7, 16. https://doi.org/10.3390/ncrna7010016   https://www.mdpi.com/journal/ncrna

2. An Overview of the Characteristics of LncRNAs

Although lncRNAs were initially assumed as transcriptional noise or genomic "junk" [9,10], studies later revealed that they play vital roles in the regulation of various cellular processes, such as cell division, proliferation, differentiation, cell cycle, cell death, and metabolism [11–15]. Recent reports indicated some lncRNAs have a small open-reading frame (sORFs/smORFs) and are associated with ribosomes, suggesting their protein-coding potential [16–20]. In fact, recent studies showed that a small number of lncRNAs are capable of encoding small proteins called micropeptides (less than 100 amino acids) that are involved in the regulation of various biological processes [21].

Initial studies suggested that lncRNAs were thought of as unstable transcripts. However, later studies demonstrated that the majority of 800 lncRNAs have half-lives greater than 16 h and are highly stable, while only a minority of lncRNAs have half-lives less than 2 h [22].

lncRNAs have been traditionally categorized according to their specific locations on the genome into five major groups, including antisense, sense, bidirectional, intronic, and intergenic RNAs [23]. In a recent report, a more detailed classification has been proposed to describe the diversity of lncRNAs. This new classification includes seven different groups of lncRNAs: (a) mRNA-like intergenic transcripts (lincRNAs), (b) anti-sense transcripts of protein coding genes (natural anti-sense transcripts -NATs-), (c) processed transcripts, (d) enhancer RNAs (eRNAs), (e) promoter upstream transcripts (PROMPTs), (f) small nucleolar RNA (snoRNA)-ended lncRNAs (sno-lncRNAs), and (g) circular intronic RNAs (ciRNAs) [24].

3. Regulatory Functions of LncRNAs Depending on Their Subcellular Location

Since lncRNAs are capable of interacting with nucleic acids (DNA, RNA) and proteins, they are involved in the regulation of diverse molecular processes such as epigenetic and (post)-transcriptional modifications, translational regulation, splicing and scaffolding [6,7,25]. These diverse functions of lncRNAs are closely associated with their cellular location. lncRNAs are predominantly found in the nucleus and cytoplasm [1], while some lncRNA transcripts can be localized in exosomes. Recent findings revealed that large quantities of lncRNAs are exported to the cytoplasm to display their vital regulatory functions in cytoplasmic processes [19,26,27]. Subcellular localization of lncRNAs is a tightly regulated process controlled by various factors, such as sequence and structural motifs [28].

Based on their location in the cell, lncRNAs are involved in different molecular processes. The nuclear lncRNAs are closely associated with chromatin structures and regulate gene expression by influencing diverse mechanisms such as transcriptional and epigenetic regulation of specific genes and pre-mRNA processing [29]. In contrast, cytoplasmic lncRNAs dominantly control the stability and translation of mRNAs [27]. For instance, lncRNAs such as MALAT1 and NEAT1 are predominantly found in the nucleus; DANCR and OIP5-AS1 are found mainly in the cytoplasm; TUG1, CasC7 and HOTAIR have both nuclear and cytoplasmic distribution [30]. Since the subcellular location determines the function of lncRNAs, in this section, we aim to highlight the regulatory functions of lncRNAs depending on their subcellular locations.

3.1. Cytoplasmic LncRNAs

Cytoplasmic lncRNAs control a wide range of cellular processes by interacting with miRNAs, mRNAs and proteins. They can reciprocally interact with miRNAs and affect the functions of miRNAs in various ways. lncRNAs can function as competing endogenous RNAs (ceRNA) to bind miRNAs and block miRNA-mRNA interactions. For instance, BACE1 (beta-secretase-1) mRNA expression has been shown to be inhibited by miR-485-5p [31]. BACE1-antisense lncRNA and miR-485-5p compete for the same binding site in the ORF of the BACE1 mRNA and BACE1-antisense lncRNA prevents the mRNA-miRNA interaction [31]. In the second mechanism of lncRNA-miRNA interaction, lncRNAs can act as miRNA sponges or decoys and attract miRNAs, competitively sequestering miRNAs away from the target mRNAs [32]. For instance, lncRNA GAS5 (Growth arrest-specific 5), a tumor suppressor, functions as a sponge by sequestering and decreasing oncogenic effects of miR-21 and inhibits the proliferation of cancer cells and induces apoptotic cell death [33,34]. Similarly, lncRNA TRPM2-AS acts as a sponge or a competitive endogenous RNA for tumor-suppressor miR-612 and consequently modulates the derepression of IGF2BP1 and FOXM1 [35]. Silencing of TRPM2-AS inhibited aggressiveness of tumors in gastric cancer patients (proliferation, metastasis, radioresistance), while its overexpression promoted progression of gastric cancer [35].

lncRNAs in cytoplasm are also involved in the modulation of turnover and translation of some specific mRNAs [27]. lncRNAs can prevent the formation of mRNA-miRNA complexes as abovementioned, or they can bind to RNA-binding proteins (RBPs) [36,37]. For instance, lncRNA LAST stabilizes mRNA levels of Cyclin D1 (CCND1) oncogene. lncRNA LAST promotes the binding of CNBP-RBP (CCHC-type zinc finger nucleic acid binding protein) to CCND1, resulting in increased expression of CCND1 by stabilizing its mRNA [38]. Morover, lncRNAs modulate protein stability by influencing to enhance or hinder access to the ubiquitin-dependent proteasomal degradation machinery [27]. A study showed that lncRNA-p21 levels were transcriptionally activated by HIF-1α (Hypoxia-inducible factor-1α) under hypoxic conditions, then lncRNA-p21 binds to both HIF-1α and VHL (von Hippel-Lindau) proteins to protect HIF-1α from VHL-mediated ubiquitination [39]. Furthermore, lncRNAs can promote the proteasomal degradation. For instance, lnc-β-Catm recruits EZH2 to catalyze K49 methylation of β-catenin which inhibits phosphorylation and ubiquitination of β-catenin and promotes its stability [40].

3.2. Nuclear LncRNAs

Some of the lncRNAs are located in the nucleus to regulate gene expression by modulating chromatin organization, RNA processing and transcription [41–45]. The modulatory roles of lncRNAs on gene expression can be either cis- or trans-acting [41] and could negatively or positively affect the expression of target gene.

4. Acting Mechanisms of LncRNAs in the Regulation of Translation

4.1. Overview of Protein Translation Process

Protein translation is a highly complex process, comprising three steps (initiation, elongation, translation) and each step requires dynamic and efficient interactions between a great number of proteins, RNAs and ribosome.

The initiation process consists of two main steps. The first step involves the formation of the pre-initiation complex, and the second step is the assembling of this complex to the large subunit of the ribosome [46]. The initiation step begins with the formation of a ternary complex (eIF2-GTP-Met-tRNA), then the complex binds to small subunit (40S) of ribosome and assembles a pre-initiation complex by binding to other initiation factors (eIF1, eIF1A, eIF3, and eIF5) [47,48]. Before the pre-initiation complex directs to the 5' end of mRNA, eIF4F complex, which is formed by eIF4E (cap-binding protein), eIF4G (scaffold protein) and eIF4A (helicase), bind to the 5' end of mRNA to unwind and activate it [46,49]. The formation of eIF4F complex is maintained by some other initiation factors, eIF4B and eIF3. The pre-initiation complex then scans the mRNA until it recognizes a start codon [50]. Once the start codon is recognized, eIF5 and eIF5B promote hydrolysis of eIF2-bound GTP, releasing of eIFs from the complex and joining to the large subunit of the ribosome [51]. Following the initiation step of translation, met-tRNA reaches the P (peptidyl)-site of the 80S ribosome awaiting amino acids for elongation of the peptide chain.

The elongation step of translation requires the recruitment of aminoacyl-tRNA to the A (aminoacyl)-site of ribosome through GTP-bound eukaryotic elongation factor 1A (eEF1A). Although there is no base-pairing between tRNA anticodon and A-site codon, tRNA generates a codon-anticodon helix by remodeling itself [52] and stabilizes the ternary complex (aa-tRNA-eIF1A-GTP) [53]. Base-pairing interactions between A-site codon and aa-tRNA anticodon induce hydrolysis of GTP by eEF1A, which is then released from the

A-site of the ribosome. eEF1A-GDP complex is recycled by eEF1B. Following the transfer of aa-tRNA to the A-site, a conformational change occurs in the ribosome which facilitates the formation of peptide bond between the aa-tRNA and the tRNA carrying the Met-tRNA at the P site. A GTPase (eEF2) binds to the A-site of the ribosome, hydrolyzes GTP and stimulates a conformational change in the ribosome resulting in movement of the ribosome one codon further. After the translocation of the ribosome, the A-site becomes empty and can accept the next aa-tRNAs to start a new cycle of elongation [52].

The last step of protein translation is termination, which begins when a stop codon (UAA, UGA, or UAG) reaches the A-site of the ribosome. Two types of release factors, eRF1 and eRF3, are involved in the termination process [54–56]. eRF1 is responsible for the recognition of stop codon and stimulation of peptide release, while eRF3 binds to eRF1 and triggers eRF1-mediated peptide release via GTPase activity [56,57]. The ternary complex (eRF1-eRF3-GTP) then binds to the ribosomal pre-termination complex and eRF3 hydrolyses GTP to release polypeptide [58].

4.2. Regulation of Translational Factors by LncRNAs

4.2.1. Inhibitory Roles of LncRNAs in Translation through Regulation of Translation Factors

A growing body of evidence demonstrates that lncRNAs can regulate each step of translation by regulating the expression and the function of translation factors. For instance, lncRNA GAS5 is involved in the regulation of apoptosis and cell proliferation. A study performed with lymphoma cells showed that GAS5 interacts with the translation initiation complex, eIF4F, by directly binding to eIF4E and decreasing the translation of c-Myc [37]. Similarly, lncRNA RP1-5O6.5 has been shown to interact with eIF4E and prevents binding of eIF4E to eIF4G, leading to inhibition of translation of p27kip1, which negatively regulates Snail levels in breast cancer cells [59]. lncRNAs SNHG1 and SNGH4 are capable of binding to eIF4E and dysregulate the function of eIF4E in mantle cell lymphoma cells [60]. In the other example, lncRNA treRNA has been shown to interact with ribonucleoproteins (RNPs) (hnRNP K, FXR1, FXR2, PUF60, and SF3B3) and form treRNA-RNP complex which suppresses the translation efficiency of E-cadherin by binding eIF4G1 [61]. A brain-specific lncRNA, BC1, has been reported to interact with eIF4A and poly(A)-binding protein (PABP) and negatively regulate translation process [62,63]. lncRNA GAPLINC is overexpressed in non-small lung cancer cells and it increases eEF2K expression (a negative regulator of eEF2) by acting as a sponge for miR-661 [64]. In the other study, lncRNA FOXD1-AS1 was shown to bind to eIF5A, however it did not change the mRNA expression levels, suggesting that FOXD1-AS1 can involve in the post-translational regulation [65]. Overall, these studies suggested that lncRNAs can play an important inhibitory roles in mRNA translation through regulation of translation factors.

4.2.2. LncRNAs Positively Regulate Protein Translation

Some lncRNAs have been reported to positively regulate protein translation. For instance, lncRNA SRA enhanced Wnt/β-catenin signaling pathway by increasing the expression of eIF4E-binding protein 1 (eIF4E-BP1) and contributed to the aggressive characteristics of endometrial cancer [66]. Another study showed that lncRNA MCM3AP-AS1 enhances the expression of eIF4E by acting as a sponge for miR15a, which supresses eIF4E expression and contributes to doxorubicin resistance in Burkitt lymphoma cells through MCM3AP-AS1/miR-15a/eIF4E axis [67]. Similarly, lncRNA SNHG12 enhanced the invasion of human vascular smooth muscle cells by serving as a sponge of miR-766-5p and influencing the miR-766-5p/eIF5A axis [68]. In the other study, a Y-linked lncRNA, LINC00278, was found to encode a micropeptide called YY1BM which led to a decrease in the expression of negative regulator of translation, eEF2K [69]. The functions of lncRNAs on translational factors are summarized in Table 1.

Table 1. The list of long non-coding RNAs (lncRNAs) involved in regulation of translational factors [37,59–69].

LncRNA	Translation Factor	Function	Reference
GAS5	Binds to eIF4E and prevents formation of initiation complex (eIF4F)	Decreases translation of c-Myc	[37]
RP1-5O6.5	Interacts with eIF4E and prevents binding to eIF4G	Promotes breast cancer metastasis by inhibiting translation of p27Kip1	[59]
SNHG1 and SNGH4	Bind to eIF4E and dysregulate it	Enhance translation and contribute aggressiveness of lymphoma cells	[60]
treRNA	Promotes the formation of a treRNA-associated protein (treRNP) complex and suppresses translation by binding to eEIF4G1	treRNP complex reduces translation efficiency of E-cadherin and decreases tumor metastasis	[61]
BC1	Interacts with eIF4A and poly(A)-binding protein (PABP)	Represses translation	[62,63]
GAPLINC	Positively regulates eEF2K expression by sponging miR-661	Promotes tumorigenesis of non-small cell lung cancer cells	[64]
SRA	Binds and increases the expression of eIF4E-binding protein 1 (eIF4E-BP1)	Increases the activity of Wnt/β-catenin signaling and promotes aggressive characteristics of endometrial cancer	[66]
MCM3AP-AS1	Positively regulates the expression of eIF4E by using miR15a as a sponge	Promotes translation and contributes doxorubicin resistance	[67]
SNGH12	Binds to miR-766-5p, which is a negative regulator of eIF5A	Targets miR-766-5p/eIF5A axis and enhances invasion of vascular smooth muscle cells	[68]
LNC00278	Decreases eEF2K expression	Micropeptide of lncRNA, YY1BM, represses the eEF2K/eEF2 axis	[69]

4.3. LncRNAs Involved in Signaling Pathways Regulating Protein Translation

The PI3K/AKT/mTOR is one of the major signaling pathways known to regulate vital cellular processes including cell proliferation, growth, survival, metabolism and protein translation. The role of PI3K/AKT/mTOR and MAPK pathways in the regulation of translational machinery are well documented and they are frequently overactivated in most types of cancer [70]. Both pathways involve the mechanistic target of rapamycin (mTOR) to regulate a variety of components of the translational machinery in homeostasis, their dysregulation results in aberrant translation which is often detected in diabetes, neurological disorders, and cancer [71–74]. The MAPK family consists of a serine/threonine kinases, that includes ERKs, JNKs and p38/SAPKs [75]. Especially the MAPK/ERK signaling pathway is amongst the most well-studied, signaling and dysregulating one-third of all human cancers [76].

PI3K/AKT/mTOR pathway regulates cell growth and proliferation by phosphorylating two downstream targets which are 4E-BP1 and ribosomal protein S6 kinase (S6Ks). mTOR complex I (mTORC1) controls translational activation by phosphorylating eIF4E inhibitor, 4E-BP1, which releases eIF4E to interact with initiation complex (eIF4F) [77]. S6K protein requires sequential phosphorylations at multiple serine/threonine sites and mTORC1 regulates its activation by phosphorylation. Once S6K is activated, it phosphorylates and activates eIF4B, which increases the recruitment of eIF4B to eEF4A and enhances translation [78]. Besides, S6K and mTORC1 signaling pathways can phosphorylate EF2-Kinase (EF2K) and decrease its sensitivity to Ca/Calmoduline for its activation [79]. Similarly, eEF2K activity is negatively regulated by MAPKs and their downstream effectors, reducing phosphorylation of eEF2, leading to increased translation by promoting peptide elongation phase of protein systhesis [80,81]. Considering the significant regulatory roles of PI3K/AKT/mTOR and MAPK signaling pathways in protein translation, regulation of their activity by lncRNAs indicate that the lncRNAs are involved in controlling protein translation through regulation of these key signaling pathways. For instance, lncRNA UASR1 promotes cell growth and migration of breast cancer cells by regulating AKT/mTOR pathway [82]. In these cells, active mediators of this pathway such as p-AKT, p-TSC2, p-4EBP1 and p-p70S6K are increased by overexpression of UASR1. Thus, UASR1 plays an oncogenic role in breast cancer cells through activation of the AKT/mTOR signaling pathway. Another lncRNA H19 is overexpressed in colorectal cancer tissues and it promotes the activity of PI3K/AKT pathway by acting as a ceRNA and regulating some components of this pathway. H19 regulates various cancer-related mRNAs (such as (AKT3, CSF1, MET, COL1A1) by competitively sponging various miRNAs. Knockdown of H19 reduced protein level of MET, ZEB1, and COL1A1 in vitro [83]. The other study showed that H19 inhibits mTORC1-mediated 4E-BP1 phosphorylation, but it does not affect the activation of S6K1 [84]. lncRNA CASC9 has been shown to suppress apoptosis and promote aggressiveness of oral squamous cell carcinoma cells by activating the AKT/mTOR pathway [85].

In contrast, some lncRNAs might negatively regulate the abovementioned pathways. For instance, lncRNA FER1L4 suppresses cell proliferation and metastasis through downregulating the expressions of PI3K and AKT in lung cancer cells [86]. Overall, lncRNAs can regulate signaling pathways involved in translational control that is an integral part of these survival adaptive pathways in normal and cancer cells. Some of these regulatory lncRNAs and their functions on signaling pathways are summarized in Table 2.

Table 2. lncRNAs in the regulation of signaling pathways and their roles in various cancers [87–96].

LncRNA	Target	Function	Reference
MALAT1	mTOR signaling	Improves glucose metabolism to contribute aggressiveness in hepatocellular carcinoma cells	[87]
HOXB-AS3	PI3K/AKT signaling	Increases proliferation, migration, and invasion of lung cancer cells	[88]
AK023391	PI3K/AKT signaling	Promotes tumorigenesis and invasion of gastric cancer	[89]
LOC101928316	PI3K/AKT/mTOR signaling	Inhibits cell proliferation, invasion and tumorigenesis of gastric cancer cells	[90]
UCA1	PI3K/AKT signaling	Promotes cell proliferation and inhibits apoptosis in retinoblastoma cells	[91]
OECC	PI3K/AKT/mTOR signaling	Increases proliferation, migration and invasion of lung cancer cells	[92]
GAS5	PTEN/PI3K/AKT signaling	Suppresses proliferation and invasion of osteosarcoma cells and promotes PTEN expression by sponging miR-23a-3p	[93]
LINC01503	MAPK/ERK signaling	Increases proliferation and tumor forming-ability of hepatocellular carcinoma cells	[94]
ST8SIA6-AS1	p38 MAPK signaling	Promotes proliferation, migration and invasion of breast cancer cells	[95]
FENDRR	p38 MAPK signaling	Inhibits cell proliferation and induces apoptosis in hepatocellular carcinoma cells	[96]

4.4. LncRNAs in Cancer

4.4.1. LncRNAs Can Contribute Hallmarks of Cancer

Deregulation of mRNA translation is commonly observed in malignant cells and is considered as a critical factor contributing to cancer initiation, tumorigenesis, and progression. Because lncRNAs play critical roles in the regulation of a wide range of cellular processes, their dysregulation is associated with cell proliferation, survival, tumorigenesis and progression of various cancers, and aberrant expression of lncRNAs can contribute to the hallmarks of cancer. Reprograming of the translation machinery in cancer cells is important function of the key oncogenic signalings, promoting cellular transformation. Increased activity of translational machinery has been shown to be critical in many cancer cells, including breast [97], pancreatic [98], liver [99], and colorectal cancer [100], and leukemia [101]. Thus, lncRNA-mediated regulation of protein translation plays an important role in promoting oncogenic signaling, and specific targeting of these lncRNAs holds promise for developing highly targeted therapies in cancer and other human diseases. Figure 1 illustrates some of the lncRNAs that are involved in tumorigenesis and cancer progression.

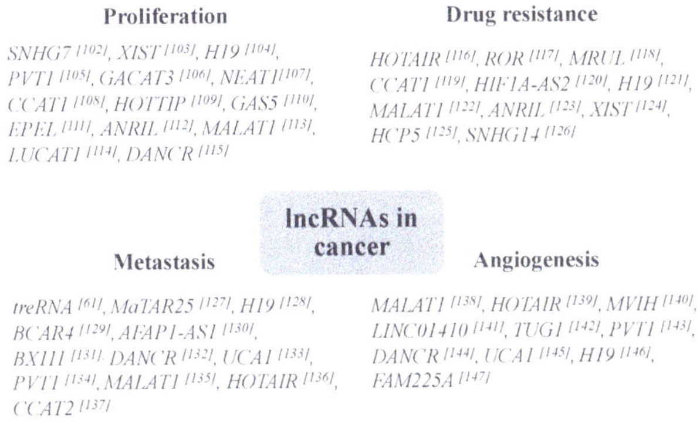

Proliferation

SNHG7 [102], XIST [103], H19 [104], PVT1 [105], GACAT3 [106], NEAT1[107], CCAT1 [108], HOTTIP [109], GAS5 [110], EPEL [111], ANRIL [112], MALAT1 [113], LUCAT1 [114], DANCR [115]

Drug resistance

HOTAIR [116], ROR [117], MRUL [118], CCAT1 [119], HIF1A-AS2 [120], H19 [121], MALAT1 [122], ANRIL [123], XIST [124], HCP5 [125], SNHG14 [126]

lncRNAs in cancer

Metastasis

treRNA [61], MaTAR25 [127], H19 [128], BCAR4 [129], AFAP1-AS1 [130], BX111 [131], DANCR [132], UCA1 [133], PVT1 [134], MALAT1 [135], HOTAIR [136], CCAT2 [137]

Angiogenesis

MALAT1 [138], HOTAIR [139], MVIH [140], LINC01410 [141], TUG1 [142], PVT1 [143], DANCR [144], UCA1 [145], H19 [146], FAM225A [147]

Figure 1. Some lncRNAs are identified to be involved in aggressive characteristics of some common types of cancers [61,102–147].

4.4.2. The Functions of LncRNAs in Regulating Translation of Cancer-Related Proteins

As mentioned above, various lncRNAs are involved in the regulation of hallmarks of cancer, suggesting that they have potential regulatory roles in cancer-related protein tranlation. Since we have already summarized the roles of some lncRNAs on PI3K/AKT/mTOR and MAPK pathways in Table 2, here we briefly focus on the interaction between lncRNAs and translation, promoting the aggressive tumor characteristics.

An example of a lncRNA that is well-known to be associated with cancer is MALAT1. MALAT1 was shown to upregulate the expression of glycolytic genes which contributes the aggressive characteristics of hepatocellular carcinoma cells. MALAT1 regulated the glucose metabolism of hepatocellular carcinoma cells by enhancing translation of metabolic transcription factor TCF7L2 through mTORC1–4EBP1 axis [87]. lncRNA NEAT1 represents another example of lncRNAs that contribute to the aggressiveness of non-small cell lung cancer by enhancing eIF4G2 via miR-582-5p sponging effects [148]. Similarly, lncRNA RP11-284P20.2 enhanced c-met mRNA translation by recruiting eIF3b to c-met and thus promoted proliferation and invasion of hepatocellular carcinoma cells [149]. In prostate cancer, lncRNA UCA1 levels were found to be positively correlated with eIF4G1 levels. UCA1 enhances eIF4G1 levels via sponging miR-331-3p, while knockingdown of UCA1 sensitizes prostate cancer cells to radiotherapy by suppressing eIF4G1 expression via miR-331-3p/eIF4G1 axis [150]. In another study, lncRNA GAPLINC increased the eEF2K expression by serving as a sponge for miR-661, thereby promoted proliferation and progression of non-small cell lung cancer [64].

lncRNAs can also regulate translation process by interacting with the ribosome or ribosome-related proteins. For example, lncRNA ZFAS1 was shown to interact with a small 40S subunit of the ribosome in breast cancer cells. The study showed that ZFAS1 did not regulate translation process directly. Instead, the lncRNA was increased during the ribosome biogenesis indicating its role in regulating the ribosome production and assembly [151]. In neuroblastoma cells, it was shown that lncNB1 enhanced E2F1 protein synthesis and N-Myc stability by binding the ribosomal protein RPL35 [152].

Overall, an emerging body of evidence suggests that lncRNAs play important roles in the regulation of protein translation process. They can enhance or suppress translation via several mechanisms, including through interacting with the ribosome-associated proteins, sponging miRNAs, and competing with endogenous RNAs. Their mechanisms of action and some examples are summarized in Figure 2.

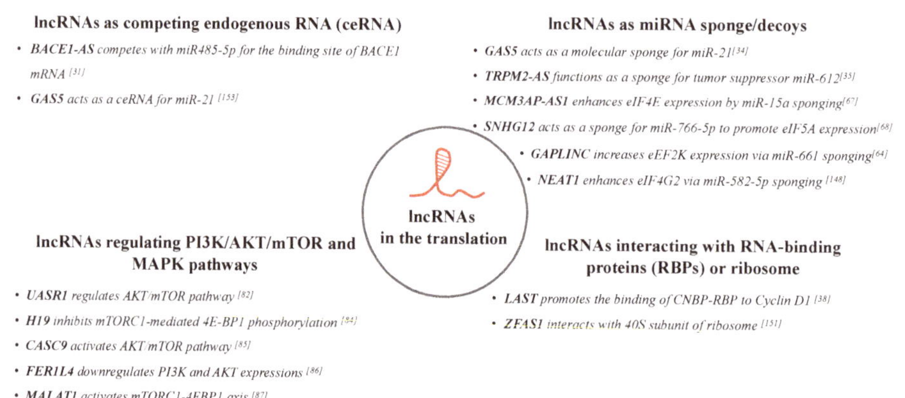

Figure 2. The mechanisms of action of lncRNAs on the regulation of cancer-related protein translation [31,34,35,38,64,67,68,82,84–87,148,151,153].

5. Conclusions

Advances in high throughput technologies resulted in the identification of a large number of lncRNAs. Although thousands of lncRNAs have been identified in the genomes of higher eukaryotes, our understanding of the mechanisms by which lncRNAs exert their precise function for most of them remains unknown. Elucidating the function of these lncRNAs is expected to provide deeper insight into the molecular mechanisms regarding their function in human diseases, including cancer and the interaction of lncRNAs with other molecules may help to design novel strategies. Accumulating evidence indicates that lncRNAs display pivotal roles in the regulation of almost every cellular process by binding to the target proteins, mRNAs, miRNA, and/or DNAs, indicating the complicated roles of lncRNAs. Recent findings revealed that lncRNAs can play important roles in the pathogenesis of human cancers, contributing to tumor growth and progression. Thefore, a better understanding of the role of lncRNAs is needed to elucidate the missing links in the molecular mechanims involved in human diseases, including cancer.

Funding: This research received no external funding.

Conflicts of Interest: The authors declare no conflict of interest.

References

1. Derrien, T.; Johnson, R.; Bussotti, G.; Tanzer, A.; Djebali, S.; Tilgner, H.; Guernec, G.; Martin, D.; Merkel, A.; Knowles, D.G.; et al. The GENCODE v7 catalog of human long noncoding RNAs: Analysis of their gene structure, evolution, and expression. *Genome Res.* **2012**, *22*, 1775–1789. [CrossRef] [PubMed]
2. Rinn, J.L.; Chang, H.Y. Genome regulation by long noncoding RNAs. *Annu. Rev. Biochem.* **2012**, *81*, 145–166. [CrossRef] [PubMed]
3. Quinn, J.J.; Chang, H.Y. Unique features of long non-coding RNA biogenesis and function. *Nat. Rev. Genet.* **2016**, *17*, 47–62. [CrossRef] [PubMed]
4. Aillaud, M.; Schulte, L.N. Emerging Roles of Long Noncoding RNAs in the Cytoplasmic Milieu. *ncRNA* **2020**, *6*, 44. [CrossRef] [PubMed]

5. Penny, G.D.; Kay, G.F.; Sheardown, S.A.; Rastan, S.; Brockdorff, N. Requirement for Xist in X chromosome inactivation. *Nature* **1996**, *379*, 131–137. [CrossRef] [PubMed]
6. Böhmdorfer, G.; Wierzbicki, A.T. Control of Chromatin Structure by Long Noncoding RNA. *Trends Cell Biol.* **2015**, *25*, 623–632. [CrossRef]
7. Connerty, P.; Lock, R.B.; de Bock, C.E. Long Non-coding RNAs: Major Regulators of Cell Stress in Cancer. *Front. Oncol.* **2020**, *10*, 285. [CrossRef]
8. DiStefano, J.K. The Emerging Role of Long Noncoding RNAs in Human Disease. *Methods Mol. Biol.* **2018**, *1706*, 91–110. [PubMed]
9. Ponjavic, J.; Ponting, C.P.; Lunter, G. Functionality or transcriptional noise? Evidence for selection within long noncoding RNAs. *Genome Res.* **2007**, *17*, 556–565. [CrossRef]
10. Struhl, K. Transcriptional noise and the fidelity of initiation by RNA polymerase II. *Nat. Struct. Mol. Biol.* **2007**, *14*, 103–105. [CrossRef]
11. Guttman, M.; Amit, I.; Garber, M.; French, C.; Lin, M.F.; Feldser, D.; Huarte, M.; Zuk, O.; Carey, B.W.; Cassady, J.P.; et al. Chromatin signature reveals over a thousand highly conserved large non-coding RNAs in mammals. *Nature* **2009**, *458*, 223–227. [CrossRef]
12. Guttman, M.; Donaghey, J.; Carey, B.W.; Garber, M.; Grenier, J.K.; Munson, G.; Young, G.; Lucas, A.B.; Ach, R.; Bruhn, L.; et al. lincRNAs act in the circuitry controlling pluripotency and differentiation. *Nature* **2011**, *477*, 295–300. [CrossRef]
13. Hung, T.; Wang, Y.; Lin, M.F.; Koegel, A.K.; Kotake, Y.; Grant, G.D.; Horlings, H.M.; Shah, N.; Umbricht, C.; Wang, P.; et al. Extensive and coordinated transcription of noncoding RNAs within cell-cycle promoters. *Nat. Genet.* **2011**, *43*, 621–629. [CrossRef] [PubMed]
14. Rossi, M.N.; Antonangeli, F. LncRNAs: New Players in Apoptosis Control. *Int. J. Cell Biol.* **2014**, *2014*, 473857. [CrossRef] [PubMed]
15. Lin, Y.H. Crosstalk of lncRNA and Cellular Metabolism and Their Regulatory Mechanism in Cancer. *Int. J. Mol. Sci.* **2020**, *21*, 2947. [CrossRef] [PubMed]
16. Chew, G.L.; Pauli, A.; Rinn, J.L.; Regev, A.; Schier, A.F.; Valen, E. Ribosome profiling reveals resemblance between long non-coding RNAs and 5′ leaders of coding RNAs. *Development* **2013**, *140*, 2828–2834. [CrossRef] [PubMed]
17. Aspden, J.L.; Eyre-Walker, Y.C.; Phillips, R.J.; Amin, U.; Mumtaz, M.A.; Brocard, M.; Couso, J.P. Extensive translation of small Open Reading Frames revealed by Poly-Ribo-Seq. *eLife* **2014**, *3*, e03528. [CrossRef] [PubMed]
18. Ruiz-Orera, J.; Messeguer, X.; Subirana, J.A.; Alba, M.M. Long noncoding RNAs as a source of new peptides. *eLife* **2014**, *3*, e03523. [CrossRef] [PubMed]
19. Van Heesch, S.; van Iterson, M.; Jacobi, J.; Boymans, S.; Essers, P.B.; de Bruijn, E.; Hao, W.; Macinnes, A.W.; Cuppen, E.; Simonis, M. Extensive localization of long noncoding RNAs to the cytosol and mono- and polyribosomal complexes. *Genome Biol.* **2014**, *15*, R6. [CrossRef]
20. Mackowiak, S.D.; Zauber, H.; Bielow, C.; Thiel, D.; Kutz, K.; Calviello, L.; Mastrobuoni, G.; Rajewsky, N.; Kempa, S.; Selbach, M.; et al. Extensive identification and analysis of conserved small ORFs in animals. *Genome Biol.* **2015**, *16*, 179. [CrossRef] [PubMed]
21. Yeasmin, F.; Yada, T.; Akimitsu, N. Micropeptides Encoded in Transcripts Previously Identified as Long Noncoding RNAs: A New Chapter in Transcriptomics and Proteomics. *Front. Genet.* **2018**, *9*, 144. [CrossRef] [PubMed]
22. Clark, M.B.; Johnston, R.L.; Inostroza-Ponta, M.; Fox, A.H.; Fortini, F.; Moscato, P.; Dinger, M.E.; Mattick, J.S. Genome-wide analysis of long noncoding RNA stability. *Genome Res.* **2012**, *22*, 885–898. [CrossRef]
23. Ponting, C.P.; Oliver, P.L.; Reik, W. Evolution and functions of long noncoding RNAs. *Cell* **2009**, *136*, 629–641. [CrossRef]
24. Yao, R.W.; Wang, Y.; Chen, L.L. Cellular functions of long noncoding RNAs. *Nat. Cell Biol.* **2019**, *21*, 542–551. [CrossRef] [PubMed]
25. López-Urrutia, E.; Bustamante Montes, L.P.; Ladrón de Guevara Cervantes, D.; Pérez-Plasencia, C.; Campos-Parra, A.D. Crosstalk Between Long Non-coding RNAs, Micro-RNAs and mRNAs: Deciphering Molecular Mechanisms of Master Regulators in Cancer. *Front. Oncol.* **2019**, *9*, 669. [CrossRef] [PubMed]
26. Rashid, F.; Shah, A.; Shan, G. Long Non-coding RNAs in the Cytoplasm. *Genom. Proteom. Bioinform.* **2016**, *14*, 73–80. [CrossRef] [PubMed]
27. Noh, J.H.; Kim, K.M.; McClusky, W.G.; Abdelmohsen, K.; Gorospe, M. Cytoplasmic functions of long noncoding RNAs. *Wiley Interdiscip. Rev. RNA* **2018**, *9*, e1471. [CrossRef]
28. Goff, L.A.; Rinn, J.L. Linking RNA biology to lncRNAs. *Genome Res.* **2015**, *25*, 1456–1465. [CrossRef]
29. Singh, D.K.; Prasanth, K.V. Functional insights into the role of nuclear-retained long noncoding RNAs in gene expression control in mammalian cells. *Chromosome Res.* **2013**, *21*, 695–711. [CrossRef]
30. Lennox, K.A.; Behlke, M.A. Cellular localization of long non-coding RNAs affects silencing by RNAi more than by antisense oligonucleotides. *Nucleic Acids Res.* **2016**, *44*, 863–877. [CrossRef]
31. Faghihi, M.A.; Zhang, M.; Huang, J.; Modarresi, F.; Van der Brug, M.P.; Nalls, M.A.; Cookson, M.R.; St-Laurent, G., 3rd; Wahlestedt, C. Evidence for natural antisense transcript-mediated inhibition of microRNA function. *Genome Biol.* **2010**, *11*, R56. [CrossRef]
32. Zhang, J.; Liu, L.; Li, J.; Le, T.D. LncmiRSRN: Identification and analysis of long non-coding RNA related miRNA sponge regulatory network in human cancer. *Bioinformatics* **2018**, *34*, 4232–4240. [CrossRef]
33. Zhang, Z.; Zhu, Z.; Watabe, K.; Zhang, X.; Bai, C.; Xu, M.; Wu, F.; Mo, Y.Y. Negative regulation of lncRNA GAS5 by miR-21. *Cell Death Differ.* **2013**, *20*, 1558–1568. [CrossRef]

34. Li, W.; Zhai, L.; Wang, H.; Li, W.; Zhai, L.; Wang, H.; Liu, C.; Zhang, J.; Chen, W.; Wei, Q. Downregulation of LncRNA GAS5 causes trastuzumab resistance in breast cancer. *Oncotarget* **2016**, *7*, 27778–27786. [CrossRef]

35. Xiao, J.; Lin, L.; Luo, D.; Shi, L.; Chen, W.; Fan, H.; Li, Z.; Ma, X.; Ni, P.; Yang, L.; et al. Long noncoding RNA TRPM2-AS acts as a microRNA sponge of miR-612 to promote gastric cancer progression and radioresistance. *Oncogenesis* **2020**, *9*, 29. [CrossRef]

36. Mercer, T.R.; Neph, S.; Dinger, M.E.; Crawford, J.; Smith, M.A.; Shearwood, A.M.; Haugen, E.; Bracken, C.P.; Rackham, O.; Stamatoyannopoulos, J.A.; et al. The human mitochondrial transcriptome. *Cell* **2011**, *146*, 645–658. [CrossRef]

37. Hu, G.; Lou, Z.; Gupta, M. The long non-coding RNA GAS5 cooperates with the eukaryotic translation initiation factor 4E to regulate c-Myc translation. *PLoS ONE* **2014**, *9*, e107016. [CrossRef]

38. Cao, L.; Zhang, P.; Li, J.; Wu, M. LAST, a c-Myc-inducible long noncoding RNA, cooperates with CNBP to promote CCND1 mRNA stability in human cells. *Elife* **2017**, *6*, e30433. [CrossRef]

39. Yang, F.; Zhang, H.; Mei, Y.; Wu, M. Reciprocal regulation of HIF-1α and lincRNA-p21 modulates the Warburg effect. *Mol. Cell.* **2014**, *53*, 88–100. [CrossRef]

40. Zhu, P.; Wang, Y.; Huang, G.; Ye, B.; Liu, B.; Wu, J.; Du, Y.; He, L.; Fan, Z. lnc-β-Catm elicits EZH2-dependent β-catenin stabilization and sustains liver CSC self-renewal. *Nat. Struct. Mol. Biol.* **2016**, *23*, 631–639. [CrossRef]

41. Bergmann, J.H.; Spector, D.L. Long non-coding RNAs: Modulators of nuclear structure and function. *Curr. Opin. Cell Biol.* **2014**, *26*, 10–18. [CrossRef] [PubMed]

42. Vance, K.W.; Ponting, C.P. Transcriptional regulatory functions of nuclear long noncoding RNAs. *Trends Genet.* **2014**, *30*, 348–355. [CrossRef]

43. Quinodoz, S.; Guttman, M. Long noncoding RNAs: An emerging link between gene regulation and nuclear organization. *Trends Cell Biol.* **2014**, *24*, 651–663. [CrossRef]

44. Chen, L.L. Linking Long Noncoding RNA Localization and Function. *Trends Biochem. Sci.* **2016**, *41*, 761–772. [CrossRef] [PubMed]

45. Schmitt, A.M.; Chang, H.Y. Long Noncoding RNAs: At the Intersection of Cancer and Chromatin Biology. *Cold Spring Harb. Perspect. Med.* **2017**, *7*, a026492. [CrossRef]

46. Jackson, R.J.; Hellen, C.U.; Pestova, T.V. The mechanism of eukaryotic translation initiation and principles of its regulation. *Nat. Rev. Mol. Cell Biol.* **2010**, *11*, 113–127. [CrossRef]

47. Dever, T.E. Gene-specific regulation by general translation factors. *Cell* **2002**, *108*, 545–556. [CrossRef]

48. Ali, M.U.; Ur Rahman, M.S.; Jia, Z.; Jiang, C. Eukaryotic translation initiation factors and cancer. *Tumor Biol.* **2017**, *39*, 1010428317709805. [CrossRef] [PubMed]

49. Grifo, J.A.; Tahara, S.M.; Morgan, M.A.; Shatkin, A.J.; Merrick, W.C. New initiation factor activity required for globin mRNA translation. *J. Biol. Chem.* **1983**, *258*, 5804–5810. [CrossRef]

50. Hinnebusch, A.G. Molecular mechanism of scanning and start codon selection in eukaryotes. *Microbiol. Mol. Biol. Rev.* **2011**, *75*, 434–467. [CrossRef] [PubMed]

51. Llácer, J.L.; Hussain, T.; Saini, A.K.; Nanda, J.S.; Kaur, S.; Gordiyenko, Y.; Kumar, R.; Hinnebusch, A.G.; Lorsch, J.R.; Ramakrishnan, V. Translational initiation factor eIF5 replaces eIF1 on the 40S ribosomal subunit to promote start-codon recognition. *eLife* **2018**, *7*, e39273. [CrossRef]

52. Voorhees, R.M.; Ramakrishnan, V. Structural basis of the translational elongation cycle. *Annu. Rev. Biochem.* **2013**, *82*, 203–236. [CrossRef] [PubMed]

53. Rodnina, M.V.; Gromadski, K.B.; Kothe, U.; Wieden, H.J. Recognition and selection of tRNA in translation. *FEBS Lett.* **2005**, *579*, 938–942. [CrossRef] [PubMed]

54. Stansfield, I.; Jones, K.M.; Kushnirov, V.V.; Dagkesamanskaya, A.R.; Poznyakovski, A.I.; Paushkin, S.V.; Nierras, C.R.; Cox, B.S.; Ter-Avanesyan, M.D.; Tuite, M.F. The products of the SUP45 (eRF1) and SUP35 genes interact to mediate translation termination in Saccharomyces cerevisiae. *EMBO J.* **1995**, *14*, 4365–4373. [CrossRef]

55. Zhouravleva, G.; Frolova, L.; Le Goff, X.; Le Guellec, R.; Inge-Vechtomov, S.; Kisselev, L.; Philippe, M. Termination of translation in eukaryotes is governed by two interacting polypeptide chain release factors, eRF1 and eRF3. *EMBO J.* **1995**, *14*, 4065–4072. [CrossRef]

56. Alkalaeva, E.Z.; Pisarev, A.V.; Frolova, L.Y.; Kisselev, L.L.; Pestova, T.V. In vitro reconstitution of eukaryotic translation reveals cooperativity between release factors eRF1 and eRF3. *Cell* **2006**, *125*, 1125–1136. [CrossRef]

57. Salas-Marco, J.; Bedwell, D.M. GTP hydrolysis by eRF3 facilitates stop codon decoding during eukaryotic translation termination. *Mol. Cell Biol.* **2004**, *24*, 7769–7778. [CrossRef]

58. Frolova, L.; Le Goff, X.; Zhouravleva, G.; Davydova, E.; Philippe, M.; Kisselev, L. Eukaryotic polypeptide chain release factor eRF3 is an eRF1- and ribosome-dependent guanosine triphosphatase. *RNA* **1996**, *2*, 334–341.

59. Jia, X.; Shi, L.; Wang, X.; Luo, L.; Ling, L.; Yin, J.; Song, Y.; Zhang, Z.; Qiu, N.; Liu, H.; et al. KLF5 regulated lncRNA RP1 promotes the growth and metastasis of breast cancer via repressing p27kip1 translation. *Cell Death Dis.* **2019**, *10*, 373. [CrossRef]

60. Hu, G.; Witzig, T.E.; Gupta, M. Novel Long Non-Coding RNA, SNHG4 Complex With Eukaryotic Initiation Factor-4E and Regulate Aberrant Protein Translation In Mantle Cell Lymphoma: Implications For Novel Biomarker. *Blood* **2013**, *122*, 81. [CrossRef]

61. Gumireddy, K.; Li, A.; Yan, J.; Setoyama, T.; Johannes, G.J.; Orom, U.A.; Tchou, J.; Liu, Q.; Zhang, L.; Speicher, D.W.; et al. Identification of a long non-coding RNA-associated RNP complex regulating metastasis at the translational step. *EMBO J.* **2013**, *32*, 2672–2684. [CrossRef]

62. Wang, H.; Iacoangeli, A.; Popp, S.; Muslimov, I.A.; Imataka, H.; Sonenberg, N.; Lomakin, I.B.; Tiedge, H. Dendritic BC1 RNA: Functional role in regulation of translation initiation. *J. Neurosci.* **2002**, *22*, 10232–10241. [CrossRef]
63. Wang, H.; Iacoangeli, A.; Lin, D.; Williams, K.; Denman, R.B.; Hellen, C.U.; Tiedge, H. Dendritic BC1 RNA in translational control mechanisms. *J. Cell Biol.* **2005**, *171*, 811–821. [CrossRef]
64. Gu, H.; Chen, J.; Song, Y.; Shao, H. Gastric Adenocarcinoma Predictive Long Intergenic Non-Coding RNA Promotes Tumor Occurrence and Progression in Non-Small Cell Lung Cancer via Regulation of the miR-661/eEF2K Signaling Pathway. *Cell Physiol. Biochem.* **2018**, *51*, 2136–2147. [CrossRef] [PubMed]
65. Gao, Y.F.; Liu, J.Y.; Mao, X.Y.; He, Z.W.; Zhu, T.; Wang, Z.B.; Li, X.; Yin, J.Y.; Zhang, W.; Zhou, H.H.; et al. LncRNA FOXD1-AS1 acts as a potential oncogenic biomarker in glioma. *CNS Neurosci. Ther.* **2020**, *26*, 66–75. [CrossRef]
66. Park, S.A.; Kim, L.K.; Kim, Y.T.; Heo, T.H.; Kim, H.J. Long non-coding RNA steroid receptor activator promotes the progression of endometrial cancer via Wnt/β-catenin signaling pathway. *Int. J. Biol. Sci.* **2020**, *16*, 99–115. [CrossRef]
67. Guo, C.; Gong, M.; Li, Z. Knockdown of lncRNA MCM3AP-AS1 Attenuates Chemoresistance of Burkitt Lymphoma to Doxorubicin Treatment via Targeting the miR-15a/EIF4E Axis. *Cancer Manag. Res.* **2020**, *12*, 5845–5855. [CrossRef] [PubMed]
68. Liu, W.; Che, J.; Gu, Y.; Song, L. LncRNA SNHG12 promotes proliferation and migration of vascular smooth muscle cells via targeting miR-766-5p/EIF5A. *Res. Square* **2020**. [CrossRef]
69. Wu, S.; Zhang, L.; Deng, J.; Guo, B.; Li, F.; Wang, Y.; Wu, R.; Zhang, S.; Lu, J.; Zhou, Y. A Novel Micropeptide Encoded by Y-Linked LINC00278 Links Cigarette Smoking and AR Signaling in Male Esophageal Squamous Cell Carcinoma. *Cancer Res.* **2020**, *80*, 2790–2803. [CrossRef]
70. Janku, F.; Yap, T.A.; Meric-Bernstam, F. Targeting the PI3K pathway in cancer: Are we making headway? *Nat. Rev. Clin. Oncol.* **2018**, *15*, 273–291. [CrossRef]
71. Kim, E.K.; Choi, E.J. Pathological roles of MAPK signaling pathways in human diseases. *Biochim. Biophys. Acta* **2010**, *1802*, 396–405. [CrossRef] [PubMed]
72. Fruman, D.A.; Chiu, H.; Hopkins, B.D.; Bagrodia, S.; Cantley, L.C.; Abraham, R.T. The PI3K Pathway in Human Disease. *Cell* **2017**, *170*, 605–635. [CrossRef]
73. Braicu, C.; Buse, M.; Busuioc, C.; Drula, R.; Gulei, D.; Raduly, L.; Rusu, A.; Irimie, A.; Atanasov, A.G.; Slaby, O.; et al. A Comprehensive Review on MAPK: A Promising Therapeutic Target in Cancer. *Cancers* **2019**, *11*, 1618. [CrossRef]
74. Xu, F.; Na, L.; Li, Y.; Chen, L. Roles of the PI3K/AKT/mTOR signalling pathways in neurodegenerative diseases and tumours. *Cell Biosci.* **2020**, *10*, 54. [CrossRef]
75. Morrison, D.K. MAP kinase pathways. *Cold Spring Harb. Perspect. Biol.* **2012**, *4*, a011254. [CrossRef]
76. Dhillon, A.S.; Hagan, S.; Rath, O.; Kolch, W. MAP kinase signalling pathways in cancer. *Oncogene* **2007**, *26*, 3279–3290. [CrossRef]
77. Heesom, K.J.; Denton, R.M. Dissociation of the eukaryotic initiation factor-4E/4E-BP1 complex involves phosphorylation of 4E-BP1 by an mTOR-associated kinase. *FEBS Lett.* **1999**, *457*, 489–493. [CrossRef]
78. Holz, M.K.; Ballif, B.A.; Gygi, S.P.; Blenis, J. mTOR and S6K1 mediate assembly of the translation preinitiation complex through dynamic protein interchange and ordered phosphorylation events. *Cell* **2005**, *123*, 569–580. [CrossRef]
79. Browne, G.J.; Proud, C.G. A novel mTOR-regulated phosphorylation site in elongation factor 2 kinase modulates the activity of the kinase and its binding to calmodulin. *Mol. Cell. Biol.* **2004**, *24*, 2986–2997. [CrossRef]
80. Knebel, A.; Morrice, N.; Cohen, P. A novel method to identify protein kinase substrates: eEF2 kinase is phosphorylated and inhibited by SAPK4/p38delta. *EMBO J.* **2001**, *20*, 4360–4369. [CrossRef] [PubMed]
81. Knebel, A.; Haydon, C.E.; Morrice, N.; Cohen, P. Stress-induced regulation of eEF2 kinase by SB203580-sensitive and -insensitive pathways. *Biochem. J.* **2002**, *367*, 525–532. [CrossRef] [PubMed]
82. Cao, Z.; Wu, P.; Su, M.; Ling, H.; Khoshaba, R.; Huang, C.; Gao, H.; Zhao, Y.; Chen, J.; Liao, Q.; et al. Long non-coding RNA UASR1 promotes proliferation and migration of breast cancer cells through the AKT/mTOR pathway. *J. Cancer* **2019**, *10*, 2025–2034. [CrossRef] [PubMed]
83. Zhong, M.E.; Chen, Y.; Zhang, G.; Xu, L.; Ge, W.; Wu, B. LncRNA H19 regulates PI3K-Akt signal pathway by functioning as a ceRNA and predicts poor prognosis in colorectal cancer: Integrative analysis of dysregulated ncRNA-associated ceRNA network. *Cancer Cell Int.* **2019**, *19*, 148. [CrossRef] [PubMed]
84. Wu, Z.R.; Yan, L.; Liu, Y.T.; Cao, L.; Guo, Y.H.; Zhang, Y.; Yao, H.; Cai, L.; Shang, H.B.; Rui, W.W.; et al. Inhibition of mTORC1 by lncRNA H19 via disrupting 4E-BP1/Raptor interaction in pituitary tumours. *Nat. Commun.* **2018**, *9*, 4624. [CrossRef]
85. Yang, Y.; Chen, D.; Liu, H.; Yang, K. Increased expression of lncRNA CASC9 promotes tumor progression by suppressing autophagy-mediated cell apoptosis via the AKT/mTOR pathway in oral squamous cell carcinoma. *Cell Death Dis.* **2019**, *10*, 41. [CrossRef]
86. Gao, X.; Wang, N.; Wu, S.; Cui, H.; An, X.; Yang, Y. Long non-coding RNA FER1L4 inhibits cell proliferation and metastasis through regulation of the PI3K/AKT signaling pathway in lung cancer cells. *Mol. Med. Rep.* **2019**, *20*, 182–190. [CrossRef]
87. Malakar, P.; Stein, I.; Saragovi, A.; Winkler, R.; Stern-Ginossar, N.; Berger, M.; Pikarsky, E.; Karni, R. Long Noncoding RNA MALAT1 Regulates Cancer Glucose Metabolism by Enhancing mTOR-Mediated Translation of TCF7L2. *Cancer Res.* **2019**, *79*, 2480–2493. [CrossRef]
88. Jiang, W.; Kai, J.; Li, D.; Wei, Z.; Wang, Y.; Wang, W. lncRNA HOXB-AS3 exacerbates proliferation, migration, and invasion of lung cancer via activating the PI3K-AKT pathway. *J. Cell Physiol.* **2020**, *235*, 7194–7203. [CrossRef]

89. Huang, Y.; Zhang, J.; Hou, L.; Wang, G.; Liu, H.; Zhang, R.; Chen, X.; Zhu, J. LncRNA AK023391 promotes tumorigenesis and invasion of gastric cancer through activation of the PI3K/Akt signaling pathway. *J. Exp. Clin. Cancer Res.* **2017**, *36*, 194. [CrossRef]

90. Li, C.; Liang, G.; Yang, S.; Sui, J.; Wu, W.; Xu, S.; Ye, Y.; Shen, B.; Zhang, X.; Zhang, Y. LncRNA-LOC101928316 contributes to gastric cancer progression through regulating PI3K-Akt-mTOR signaling pathway. *Cancer Med.* **2019**, *8*, 4428–4440. [CrossRef]

91. Yuan, Z.; Li, Z. Long noncoding RNA UCA1 facilitates cell proliferation and inhibits apoptosis in retinoblastoma by activating the PI3K/Akt pathway. *Transl. Cancer Res.* **2020**, *9*. [CrossRef]

92. Zou, Y.; Zhang, B.; Mao, Y.; Zhang, H.; Hong, W. Long non-coding RNA OECC promotes cell proliferation and metastasis through the PI3K/Akt/mTOR signaling pathway in human lung cancer. *Oncol. Lett.* **2019**, *18*, 3017–3024. [CrossRef]

93. Liu, J.; Chen, M.; Ma, L.; Dang, X.; Du, G. LncRNA GAS5 Suppresses the Proliferation and Invasion of Osteosarcoma Cells via the miR-23a-3p/PTEN/PI3K/AKT Pathway. *Cell Transplant.* **2020**, *29*, 963689720953093. [CrossRef]

94. Wang, M.R.; Fang, D.; Di, M.P.; Guan, J.L.; Wang, G.; Liu, L.; Sheng, J.Q.; Tian, D.A.; Li, P.Y. Long non-coding RNA LINC01503 promotes the progression of hepatocellular carcinoma via activating MAPK/ERK pathway. *Int. J. Med. Sci.* **2020**, *17*, 1224–1234. [CrossRef]

95. Fang, K.; Hu, C.; Zhang, X.; Hou, Y.; Gao, D.; Guo, Z.; Li, L. LncRNA ST8SIA6-AS1 promotes proliferation, migration and invasion in breast cancer through the p38 MAPK signalling pathway. *Carcinogenesis* **2020**, *41*, 1273–1281. [CrossRef] [PubMed]

96. Qian, G.; Jin, X.; Zhang, L. LncRNA FENDRR Upregulation Promotes Hepatic Carcinoma Cells Apoptosis by Targeting miR-362-5p Via NPR3 and p38-MAPK Pathway. *Cancer Biother. Radiopharm.* **2020**, *35*, 629–639. [CrossRef]

97. Zhang, T.; Hu, H.; Yan, G.; Wu, T.; Liu, S.; Chen, W.; Ning, Y.; Lu, Z. Long Non-Coding RNA and Breast Cancer. *Technol. Cancer Res. Treat.* **2019**, *18*, 1533033819843889. [CrossRef]

98. Zhou, W.; Chen, L.; Li, C.; Huang, R.; Guo, M.; Ning, S.; Ji, J.; Guo, X.; Lou, G.; Jia, X.; et al. The multifaceted roles of long noncoding RNAs in pancreatic cancer: An update on what we know. *Cancer Cell Int.* **2020**, *20*, 41. [CrossRef]

99. Huang, Z.; Zhou, J.K.; Peng, Y.; He, W.; Huang, C. The role of long noncoding RNAs in hepatocellular carcinoma. *Mol. Cancer* **2020**, *19*, 77. [CrossRef]

100. Siddiqui, H.; Al-Ghafari, A.; Choudhry, H.; Al Doghaither, H. Roles of long non-coding RNAs in colorectal cancer tumorigenesis: A Review. *Mol. Clin. Oncol.* **2019**, *11*, 167–172. [CrossRef]

101. Gao, J.; Wang, F.; Wu, P.; Chen, Y.; Jia, Y. Aberrant LncRNA Expression in Leukemia. *J. Cancer* **2020**, *11*, 4284–4296. [CrossRef] [PubMed]

102. Cheng, D.; Fan, J.; Ma, Y.; Zhou, Y.; Qin, K.; Shi, M.; Yang, J. LncRNA SNHG7 promotes pancreatic cancer proliferation through ID4 by sponging miR-342-3p. *Cell Biosci.* **2019**, *9*, 28. [CrossRef]

103. Wei, W.; Liu, Y.; Lu, Y.; Yang, B.; Tang, L. LncRNA XIST Promotes Pancreatic Cancer Proliferation Through miR-133a/EGFR. *J. Cell Biochem.* **2017**, *118*, 3349–3358. [CrossRef] [PubMed]

104. Berteaux, N.; Lottin, S.; Monté, D.; Pinte, S.; Quatannens, B.; Coll, J.; Hondermarck, H.; Curgy, J.J.; Dugimont, T.; Adriaenssens, E. H19 mRNA-like noncoding RNA promotes breast cancer cell proliferation through positive control by E2F1. *J. Biol. Chem.* **2005**, *280*, 29625–29636. [CrossRef] [PubMed]

105. Tang, J.; Li, Y.; Sang, Y.; Yu, B.; Lv, D.; Zhang, W.; Feng, H. LncRNA PVT1 regulates triple-negative breast cancer through KLF5/beta-catenin signaling. *Oncogene* **2018**, *37*, 4723–4734. [CrossRef]

106. Zhong, H.; Yang, J.; Zhang, B.; Wang, X.; Pei, L.; Zhang, L.; Lin, Z.; Wang, Y.; Wang, C. LncRNA GACAT3 predicts poor prognosis and promotes cell proliferation in breast cancer through regulation of miR-497/CCND2. *Cancer Biomark.* **2018**, *22*, 787–797. [CrossRef] [PubMed]

107. Zhang, M.; Wu, W.B.; Wang, Z.W.; Wang, X.H. lncRNA NEAT1 is closely related with progression of breast cancer via promoting proliferation and EMT. *Eur. Rev. Med. Pharmacol. Sci.* **2017**, *21*, 1020–1026.

108. You, Z.; Liu, C.; Wang, C.; Ling, Z.; Wang, Y.; Wang, Y.; Zhang, M.; Chen, S.; Xu, B.; Guan, H.; et al. LncRNA CCAT1 Promotes Prostate Cancer Cell Proliferation by Interacting with DDX5 and MIR-28-5P. *Mol. Cancer Ther.* **2019**, *18*, 2469–2479. [CrossRef] [PubMed]

109. Yang, B.; Gao, G.; Wang, Z.; Sun, D.; Wei, X.; Ma, Y.; Ding, Y. Long non-coding RNA HOTTIP promotes prostate cancer cells proliferation and migration by sponging miR-216a-5p. *Biosci. Rep.* **2018**, *38*, BSR20180566. [CrossRef]

110. Zhang, Y.; Su, X.; Kong, Z.; Fu, F.; Zhang, P.; Wang, D.; Wu, H.; Wan, X.; Li, Y. An androgen reduced transcript of LncRNA GAS5 promoted prostate cancer proliferation. *PLoS ONE* **2017**, *12*, e0182305. [CrossRef]

111. Park, S.M.; Choi, E.Y.; Bae, D.H.; Sohn, H.A.; Kim, S.Y.; Kim, Y.J. The LncRNA EPEL Promotes Lung Cancer Cell Proliferation Through E2F Target Activation. *Cell Physiol. Biochem.* **2018**, *45*, 1270–1283. [CrossRef] [PubMed]

112. Nie, F.Q.; Sun, M.; Yang, J.S.; Xie, M.; Xu, T.P.; Xia, R.; Liu, Y.W.; Liu, X.H.; Zhang, E.B.; Lu, K.H.; et al. Long noncoding RNA ANRIL promotes non-small cell lung cancer cell proliferation and inhibits apoptosis by silencing KLF2 and P21 expression. *Mol. Cancer Ther.* **2015**, *14*, 268–277. [CrossRef]

113. Feng, C.; Zhao, Y.; Li, Y.; Zhang, T.; Ma, Y.; Liu, Y. LncRNA MALAT1 Promotes Lung Cancer Proliferation and Gefitinib Resistance by Acting as a miR-200a Sponge. *Arch. Bronconeumol.* **2019**, *55*, 627–633. [CrossRef] [PubMed]

114. Wu, R.; Li, L.; Bai, Y.; Yu, B.; Xie, C.; Wu, H.; Zhang, Y.; Huang, L.; Yan, Y.; Li, X.; et al. The long noncoding RNA LUCAT1 promotes colorectal cancer cell proliferation by antagonizing Nucleolin to regulate MYC expression. *Cell Death Dis.* **2020**, *11*, 908. [CrossRef] [PubMed]

115. Wang, Y.; Lu, Z.; Wang, N.; Feng, J.; Zhang, J.; Luan, L.; Zhao, W.; Zeng, X. Long noncoding RNA DANCR promotes colorectal cancer proliferation and metastasis via miR-577 sponging. *Exp. Mol. Med.* **2018**, *50*, 1–17. [CrossRef]

116. Liu, Z.; Sun, M.; Lu, K.; Liu, J.; Zhang, M.; Wu, W.; De, W.; Wang, Z.; Wang, R. The long noncoding RNA HOTAIR contributes to cisplatin resistance of human lung adenocarcinoma cells via downregualtion of p21(WAF1/CIP1) expression. *PLoS ONE* **2013**, *8*, e77293.

117. Pan, Y.; Chen, J.; Tao, L.; Zhang, K.; Wang, R.; Chu, X.; Chen, L. Long noncoding RNA ROR regulates chemoresistance in docetaxel-resistant lung adenocarcinoma cells via epithelial mesenchymal transition pathway. *Oncotarget* **2017**, *8*, 33144–33158. [CrossRef]

118. Wang, Y.; Zhang, D.; Wu, K.; Zhao, Q.; Nie, Y.; Fan, D. Long noncoding RNA MRUL promotes ABCB1 expression in multidrug-resistant gastric cancer cell sublines. *Mol. Cell Biol.* **2014**, *34*, 3182–3193. [CrossRef]

119. Chen, J.; Zhang, K.; Song, H.; Wang, R.; Chu, X.; Chen, L. Long noncoding RNA CCAT1 acts as an oncogene and promotes chemoresistance in docetaxel-resistant lung adenocarcinoma cells. *Oncotarget* **2016**, *7*, 62474–62489. [CrossRef]

120. Jiang, Y.Z.; Liu, Y.R.; Xu, X.E.; Jin, X.; Hu, X.; Yu, K.D.; Shao, Z.M. Transcriptome Analysis of Triple-Negative Breast Cancer Reveals an Integrated mRNA-lncRNA Signature with Predictive and Prognostic Value. *Cancer Res.* **2016**, *76*, 2105–2114. [CrossRef]

121. Ren, J.; Ding, L.; Zhang, D.; Shi, G.; Xu, Q.; Shen, S.; Wang, Y.; Wang, T.; Hou, Y. Carcinoma-associated fibroblasts promote the stemness and chemoresistance of colorectal cancer by transferring exosomal lncRNA H19. *Theranostics* **2018**, *8*, 3932–3948. [CrossRef] [PubMed]

122. Fang, Z.; Chen, W.; Yuan, Z.; Liu, X.; Jiang, H. LncRNA-MALAT1 contributes to the cisplatin-resistance of lung cancer by upregulating MRP1 and MDR1 via STAT3 activation. *Biomed. Pharmacother.* **2018**, *101*, 536–542. [CrossRef]

123. Lan, W.G.; Xu, D.H.; Xu, C.; Ding, C.L.; Ning, F.L.; Zhou, Y.L.; Ma, L.B.; Liu, C.M.; Han, X. Silencing of long non-coding RNA ANRIL inhibits the development of multidrug resistance in gastric cancer cells. *Oncol. Rep.* **2016**, *36*, 263–270. [CrossRef]

124. Zhu, J.; Zhang, R.; Yang, D.; Li, J.; Yan, X.; Jin, K.; Li, W.; Liu, X.; Zhao, J.; Shang, W.; et al. Knockdown of Long Non-Coding RNA XIST Inhibited Doxorubicin Resistance in Colorectal Cancer by Upregulation of miR-124 and Downregulation of SGK1. *Cell Physiol. Biochem.* **2018**, *51*, 113–128. [CrossRef]

125. Liu, Y.; Wang, J.; Dong, L.; Xia, L.; Zhu, H.; Li, Z.; Yu, X. Long Noncoding RNA HCP5 Regulates Pancreatic Cancer Gemcitabine (GEM) Resistance By Sponging Hsa-miR-214-3p To Target HDGF. *Oncol. Ther.* **2019**, *12*, 8207–8216. [CrossRef]

126. Zhang, X.; Zhao, P.; Wang, C.; Xin, B. SNHG14 enhances gemcitabine resistance by sponging miR-101 to stimulate cell autophagy in pancreatic cancer. *Biochem. Biophys. Res. Commun.* **2019**, *510*, 508–514. [CrossRef]

127. Chang, K.C.; Diermeier, S.D.; Yu, A.T.; Brine, L.D.; Russo, S.; Bhatia, S.; Alsudani, H.; Kostroff, K.; Bhuiya, T.; Brogi, E.; et al. MaTAR25 lncRNA regulates the Tensin1 gene to impact breast cancer progression. *Nat. Commun.* **2020**, *11*, 6438. [CrossRef]

128. Zhang, Y.; Huang, W.; Yuan, Y.; Li, J.; Wu, J.; Yu, J.; He, Y.; Wei, Z.; Zhang, C. Long non-coding RNA H19 promotes colorectal cancer metastasis via binding to hnRNPA2B1. *J. Exp. Clin. Cancer Res.* **2020**, *39*, 141. [CrossRef]

129. Xing, Z.; Lin, A.; Li, C.; Liang, K.; Wang, S.; Liu, Y.; Park, P.K.; Qin, L.; Wei, Y.; Hawke, D.H.; et al. lncRNA directs cooperative epigenetic regulation downstream of chemokine signals. *Cell* **2014**, *159*, 1110–1125. [CrossRef]

130. Leng, X.; Ding, X.; Wang, S.; Fang, T.; Shen, W.; Xia, W.; You, R.; Xu, K.; Yin, R. Long noncoding RNA AFAP1-AS1 is upregulated in NSCLC and associated with lymph node metastasis and poor prognosis. *Oncol. Lett.* **2018**, *16*, 727–732. [CrossRef]

131. Deng, S.J.; Chen, H.Y.; Ye, Z.; Deng, S.C.; Zhu, S.; Zeng, Z.; He, C.; Liu, M.L.; Huang, K.; Zhong, J.X.; et al. Hypoxia-induced LncRNA-BX111 promotes metastasis and progression of pancreatic cancer through regulating ZEB1 transcription. *Oncogene* **2018**, *37*, 5811–5828. [CrossRef] [PubMed]

132. Luo, Y.; Wang, Q.; Teng, L.; Zhang, J.; Song, J.; Bo, W.; Liu, D.; He, Y.; Tan, A. LncRNA DANCR promotes proliferation and metastasis in pancreatic cancer by regulating miRNA-33b. *FEBS Open Bio* **2020**, *10*, 18–27. [CrossRef] [PubMed]

133. Zhang, M.; Zhao, Y.; Zhang, Y.; Wang, D.; Gu, S.; Feng, W.; Peng, W.; Gong, A.; Xu, M. LncRNA UCA1 promotes migration and invasion in pancreatic cancer cells via the Hippo pathway. *Biochim. Biophys. Acta Mol. Basis Dis.* **2018**, *1864*, 1770–1782. [CrossRef]

134. Sun, F.; Wu, K.; Yao, Z.; Mu, X.; Zheng, Z.; Sun, M.; Wang, Y.; Liu, Z.; Zhu, Y. Long Noncoding RNA PVT1 Promotes Prostate Cancer Metastasis by Increasing NOP2 Expression via Targeting Tumor Suppressor MicroRNAs. *OncoTargets Ther.* **2020**, *13*, 6755–6765. [CrossRef]

135. Yang, M.H.; Hu, Z.Y.; Xu, C.; Xie, L.Y.; Wang, X.Y.; Chen, S.Y.; Li, Z.G. MALAT1 promotes colorectal cancer cell proliferation/migration/invasion via PRKA kinase anchor protein 9. *Biochim. Biophys. Acta* **2015**, *1852*, 166–174. [CrossRef]

136. Gupta, R.A.; Shah, N.; Wang, K.C.; Kim, J.; Horlings, H.M.; Wong, D.J.; Tsai, M.C.; Hung, T.; Argani, P.; Rinn, J.L.; et al. Long non-coding RNA HOTAIR reprograms chromatin state to promote cancer metastasis. *Nature* **2010**, *464*, 1071–1076. [CrossRef] [PubMed]

137. Ling, H.; Spizzo, R.; Atlasi, Y.; Nicoloso, M.; Shimizu, M.; Redis, R.S.; Nishida, N.; Gafà, R.; Song, J.; Guo, Z.; et al. CCAT2, a novel noncoding RNA mapping to 8q24, underlies metastatic progression and chromosomal instability in colon cancer. *Genome Res.* **2013**, *23*, 1446–1461. [CrossRef] [PubMed]

138. Tee, A.E.; Liu, B.; Song, R.; Li, J.; Pasquier, E.; Cheung, B.B.; Jiang, C.; Marshall, G.M.; Haber, M.; Norris, M.D.; et al. The long noncoding RNA MALAT1 promotes tumor-driven angiogenesis by up-regulating pro-angiogenic gene expression. *Oncotarget* **2016**, *7*, 8663–8675. [CrossRef]

139. Fu, W.M.; Lu, Y.F.; Hu, B.G.; Liang, W.C.; Zhu, X.; Yang, H.D.; Li, G.; Zhang, J.F. Long noncoding RNA Hotair mediated angiogenesis in nasopharyngeal carcinoma by direct and indirect signaling pathways. *Oncotarget* **2016**, *7*, 4712–4723. [CrossRef]

140. Yuan, S.X.; Yang, F.; Yang, Y.; Tao, Q.F.; Zhang, J.; Huang, G.; Yang, Y.; Wang, R.Y.; Yang, S.; Huo, X.S.; et al. Long noncoding RNA associated with microvascular invasion in hepatocellular carcinoma promotes angiogenesis and serves as a predictor for hepatocellular carcinoma patients' poor recurrence-free survival after hepatectomy. *Hepatology* **2012**, *56*, 2231–2241. [CrossRef]

141. Zhang, J.X.; Chen, Z.H.; Chen, D.L.; Tian, X.P.; Wang, C.Y.; Zhou, Z.W.; Gao, Y.; Xu, Y.; Chen, C.; Zheng, Z.S.; et al. LINC01410-miR-532-NCF2-NF-kB feedback loop promotes gastric cancer angiogenesis and metastasis. *Oncogene* **2018**, *37*, 2660–2675. [CrossRef] [PubMed]

142. Cai, H.; Liu, X.; Zheng, J.; Xue, Y.; Ma, J.; Li, Z.; Xi, Z.; Li, Z.; Bao, M.; Liu, Y. Long non-coding RNA taurine upregulated 1 enhances tumor-induced angiogenesis through inhibiting microRNA-299 in human glioblastoma. *Oncogene* **2017**, *36*, 318–331. [CrossRef] [PubMed]

143. Zhao, J.; Du, P.; Cui, P.; Qin, Y.; Hu, C.; Wu, J.; Zhou, Z.; Zhang, W.; Qin, L.; Huang, G. LncRNA PVT1 promotes angiogenesis via activating the STAT3/VEGFA axis in gastric cancer. *Oncogene* **2018**, *37*, 4094–4109. [CrossRef]

144. Lin, X.; Yang, F.; Qi, X.; Li, Q.; Wang, D.; Yi, T.; Yin, R.; Zhao, X.; Zhong, X.; Bian, C. LncRNA DANCR promotes tumor growth and angiogenesis in ovarian cancer through direct targeting of miR-145. *Mol. Carcinog.* **2019**, *58*, 2286–2296. [CrossRef]

145. Guo, Z.; Wang, X.; Yang, Y.; Chen, W.; Zhang, K.; Teng, B.; Huang, C.; Zhao, Q.; Qiu, Z. Hypoxic Tumor-Derived Exosomal Long Noncoding RNA UCA1 Promotes Angiogenesis via miR-96-5p/AMOTL2 in Pancreatic Cancer. *Mol. Ther. Nucl. Acids* **2020**, *22*, 179–195. [CrossRef]

146. Jiang, X.; Yan, Y.; Hu, M.; Chen, X.; Wang, Y.; Dai, Y.; Wu, D.; Wang, Y.; Zhuang, Z.; Xia, H. Increased level of H19 long noncoding RNA promotes invasion, angiogenesis, and stemness of glioblastoma cells. *J. Neurosurg.* **2016**, *124*, 129–136. [CrossRef] [PubMed]

147. Zhang, C.; Luo, Y.; Cao, J.; Wang, X.; Miao, Z.; Shao, G. Exosomal lncRNA FAM225A accelerates esophageal squamous cell carcinoma progression and angiogenesis via sponging miR-206 to upregulate NETO2 and FOXP1 expression. *Cancer Med.* **2020**, *9*, 8600–8611. [CrossRef] [PubMed]

148. Zhang, X.; Sun, Z.; Zou, Y. LncRNA NEAT1 exacerbates non-small cell lung cancer by upregulating EIF4G2 via miR-582-5p sponging. *Arch. Biol. Sci.* **2020**, *72*, 243–252. [CrossRef]

149. Fang, Q.L.; Zhou, J.Y.; Xiong, Y.; Xiong, Y.; Xie, C.R.; Wang, F.Q.; Li, Y.T.; Yin, Z.Y.; Luo, G.H. Long non-coding RNA RP11-284P20.2 promotes cell proliferation and invasion in hepatocellular carcinoma by recruiting EIF3b to induce c-met protein synthesis. *Biosci. Rep.* **2020**, *40*, BSR20200297. [CrossRef]

150. Hu, M.; Yang, J. Down-regulation of lncRNA UCA1 enhances radiosensitivity in prostate cancer by suppressing EIF4G1 expression via sponging miR-331-3p. *Cancer Cell Int.* **2020**, *20*, 449. [CrossRef]

151. Hansji, H.; Leung, E.Y.; Baguley, B.C.; Finlay, G.J.; Cameron-Smith, D.; Figueiredo, V.C.; Askarian-Amiri, M.E. ZFAS1: A long noncoding RNA associated with ribosomes in breast cancer cells. *Biol. Direct.* **2016**, *11*, 62. [CrossRef] [PubMed]

152. Liu, P.Y.; Tee, A.E.; Milazzo, G.; Hannan, K.M.; Maag, J.; Mondal, S.; Atmadibrata, B.; Bartonicek, N.; Peng, H.; Ho, N.; et al. The long noncoding RNA lncNB1 promotes tumorigenesis by interacting with ribosomal protein RPL35. *Nat. Commun.* **2019**, *10*, 5026. [CrossRef] [PubMed]

153. Song, J.; Ahn, C.; Chun, C.H.; Jin, E.J. A long non-coding RNA, GAS5, plays a critical role in the regulation of miR-21 during osteoarthritis. *J. Orthop. Res.* **2014**, *32*, 1628–1635. [CrossRef] [PubMed]

non-coding **RNA**

MDPI

Review

The Role of lncRNAs in Gene Expression Regulation through mRNA Stabilization

Maialen Sebastian-delaCruz [1,2], Itziar Gonzalez-Moro [2,3], Ane Olazagoitia-Garmendia [1,2], Ainara Castellanos-Rubio [1,2,4,5] and Izortze Santin [2,3,5,*]

1 Department of Genetics, Physical Anthropology and Animal Physiology, University of the Basque Country, 48940 Leioa, Spain; maialen.sebastian@ehu.eus (M.S.-d.); ane.olazagoitia@ehu.eus (A.O.-G.); ainara.castellanos@ehu.eus (A.C.-R.)
2 Biocruces Bizkaia Health Research Institute, 48903 Barakaldo, Spain; itziar.gonzalezm@ehu.eus
3 Department of Biochemistry and Molecular Biology, University of the Basque Country, 48940 Leioa, Spain
4 Ikerbasque, Basque Foundation for Science, 48009 Bilbao, Spain
5 CIBER de Diabetes y Enfermedades Metabólicas Asociadas (CIBERDEM), Instituto de Salud Carlos III, 28029 Madrid, Spain
* Correspondence: izortze.santin@ehu.eus; Tel.: +34-94-601-32-09

Abstract: mRNA stability influences gene expression and translation in almost all living organisms, and the levels of mRNA molecules in the cell are determined by a balance between production and decay. Maintaining an accurate balance is crucial for the correct function of a wide variety of biological processes and to maintain an appropriate cellular homeostasis. Long non-coding RNAs (lncRNAs) have been shown to participate in the regulation of gene expression through different molecular mechanisms, including mRNA stabilization. In this review we provide an overview on the molecular mechanisms by which lncRNAs modulate mRNA stability and decay. We focus on how lncRNAs interact with RNA binding proteins and microRNAs to avoid mRNA degradation, and also on how lncRNAs modulate epitranscriptomic marks that directly impact on mRNA stability.

Keywords: long non-coding RNA; mRNA stability; RNA binding protein; microRNA; gene expression

check for **updates**

Citation: Sebastian-delaCruz, M.; Gonzalez-Moro, I.; Olazagoitia-Garmendia, A.; Castellanos-Rubio, A.; Santin, I. The Role of lncRNAs in Gene Expression Regulation through mRNA Stabilization. *Non-coding RNA* **2021**, 7, 3. http://doi.org/10.3390/ncrna 7010003

Received: 9 December 2020
Accepted: 30 December 2020
Published: 5 January 2021

Publisher's Note: MDPI stays neutral with regard to jurisdictional claims in published maps and institutional affiliations.

1. Introduction

Gene expression and translation is influenced by messenger RNA (mRNA) stability in almost all living organisms. mRNA from bacterial cells can last from seconds to more than one hour, but on average it stays functional between 1 and 3 min [1,2]. Conversely, the lifetime of mammalian mRNA ranges from a couple of minutes to even days, making eukaryotic mRNA more stable than bacterial mRNA. However, from bacteria to mammals, mRNA lifetime needs to be finely regulated in order to enable correct cell homeostasis [1]. The control of the abundance of a particular mRNA fluctuates to adapt to environmental changes, cell growth, differentiation, or to adjust to an unfamiliar situation [3,4]. In this line, the regulation of mRNA stability is essential for tissues and organs exposed to stress signals, such as starvation, infection, inflammation, toxins, or tissue invasion by immune cells [5,6].

The levels of mRNA molecules in the cell are determined by a balance between production and decay [7,8]. Maintaining an accurate balance is crucial for the correct function of a wide variety of biological processes and for the maintenance of an appropriate cellular homeostasis. Many variables such as primary and secondary structure, translation rate and location, among others, influence mRNA stability [5,9,10], and thus minor changes in the structure or the sequence of mRNA molecules might directly influence their half-life.

Eukaryotic mRNAs are transcribed in the nucleus, they are capped (7-methylguanosine cap in 5′end), spliced, polyadenylated (poly(A) tail in 3′end), and lastly, mature mRNAs are exported to the cytoplasm where they are translated into the corresponding polypeptides [1]. Once in the cytoplasm, the 5′cap and the 3′tail serve to attract specific protein complexes that regulate mRNA stability, via protecting mRNA molecules from the attack of ribonucleases and decapping enzymes [1,9,11].

In the last few years, significant progress has been made towards the understanding of mRNA degradation and stability. In general, the decay of mRNA molecules in eukaryotic cells starts with the deadenylation and/or decapping of the mature mRNA, followed by degradation carried out by exonucleases [12–15]. However, the regulation of mRNA stability depends largely on how a three step process (deadenylation, decapping, and degradation) is modulated by regulatory factors, and thus these factors should be taken into account when analyzing the regulation of mRNA stability. Indeed, several studies have pointed out the key role of RNA-binding proteins and miRNAs in the regulation of this process [7,9,16,17]. In addition, long non-coding RNAs (lncRNAs) are emerging as prominent regulators of mRNA stability and decay [3,4,18–21]. LncRNAs are RNA molecules without protein coding potential with lengths exceeding 200 nucleotides [22]. They play important roles in biological processes such as chromatin remodeling, transcriptional activation and interference, RNA processing, and mRNA translation [23]. Regarding their mechanisms of action, different models have been proposed, including functioning as signal, decoy, scaffold, guide, and enhancer RNAs [24]. Importantly, the expression of lncRNAs occurs in a cell-, tissue-, and species-specific manner, and accumulating evidence suggests that different splice variants of individual lncRNAs are also expressed in a cell-, tissue-, and species-specific way [25].

In this review we provide an overview of the main molecular mechanisms by which lncRNAs modulate mRNA stability and gene expression (Figure 1). A detailed description of how lncRNAs interact with target mRNAs, RNA binding proteins or miRNAs to avoid mRNA degradation is provided and a brief explanation on how lncRNAs modulate epitranscriptomic changes to impact on mRNA stability is also described.

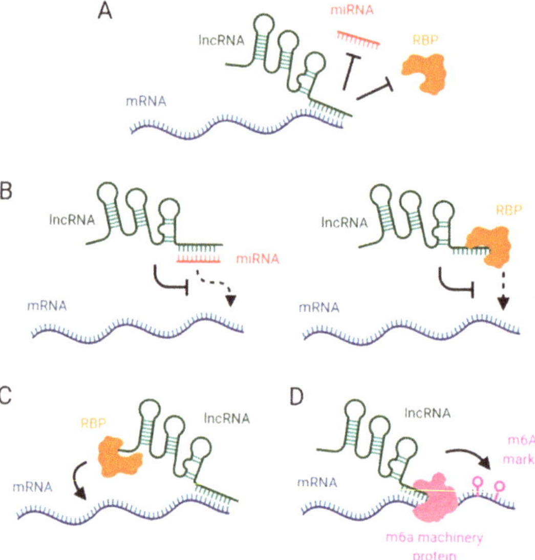

Figure 1. Mechanisms of action of lncRNA-mediated mRNA stability regulation. LncRNAs can modulate mRNA stability through different mechanisms: (**A**) Direct interaction with miRNA or RBP binding sites in target mRNA; (**B**) Sequestration of miRNAs or RBPs to avoid their interaction with mRNA molecules; (**C**) Acting as scaffolds to enhance RBP-mRNA interactions; (**D**) Interaction with m6A machinery to modulate m6A levels of target mRNAs.

2. LncRNAs Affecting mRNA Stability via miRNA Blockage

Recent several studies have been focused on the analysis of the cross-talk between non-coding and coding RNAs to characterize the implication of these interactions in several processes that include chromatin remodeling, mRNA and protein stability, transcription, and mRNA turnover [26]. Accumulating evidence has demonstrated that microRNAs (miR-NAs), which are non-coding RNA molecules of around 22 nucleotides, and long non-coding RNAs, which are longer than 200 nucleotides, interact to regulate their own expression and the expression of mRNAs through several molecular mechanisms [27–30]. MicroRNAs can silence cytoplasmic mRNAs by triggering an endonuclease cleavage, by promoting translation repression or by accelerating mRNA deadenylation and decapping [31,32]. Thus, miRNA blocking by lncRNAs can directly inhibit these processes, promoting mRNA stabilization and inducing gene expression.

In this section, we will provide an overview of lncRNAs that prevent interaction of miRNAs with target mRNAs to protect them from miRNA-driven degradation. These lncR-NAs are referred to as competitive endogenous RNAs (ceRNAs), decoys or sponges [33]. LncRNAs can act as ceRNAs by two different mechanisms. On the one hand, they are able to sequester miRNAs, avoiding their binding to target mRNAs, and on the other hand, they can directly interact with target mRNA transcripts to block miRNA binding sites in mRNA molecules. In this case, lncRNAs and miRNAs share common binding sites in the target mRNA. The interactions between miRNAs and ceRNAs are crucial for the regulation of several basal biological processes but have also been described to participate in different pathogenic conditions.

Most lncRNAs that block miRNA activity to enhance mRNA stability are transcribed from the opposite DNA strand to their paired (sometimes complementary) sense protein coding genes and are known as natural antisense transcripts (NATs). Even though there are some examples of NATs that code for proteins, such as Wrap53 [34] and DHPS [35], NATs usually lack protein coding potential [36] and are generally classified as lncRNAs [37]. NATs can alter their paired sense gene expression by exerting their effect at different levels, including transcription, mRNA processing, splicing, stability and translation [36,38–41]. Regarding the mechanisms by which NATs alter mRNA stability, the "Recycling hypothesis" suggests that reversible RNA duplex formation might trigger conformational changes in mRNA molecules, hindering the accessibility to RNA binding proteins (RBPs), both stabilizing and destabilizing RBPs, and miRNAs [37].

The best characterized NAT is probably the lncRNA *BACE1 antisense RNA (BACE1-AS)*. This lncRNA is partially antisense to *BACE1*, a gene encoding the β-site amyloid precursor protein cleaving enzyme 1, which plays a crucial role in the pathophysiology of Alzheimer's disease [42]. Interestingly, it has been shown that lncRNA *BACE1-AS* is markedly up-regulated in brain samples from patients with Alzheimer's disease and promotes the stability of the *BACE1* transcript [43]. The lncRNA *BACE1-AS* regulates the expression of its sense partner through a synergistic mechanism that includes prevention of miRNA-induced mRNA decay and translational repression. Specifically, *miR-485-5p* and *BACE1-AS* share a common binding site in the sixth exon of *BACE1* mRNA transcript, and thus binding of *BACE1-AS* to this site avoids the interaction of *miR-485-5p*, hindering the translational repression and destabilization of *BACE1* mRNA by *miR-485-5p*, and eventually elevating *BACE1* levels [43].

Similar to *BACE1-AS*, *PTB antisense RNA (PTB-AS)* also modulates its sense mRNA stability by masking miRNA binding sites [44]. *PTB-AS* binds to the 3′ untranslated region (UTR) of *PTBP1*, a RBP that promotes gliomagenesis [45], and prevents *miR-9* binding, a neural-specific miRNA known to target the 3′ UTR of *PTBP1* for degradation [46].

In addition, a lncRNA named *FGFR3 antisense transcript 1 (FGFR3-AS1)* which is antisense to *FGFR3* gene, was shown to be upregulated in an expression analysis performed in tumorigenic tissue from patients with osteosarcoma, when compared to non-cancerous tissue [30]. Bioinformatic analysis indicated that *FGFR3-AS1* and *FGFR3* formed a "tail-to-tail" fully complementarity pairing pattern composed of 1053 nucleotides, suggesting a potential regulatory effect of *FGFR3-AS1* in the expression of the *FGFR3* gene. In silico results were confirmed by RNA protection assays that showed that the non-overlapping part of *FGFR3* mRNA was totally digested, but the overlapping 3′UTR of *FGFR3* mRNA was protected from RNase digestion. Moreover, the authors showed that this antisense pairing between *FGFR3-AS1* and *FGFR3* mRNA upregulated *FGFR3* expression by increasing *FGFR3* mRNA stability. Interestingly, many miRNAs have been reported to bind to the 3′UTR of *FGFR3*, inducing *FGFR3* mRNA degradation [47]. Thus, the antisense pairing between *FGFR3-AS1* and the 3′UTR of *FGFR3* might block potential miRNA binding sites, protecting *FGFR3* from miRNA-induced degradation and/or translation inhibition. However, whether *FGFR3-AS1*-driven *FGFR3* mRNA stabilization occurs through this mechanism remains to be clarified.

Paxillin antisense RNA 1 (*PXN-AS1*), a lncRNA overlapping *PXN* mRNA, was identified after discovering alternative splicing events on a transcriptome sequencing analysis of a hepatocellular carcinoma (HCC) cell line with stable deletion of *Muscleblind-like-3* (*MBNL3*), an oncofetal splicing factor. Two main transcripts were identified: lncRNA *PXN-AS1-L* (containing exon 4) and lncRNA *PXN-AS1-S* (lacking exon 4). Both, *PXN-AS1-L* and *PXN-AS1-S*, were preferentially expressed in the cytoplasm, but had different regulatory effects on the expression of the *PXN* transcript. While *PXN-AS1-L* upregulated PXN protein, *PXN-AS1-S* downregulated it. Interestingly, *PXN-AS1-L* upregulated *PXN* mRNA by preventing *miRNA-24-AGO2* complex binding to the 3′UTR of *PXN* mRNA [48].

The lncRNA *Sirt1 antisense* (*Sirt1-AS*) is transcribed from the *Sirt1* antisense strand and has been shown to interact with *Sirt1* mRNA, forming an RNA duplex that increases stability of its paired transcript, prolonging its half-life up to 10 h and eventually augmenting SIRT1 protein expression [49]. Using luciferase assay experiments it was shown that *Sirt1-AS* lncRNA interacted with the 3′UTR of the *Sirt1* mRNA transcript. This interaction masked *miR-3a* binding sites, avoiding *miR-3a*-driven *Sirt1* mRNA degradation. Interestingly, SIRT1 is a NAD-dependent class III protein deacetylase, which regulates the balance between myoblast proliferation and differentiation, and plays a crucial role in muscle formation [50]. Thus, lncRNA *Sirt1-AS* might participate in myogenesis by blocking *miR-34a* binding to *Sirt1* mRNA which turns in increased SIRT1 protein and increased myoblast proliferation.

Similar to *Sirt1-AS*, a lncRNA named *Urothelial Cancer Associated 1* (*UCA1*) has been shown to regulate mRNA stabilization through directly binding to 3′UTRs of target mRNAs to protect them from miRNA-mediated degradation [51].

Other mechanisms by which lncRNAs block the effect of miRNAs on mRNA degradation are the ones described in the "Competing endogenous RNA" hypothesis, in which lncRNAs compete with miRNAs or RBPs to bind the same common target sequence in mRNAs. Some examples of lncRNAs that act as ceRNAs are described in the following paragraphs.

In addition to the ability to prevent miRNA-induced degradation by binding to mRNA transcripts, it has been described that *UCA1* can also control mRNA stabilization and gene expression by sponging miRNAs that negatively regulate gene expression [51]. In this specific case, lncRNA *UCA1* was implicated in the progression of colorectal cancer through its capacity to control a ceRNA network that fostered upregulation of several genes, including *ANLN*, *BIRC5*, *IPO7*, *KIF2A*, and *KIF23*.

OIP5 Antisense RNA 1 (*OIP5-AS1*) is the mammalian homolog of *Cyrano* gene in zebrafish and it is important for controlling neurogenesis during development [52]. It is located upstream of the *OIP5* sense gene, but they do not overlap. It is known to act as a ceRNA for *miR-143-3p* in cervical cancer (CC) cells, sustaining the expression of *miR-143-3p*-targets, *ITGA6* [53] and *SMAD3* [54], and promoting proliferation, migration and invasion of CC cells [53,54].

Another lncRNA acting as a ceRNA that affects mRNA stability is *MACC1 Antisense RNA 1* (*MACC1-AS1*), an intronic antisense lncRNA located between the fourth and fifth exon of *MACC1*, a transcriptional regulator of epithelial-mesenchymal transition (EMT) [55] that enhances gastric tumor progression [56]. It shares binding sites for *miR-384* and *miR-145-3p* within *PTN* and *c-Myc* transcripts respectively, which are two well-known oncogenic genes [57]. Similar to lncRNA *OIP5-AS1*, *MACC1-AS1* has the capacity to sequester *miR-384* and *miR-145-3p*, sustaining the stability of *PTN* and *c-Myc* mRNAs, and promoting cell proliferation and tumorigenesis.

Other lncRNA that also acts as a miRNA sponge is lncRNA *PTENP1pg1* [58]. *PTENpg1* controls the expression of the tumor suppressor gene *PTEN*, and thus, plays a crucial role in tumorigenesis processes. Interestingly, antisense to this *PTENP1pg1*, there is another lncRNA named *PTENP1pg1-AS*, which has two isoforms, alpha and beta. While the alpha isoform functions in *trans* and epigenetically modulates *PTEN* transcription by the recruitment of DNMT3a and EZH2, the beta isoform interacts with *PTENpg1* through an RNA:RNA pairing interaction, affecting PTEN protein output via changes of *PTENpg1* stability and microRNA sponge activity.

It is also worth mentioning the lncRNA *uc.173* that inhibits miRNA function through a molecular mechanism that implies posttranscriptional reduction of a pri-miRNA. This lncRNA is transcribed from an ultraconserved region (UCR) in human chromosome 3. UCRs represent conserved sequences of the human genome that are likely to be functional but do not have coding potential [59]. RNA molecules transcribed from UCRs originate from genomic regions located in both intra- and intergenic regions with almost perfect evolutionary conservation in most of the mammalian genomes, suggesting that may have a key function in cell physiology and pathogenic processes [59]. Indeed, lncRNA *uc.173* has been described to be implicated in intestinal mucosal cell growth and renewal [60]. This lncRNA, which is highly expressed in intestinal mucosa, stimulates intestinal epithelial cell renewal by downregulating *miRNA195* expression through posttranscriptional reduction of *pri-miR-195*. Although the precise molecular mechanisms by which this lncRNA destabilizes the *pri-miR-195* transcript are unknown, it seems that the process is achieved through a direct lncRNA-mRNA interaction that enhances the degradation of *pri-miR-195* transcript. Downregulation of *miRNA195* by lncRNA *uc.173* results in upregulation of genes implicated in intestinal epithelium growth [60].

Finally, another interesting example is the tumor-promoting lncRNA *ncNRFR* (*noncoding Nras functional RNA*). This lncRNA contains a 22-nucleotide sequence that is identical to miRNA *let-7a* and differs from other miRNAs (*let-7b, let-7c, let-7d, let-7e, let-7f, let-7g, let-7i*, and *miR-98*) in only 1–4 nucleotides [61]. Overexpression of *ncNRFR* in a cell line of colon epithelial cells increased the activity of a heterologous reporter bearing a miRNA *let-7* target site, suggesting that *ncNRFR* lowered miRNA *let-7* function. The miRNA *let-7* is a tumor suppressor that inhibits the expression of several oncogenes, and thus tumorigenic function of *ncNRFR* might be linked to its ability to suppress the action of miRNA *let-7* upon endogenous target mRNAs. The molecular mechanisms by which *ncNRFR* blocks *let-7* remain to be clarified, although taking into account the high homology in the sequence of these two ncRNAs, it is plausible to think that *ncNRFR* might directly compete with *let-7* to bind target mRNA transcripts and inhibit *let-7*-mediated mRNA degradation.

In summary, during the last few years it has become apparent that there is a significant crosstalk between miRNAs and lncRNAs in the regulation of gene expression. Indeed, various ceRNAs have already been identified and their capacity either to sequester miRNAs or to block miRNA binding to target mRNAs has been widely described. Sequestration or blocking of miRNAs by lncRNAs implies a reduced interaction of the miRNAs with their target mRNAs, which eventually turns into increased mRNA stability and expression. Although this field of research has just started to emerge, future studies analyzing the interaction between these two non-coding molecules will explain many of the "unknowns" that still linger regarding the regulation of gene expression, both in basal and pathogenic conditions.

3. Interaction between lncRNAs and RBPs in mRNA Stabilization

It is well established that AU-rich elements (AREs) [62] and GU-rich elements (GREs) [63] are distinct sequence elements in the 3′-UTR of mRNAs. These regions are among the most common determinants of RNA stability in mammalian cells by which various RNA binding proteins (RBPs), including both stabilizing and destabilizing factors bind to, thereby modulating mRNA stability and/or translational efficiency [64]. There exist hundreds of different RBPs with a diverse number of functions through distinct RNA binding domains to which proteins bind and affect RNA fate [65].

A wide variety of research works have shown how RBPs directly bind to mRNA to accelerate mRNA decay or affect translation (increasing or blocking the processes) [64,66]. Interestingly, more and more lncRNAs are being described to also bind RBPs [67–69]. One of the best studied lncRNAs, *Xist*, can form ribonucleoprotein complexes (RNPs) in the nucleus to affect target gene transcription regulation [70]. It has been shown that lncRNAs can also be cytoplasmic and bind RBPs to affect other mRNA metabolism processes such as mRNA stability and turnover [71–73]. Depending on which factors interact with a given lncRNA, this could increase or decrease the targeted mRNA.

In some cases, lncRNAs bind to mRNA transcripts and help to recruit RBPs (stabilizing or destabilizing) affecting mRNA levels. For example, the *LncRNA-assisted stabilization of transcripts (LAST)* can stabilize *CCND1* mRNA through protection against nuclease activity by promoting the interaction between the RBP named CNBP and the 5′UTR of *CCND1* mRNA [74]. In other cases, lncRNAs prevent RBP and target mRNAs interaction by binding the mRNA transcript. This is the case of the lncRNA *Sros1*, which blocks the binding of *Stat1* mRNA to the RBP CAPRIN1, stabilizing the *Stat1* mRNA [75], and of lncRNA *7SL*, which interacts with the 3′UTR of *TP53* mRNA, thereby preventing HuR binding and repressing *TP53* translation [76].

Another example is *PDCD4 Antisense RNA 1 (PDCD4-AS1)*, a NAT affecting stability of *PDCD4*, which is a tumor suppressor coding gene implicated in breast cancer (BC) [77]. In a study by Jadaliha et al. [78], overlapping regions between *PDCD4-AS1* and *PDCD4* were reported. Thus, *PDCD4-AS1* and *PDCD4* mRNA were found to form an RNA duplex, inducing an increase in *PDCD4* mRNA stability. In this case, RNA duplex formation prevented the interaction between *PDCD4* mRNA and HuR [79]. Although HuR usually acts as a stabilizing protein, it has been shown that HuR can form a complex with KSRP to destabilize mRNA molecules and induce a significant reduction in specific protein levels [79].

It is also possible that lncRNA-RBPs interactions influence downstream target mRNA. In turn, *LINC00324* [80], *TRPM2 Antisense RNA (TRPM2-AS)* [81], lncRNA *MY* [82], lncRNA *MEG3* [83], lncRNA *Gadd7* [84], lncRNA *FIRRE* [85], or lncRNA *H19* [86] among others, can bind different RBPs (both stabilizing and destabilizing), and thus affect target mRNA decay.

There are also lncRNA-RBP interactions that would indirectly affect mRNA. There are some lncRNAs that modulate RBP activity and hence, will affect downstream mRNA levels. LncRNA *NORAD* has been described to sequester PUMILIO proteins, which are key regulators for many mRNA stability and translation processes. Thus, *NORAD*-PUMILIO interaction represses mRNA stability and translation of target mRNAs [87]. LncRNAs *OCC1* and *OIP5-AS1* have also been described to bind HuR (a RBP that binds to thousands of mRNAs). While *OCC1* enhances HuR degradation [88], *OIP5-AS1* functions as a sponge for HuR and prevents binding to its targets [89]. In the case of *treRNA1*, this lncRNA downregulates the expression of E-cadherin by suppressing the translation of its mRNA. *TreRNA1* forms an RNP complex that, in turn, binds to eIF4G1 (an initiation factor of protein synthesis) affecting translation of the target mRNA [90].

It is also known that RBPs can influence lncRNA stability, that could also affect lncRNA function and target mRNAs at different levels. *LincRNA-p21* interacts with target *CTNNB1* and *JUNB* mRNAs and inhibits their translation efficiency. However, HuR RBP can inhibit the expression of *lincRNA-p21* by inducing its degradation, which promotes the binding of HuR to *CTNNB1* and *JUNB* mRNAs enhancing their translation, thus increasing the levels of these proteins [91]. The cytoplasmic RBP HuD can also increase the stability of *BACE1-AS* to further influence target mRNA stability [92].

However, lncRNAs can also reduce mRNA stability, making transcripts prone to degradation. One mechanism by which this happens is the Staufen 1 mediated decay. Staufen 1 (STAU1) protein binds the 3′UTR that contain duplex RNA structures to mediate mRNA decay and regulate gene expression [93]. LncRNAs have been found to form STAU1 binding sites by interacting with the 3′UTR of coding genes, thus downregulating their expression [94]. *TINCR* lncRNA was first found to bind STAU1 protein in the context of epidermal differentiation [95]. Further studies performed in gastric cancer, confirmed the binding of *TINCR* to STAU1 protein and found that this interaction induced the STAU1 mediated decay of *KLF2* mRNA. *KLF2*, which induces apoptosis, was described to be reduced in the cancer tissues, opposite to what happens to *TINCR*. Thus, interaction of *TINCR* lncRNA with STAU1 in cancer cells induces the degradation of *KLF2*, preventing apoptosis and contributing to the oncogenic potential of gastric carcinoma [96]. Additionally, other mechanisms of lncRNA-mediated mRNA degradation have also been described. As it is the case of *aHIF* antisense lncRNA, which overlaps the 3′UTR of *HIF1a* coding gene, and has the ability to destabilize *HIF1a* mRNA, subsequently decreasing HIF-1α protein expression in response to chronic hypoxia [97]. Rossignol F. et al. hypothesized that this destabilization occurs via *aHIF*-mediated exposure of AU rich elements present in the 3′ UTR of *HIF1a* mRNA, although the molecular mechanisms by which the mRNA is degraded have not been described yet [98].

All these RNA-protein complexes rely mostly in RNA secondary and tertiary structures that allow the direct interaction between molecules. Therefore, impairment of RNA structure will affect binding and function of the complex, leading to dysregulation of the related pathways [65,99]. There is growing evidence about disease-associated SNPs affecting lncRNA structure [100,101]. Taking into account that many complex disease-associated SNPs are enriched within lncRNAs [102], identifying those lncRNAs and how their binding to RBPs is affected could help find key targets in the associated diseases. One example is *lnc13*, which regulates the stability of *STAT1* mRNA in pancreatic beta cells [103]. *Lnc13* was first discovered in the context of celiac disease, a chronic inflammatory disorder of the intestine, where it has a stability-unrelated function and it regulates gene expression in the chromatin [104]. However, Gonzalez-Moro I. et al. recently related *lnc13* with other autoimmune disorder, type 1 diabetes (T1D), as they found that upregulation of *lnc13* in pancreatic beta cells induces the activation of the pro-inflammatory STAT1 pathway promoting the production of downstream inflammatory chemokines. *Lnc13* was found to enhance STAT1 protein levels by stabilizing its mRNA via interaction with the protein PCBP2 (Poly(rC)-binding protein 2) in the cytoplasm. Viral infections, which have been proposed as triggering factors for T1D [105], were found to induce *lnc13* translocation from the nucleus to the cytoplasm, enabling the interaction of *STAT1* mRNA with PCBP2, which promotes the signaling events that will ultimately lead to pancreatic beta cell destruction and T1D development [103].

Finally, the cell specific expression and functions of lncRNAs should be taken into consideration as this broadens the pathways that can be affected by lncRNA function. For example, *Linc-RoR* interacts with both hnRNP I (stabilizing factor) and AUF1 (destabilizing factor), with an opposite consequence in their interaction with *c-Myc* mRNA [106]. Alternatively, lncRNA *Epr* changes *Cdkn1a* gene expression by affecting both its transcription and mRNA decay through its association with the transcription factor SMAD3 or the RBP KHSRP, respectively. KHSRP is predominantly an mRNA decay promoting factor in epithelial cells and the interaction with *Epr* blocks its ability to induce decay of *Cdkn1a* mRNA [107,108].

All these mechanisms show the importance of studying lncRNA regulatory roles in mRNA stability and turnover, but also demonstrate the intricate work beyond studies for lncRNA functional characterization.

4. LncRNAs, Epitranscriptomic Changes and mRNA Stability

RNA modifications have been recently involved in the regulation of mRNA stability and it has been stated that the regulation of mRNA stability through RNA modification is a crucial step for the tight regulation of gene expression [109]. N6-methyladenosine (m^{6}A) methylation is the most prevalent RNA modification in mRNAs and noncoding RNAs, and it has been involved in a wide range of RNA metabolic processes, including stability [110].

YTHDF2, an m^{6}A reader protein, has been described to selectively bind to m^{6}A-containing mRNAs, resulting in the localization of bound mRNAs from the translatable pool to cellular mRNA decay sites, such as processing bodies [110]. In contrast to the mRNA-decay-promoting function of YTHDF2, insulin-like growth factor 2 mRNA-binding proteins (IGF2BPs) promote the stability and storage of target mRNAs in an m^{6}A-dependent manner [111]. The opposite role of IGF2BPs versus YTHDF2 imposes an additional layer of complexity on m^{6}A function. IGF2BPs and YTHDF2 may recognize different targets or compete for the same m^{6}A sites to fine-tune expression of shared targets through controlling the balance between mRNA stabilization and decay [111]. On the other hand, another m^{6}A reader protein, YTHDF1, actively promotes protein synthesis by interacting with the translation machinery [112]. Altogether, YTHDF2 and IGF2BPs control the lifetime of target transcripts, whereas YTHDF1-mediated translation promotion increases translation efficiency.

In this context, there are few works describing lncRNAs influencing m^{6}A-mediated mRNA stability. *GAS5-AS1* interacts with the tumor suppressor *GAS5* and increases its stability by influencing the interaction between *GAS5* mRNA and the RNA demethylase ALKBH5 leading to a decreased *GAS5* m^{6}A methylation. Moreover, it was shown that m^{6}A-mediated *GAS5* mRNA degradation relies on YTHDF2-dependent pathway [113]. *LINC00470* associates with *PTEN* mRNA and suppresses its stability through interaction with the m^{6}A writer METTL3. In addition, *LINC00470*-METTL3-mediated *PTEN* mRNA degradation also relies on YTHDF2 [114]. Lastly, *LIN28B-AS1* is able to regulate mRNA stability of *LIN28B* by directly interacting with IGF2BP1 but not with *LIN28B*, as IGF2BP1 affects *LIN28B* mRNA stabilization [115].

In the context of mRNA-lncRNA interactions, lncRNA *LNC942*, upregulated in breast cancer, has been described to interact with the methylase METTL4 driving it to the mRNA of target genes *CXCR4* and *CYP1B1*. These two genes are involved in breast cancer initiation and progression, and their methylation augments the stability of the mRNA molecules, which results on higher protein levels and induction of tumorigenesis [115] probably due to an increased interaction with IGF2BP and YTHDF1 readers.

5. Concluding Remarks and Future Prospects

Correct tuning of mRNA stability is a crucial process to maintain appropriate homeostasis, and thus its dysregulation may lead to the development of several pathologies, including cancer. Stability of mRNA molecules is tightly regulated by several mechanisms, including the action of lncRNA molecules. During the last few years, lncRNAs have been implicated in the modulation of mRNA stability and several mechanisms of action have been described. On the one hand, they can prevent miRNA- and RBP-binding to target mRNAs by blocking target binding sites through direct lncRNA-mRNA interaction. On the other hand, they can sequester miRNAs and RBPs to avoid their interaction with target mRNAs, or to inhibit RBP-driven posttranscriptional modifications that affect mRNA stability. Thus, lncRNAs have emerged as crucial regulators of mRNA stability, another molecular mechanism by which these non-coding molecules participate in the regulation of gene expression.

Taking into account that lncRNAs play important roles in the regulation of mRNA stability, the functional characterization of the molecular mechanisms by which these non-coding molecules participate on mRNA equilibrium maintenance will open the door to the development of new lncRNA-based strategies to modify mRNA half-life and subsequent protein expression. Additionally, the functional understanding of lncRNAs that regulate mRNA stability in non-mammalian organisms as *Drosophila melanogaster* or zebrafish, which are easier to genetically manipulate, will help find human orthologous lncRNAs important in mRNA biology. As described formerly in this review, lncRNA-driven mRNA stability changes might impact several biological processes which are important, in both, health and disease. Thus, a better understanding of how lncRNAs act on mRNA stability will provide useful information for the development of new therapeutic strategies to treat and/or cure several diseases in which a dysregulated gene expression pattern is responsible of their development.

Author Contributions: M.S.-d., I.G.-M., A.O.-G., A.C.-R. and I.S. performed the bibliographic research, wrote, revised, and approved the final version of the manuscript. I.S. is the guarantor of this work. All authors have read and agreed to the published version of the manuscript.

Funding: LncRNA work in author's laboratory is supported by European Foundation for the Study of Diabetes (EFSD)-EFSD/JDRF/Lilly Programme on Type 1 Diabetes Research and the Spanish Ministry of Science, Innovation and Universities (PID2019-104475GA-I00) to I.S; and Spanish Ministry of Science, Innovation and Universities (SAF2017-91873-EXP and PGC2018-097573-A-I00) to ACR. M.S.D. and I.G.M. are supported by a Predoctoral Fellowship Grant from the UPV/EHU (Universidad del País Vasco/Euskal Herriko Unibertsitatea) and A.O.G. is supported by a Predoctoral Fellowship Grant from the Education Department of Basque Government.

Institutional Review Board Statement: Not applicable.

Informed Consent Statement: Not applicable.

Data Availability Statement: Not applicable.

Conflicts of Interest: The authors declare no conflict of interest.

References

1. Bicknell, A.A.; Ricci, E.P. When mRNA translation meets decay. *Biochem. Soc. Trans.* **2017**, *45*, 339–351. [CrossRef] [PubMed]
2. Laalami, S.; Zig, L.; Putzer, H. Initiation of mRNA decay in bacteria. *Cell. Mol. Life Sci.* **2014**, *71*, 1799–1828. [CrossRef] [PubMed]
3. Hombach, S.; Kretz, M. Non-coding RNAs: Classification, biology and functioning. *Adv. Exp. Med. Biol.* **2016**, *937*, 3–17. [PubMed]
4. Yao, R.W.; Wang, Y.; Chen, L.L. Cellular functions of long noncoding RNAs. *Nat. Cell Biol.* **2019**, *21*, 542–551. [CrossRef] [PubMed]
5. Ross, J. mRNA Stability in Mammalian Cells. *Microbiol. Mol. Biol. Rev.* **1995**, *59*, 423–450. [CrossRef]
6. Pérez-Ortín, J.E.; Alepuz, P.; Chávez, S.; Choder, M. Eukaryotic mRNA decay: Methodologies, pathways, and links to other stages of gene expression. *J. Mol. Biol.* **2013**, *425*, 3750–3775. [CrossRef]
7. Radhakrishnan, A.; Green, R. Connections Underlying Translation and mRNA Stability. *J. Mol. Biol.* **2016**, *428*, 3558–3564. [CrossRef]
8. Nilsen, T.W. Mechanisms of microRNA-mediated gene regulation in animal cells. *Trends Genet.* **2007**, *23*, 243–249. [CrossRef]
9. Eulalio, A.; Huntzinger, E.; Nishihara, T.; Rehwinkel, J.; Fauser, M.; Izaurralde, E. Deadenylation is a widespread effect of miRNA regulation. *RNA* **2009**, *15*, 21–32. [CrossRef]
10. Karousis, E.D.; Mühlemann, O. Nonsense-mediated mRNA decay begins where translation ends. *Cold Spring Harb. Perspect. Biol.* **2019**, *11*, a032862. [CrossRef]
11. Schoenberg, D.R.; Maquat, L.E. Regulation of cytoplasmic mRNA decay. *Nat. Rev. Genet.* **2012**, *13*, 246–259. [CrossRef] [PubMed]
12. Łabno, A.; Tomecki, R.; Dziembowski, A. Cytoplasmic RNA decay pathways—Enzymes and mechanisms. *Biochim. Biophys. Acta Mol. Cell Res.* **2016**, *1863*, 3125–3147. [CrossRef] [PubMed]
13. Garneau, N.L.; Wilusz, J.; Wilusz, C.J. The highways and byways of mRNA decay. *Nat. Rev. Mol. Cell Biol.* **2007**, *8*, 113–126. [CrossRef] [PubMed]
14. Lykke-Andersen, S.; Jensen, T.H. Nonsense-mediated mRNA decay: An intricate machinery that shapes transcriptomes. *Nat. Rev. Mol. Cell Biol.* **2015**, *16*, 665–677. [CrossRef] [PubMed]
15. Moretti, F.; Kaiser, C.; Zdanowicz-Specht, A.; Hentze, M.W. PABP and the poly(A) tail augment microRNA repression by facilitated miRISC binding. *Nat. Struct. Mol. Biol.* **2012**, *19*, 603–608. [CrossRef]
16. Iwakawa, H.O.; Tomari, Y. The Functions of MicroRNAs: mRNA Decay and Translational Repression. *Trends Cell Biol.* **2015**, *25*, 651–665. [CrossRef]
17. Yoon, J.H.; Abdelmohsen, K.; Gorospe, M. Posttranscriptional gene regulation by long noncoding RNA. *J. Mol. Biol.* **2013**, *425*, 3723–3730. [CrossRef]
18. Grudzien-Nogalska, E.; Kiledjian, M. New insights into decapping enzymes and selective mRNA decay. *Wiley Interdiscip. Rev. RNA.* **2017**, *8*, e1379. [CrossRef]

19. Kondo, Y.; Shinjo, K.; Katsushima, K. Long non-coding RNAs as an epigenetic regulator in human cancers. *Cancer Sci.* **2017**, *108*, 1927–1933. [CrossRef]

20. Akhade, V.S.; Pal, D.; Kanduri, C. Long Noncoding RNA: Genome organization and mechanism of action. In *Advances in Experimental Medicine and Biology*; Springer New York LLC: New York, NY, USA, 2017; Volume 1008, pp. 47–74.

21. Esteller, M. Non-coding RNAs in human disease. *Nat. Rev. Genet.* **2011**, *19*, 861–874. [CrossRef]

22. Zhang, X.; Wang, W.; Zhu, W.; Dong, J.; Cheng, Y.; Yin, Z.; Shen, F. Mechanisms and Functions of Long Non-Coding RNAs at Multiple Regulatory Levels. *Int. J. Mol. Sci.* **2019**, *20*, 5573. [CrossRef] [PubMed]

23. Fang, Y.; Fullwood, M.J. Roles, Functions, and Mechanisms of Long Non-coding RNAs in Cancer. *Genom. Proteom. Bioinform.* **2016**, *14*, 42–54. [CrossRef] [PubMed]

24. Chowdhury, I.H.; Narra, H.P.; Sahni, A.; Khanipov, K.; Schroeder, C.L.C.; Patel, J.; Fofanov, Y.; Sahni, S.K. Expression Profiling of Long Noncoding RNA Splice Variants in Human Microvascular Endothelial Cells: Lipopolysaccharide Effects In Vitro. *Mediat. Inflamm.* **2017**, *2017*, 3427461. [CrossRef] [PubMed]

25. Fernandes, J.C.R.; Acuña, S.M.; Aoki, J.I.; Floeter-Winter, L.M.; Muxel, S.M. Long non-coding RNAs in the regulation of gene expression: Physiology and disease. *Non-Coding RNA* **2019**, *5*, 17. [CrossRef]

26. Braconi, C.; Kogure, T.; Valeri, N.; Huang, N.; Nuovo, G.; Costinean, S.; Negrini, M.; Miotto, E.; Croce, C.M.; Patel, T. MicroRNA-29 can regulate expression of the long non-coding RNA gene MEG3 in hepatocellular cancer. *Oncogene* **2011**, *30*, 4750–4756. [CrossRef]

27. Fan, M.; Li, X.; Jiang, W.; Huang, Y.; Li, J.; Wang, Z. A long non-coding RNA, PTCSC3, as a tumor suppressor and a target of miRNAs in thyroid cancer cells. *Exp. Ther. Med.* **2013**, *5*, 1143–1146. [CrossRef]

28. Liu, C.; Zhang, Y.H.; Deng, Q.; Li, Y.; Huang, T.; Zhou, S.; Cai, Y.D. Cancer-Related Triplets of mRNA-lncRNA-miRNA Revealed by Integrative Network in Uterine Corpus Endometrial Carcinoma. *BioMed Res. Int.* **2017**, *2017*, 3859582. [CrossRef]

29. Sun, J.; Wang, X.; Fu, C.; Wang, X.; Zou, J.; Hua, H.; Bi, Z. Long noncoding RNA FGFR3-AS1 promotes osteosarcoma growth through regulating its natural antisense transcript FGFR3. *Mol. Biol. Rep.* **2016**, *43*, 427–436. [CrossRef]

30. Valencia-Sanchez, M.A.; Liu, J.; Hannon, G.J.; Parker, R. Control of translation and mRNA degradation by miRNAs and siRNAs. *Genes Dev.* **2006**, *20*, 515–524. [CrossRef]

31. Chen, C.Y.; Zheng, D.; Xia, Z.; Shyu, A.B. Ago-TNRC6 triggers microRNA-mediated decay by promoting two deadenylation steps. *Nat. Struct. Mol. Biol.* **2009**, *16*, 1160–1166. [CrossRef]

32. Salmena, L.; Poliseno, L.; Tay, Y.; Kats, L.; Pandolfi, P.P. A ceRNA hypothesis: The rosetta stone of a hidden RNA language? *Cell* **2011**, *146*, 353–358. [CrossRef] [PubMed]

33. Mahmoudi, S.; Henriksson, S.; Corcoran, M.; Méndez-Vidal, C.; Wiman, K.G.; Farnebo, M. Wrap53, a Natural p53 Antisense Transcript Required for p53 Induction upon DNA Damage. *Mol. Cell* **2009**, *33*, 462–471. [CrossRef] [PubMed]

34. Su, W.Y.; Li, J.T.; Cui, Y.; Hong, J.; Du, W.; Wang, Y.C.; Lin, Y.W.; Xiong, H.; Wang, J.L.; Kong, X.; et al. Bidirectional regulation between WDR83 and its natural antisense transcript DHPS in gastric cancer. *Cell Res.* **2012**, *22*, 1374–1389. [CrossRef] [PubMed]

35. Katayama, S.; Tomaru, Y.; Kasukawa, T.; Waki, K.; Nakanishi, M.; Nakamura, M.; Nishida, H.; Yap, C.C.; Suzuki, M.; Kawai, J.; et al. Molecular biology: Antisense transcription in the mammalian transcriptome. *Science* **2005**, *309*, 1564–1566. [PubMed]

36. Nishizawa, M.; Ikeya, Y.; Okumura, T.; Kimura, T. Post-transcriptional inducible gene regulation by natural antisense RNA. *Front. Biosci.* **2015**, *20*, 1–36. [CrossRef] [PubMed]

37. Wanowska, E.; Kubiak, M.R.; Rosikiewicz, W.; Makałowska, I.; Szcześniak, M.W. Natural antisense transcripts in diseases: From modes of action to targeted therapies. *Wiley Interdiscip. Rev. RNA* **2018**, *9*, e1461. [CrossRef]

38. Pelechano, V.; Steinmetz, L.M. Gene regulation by antisense transcription. *Nat. Rev. Genet.* **2013**, *14*, 880–893. [CrossRef]

39. Zinad, H.S.; Natasya, I.; Werner, A. Natural antisense transcripts at the interface between host genome and mobile genetic elements. *Front. Microbiol.* **2017**, *8*, 2292. [CrossRef]

40. Faghihi, M.A.; Wahlestedt, C. Regulatory roles of natural antisense transcripts. *Nat. Rev. Mol. Cell Biol.* **2009**, *10*, 637–643. [CrossRef]

41. Roberson, E.D.; Mucke, L. 100 Years and counting: Prospects for defeating Alzheimer's disease. *Science* **2006**, *314*, 781–784. [CrossRef]

42. Faghihi, M.A.; Zhang, M.; Huang, J.; Modarresi, F.; van der Brug, M.P.; Nalls, M.A.; Cookson, M.R.; St-Laurent, G.; Wahlestedt, C. Evidence for natural antisense transcript-mediated inhibition of microRNA function. *Genome Biol.* **2010**, *11*, R56. [CrossRef] [PubMed]

43. Zhu, L.; Wei, Q.; Qi, Y.; Ruan, X.; Wu, F.; Li, L.; Zhou, J.; Liu, W.; Jiang, T.; Zhang, J.; et al. PTB-AS, a Novel Natural Antisense Transcript, Promotes Glioma Progression by Improving PTBP1 mRNA Stability with SND1. *Mol. Ther.* **2019**, *27*, 1621–1637. [CrossRef] [PubMed]

44. Izaguirre, D.I.; Zhu, W.; Hai, T.; Cheung, H.C.; Krahe, R.; Cote, G.J. PTBP1-dependent regulation of USP5 alternative RNA splicing plays a role in glioblastoma tumorigenesis. *Mol. Carcinog.* **2012**, *51*, 895–906. [CrossRef]

45. Xue, Y.; Ouyang, K.; Huang, J.; Zhou, Y.; Ouyang, H.; Li, H.; Wang, G.; Wu, Q.; Wei, C.; Bi, Y.; et al. Direct conversion of fibroblasts to neurons by reprogramming PTB-regulated MicroRNA circuits. *Cell* **2013**, *152*, 82–96. [CrossRef] [PubMed]

46. Bi, Y.; Jing, Y.; Cao, Y. Overexpression of miR-100 inhibits growth of osteosarcoma through FGFR3. *Tumor Biol.* **2015**, *36*, 8405–8411. [CrossRef]

47. Yuan, J.H.; Liu, X.N.; Wang, T.T.; Pan, W.; Tao, Q.F.; Zhou, W.P.; Wang, F.; Sun, S.H. The MBNL3 splicing factor promotes hepatocellular carcinoma by increasing PXN expression through the alternative splicing of lncRNA-PXN-AS1. *Nat. Cell Biol.* **2017**, *19*, 820–832. [CrossRef]

48. Wang, G.Q.; Wang, Y.; Xiong, Y.; Chen, X.C.; Ma, M.L.; Cai, R.; Gao, Y.; Sun, Y.M.; Yang, G.S.; Pang, W.J. Sirt1 AS lncRNA interacts with its mRNA to inhibit muscle formation by attenuating function of miR-34a. *Sci. Rep.* **2016**, *6*, 21865. [CrossRef]

49. Pardo, P.S.; Boriek, A.M. The physiological roles of Sirt1 in skeletal muscle. *Aging* **2011**, *3*, 430–437. [CrossRef]

50. Barbagallo, C.; Brex, D.; Caponnetto, A.; Cirnigliaro, M.; Scalia, M.; Magnano, A.; Caltabiano, R.; Barbagallo, D.; Biondi, A.; Cappellani, A.; et al. LncRNA UCA1, Upregulated in CRC Biopsies and Downregulated in Serum Exosomes, Controls mRNA Expression by RNA-RNA Interactions. *Mol. Ther. Nucleic Acids* **2018**, *12*, 229–241. [CrossRef]

51. Ulitsky, I.; Shkumatava, A.; Jan, C.H.; Sive, H.; Bartel, D.P. Conserved function of lincRNAs in vertebrate embryonic development despite rapid sequence evolution. *Cell* **2011**, *147*, 1537–1550. [CrossRef]

52. Yang, J.; Jiang, B.; Hai, J.; Duan, S.; Dong, X.; Chen, C. Long noncoding RNA opa-interacting protein 5 antisense transcript 1 promotes proliferation and invasion through elevating integrin α6 expression by sponging miR-143-3p in cervical cancer. *J. Cell. Biochem.* **2019**, *120*, 907–916. [CrossRef] [PubMed]

53. Chen, X.; Xiong, D.; Yang, H.; Ye, L.; Mei, S.; Wu, J.; Chen, S.; Shang, X.; Wang, K.; Huang, L. Long noncoding RNA OPA-interacting protein 5 antisense transcript 1 upregulated SMAD3 expression to contribute to metastasis of cervical cancer by sponging miR-143-3p. *J. Cell. Physiol.* **2019**, *234*, 5264–5275. [CrossRef] [PubMed]

54. Radhakrishnan, H.; Walther, W.; Zincke, F.; Kobelt, D.; Imbastari, F.; Erdem, M.; Kortüm, B.; Dahlmann, M.; Stein, U. MACC1—the first decade of a key metastasis molecule from gene discovery to clinical translation. *Cancer Metastasis Rev.* **2018**, *37*, 805–820. [CrossRef]

55. Zhao, Y.; Liu, Y.; Lin, L.; Huang, Q.; He, W.; Zhang, S.; Dong, S.; Wen, Z.; Rao, J.; Liao, W.; et al. The lncRNA MACC1-AS1 promotes gastric cancer cell metabolic plasticity via AMPK/Lin28 mediated mRNA stability of MACC1. *Mol. Cancer* **2018**, *17*, 69. [CrossRef] [PubMed]

56. Zhang, X.; Zhou, Y.; Chen, S.; Li, W.; Chen, W.; Gu, W. LncRNA MACC1-AS1 sponges multiple miRNAs and RNA-binding protein PTBP1. *Oncogenesis* **2019**, *8*, 73. [CrossRef]

57. Johnsson, P.; Ackley, A.; Vidarsdottir, L.; Lui, W.O.; Corcoran, M.; Grandér, D.; Morris, K.V. A pseudogene long-noncoding-RNA network regulates PTEN transcription and translation in human cells. *Nat. Struct. Mol. Biol.* **2013**, *20*, 440–446. [CrossRef]

58. Bejerano, G.; Pheasant, M.; Makunin, I.; Stephen, S.; Kent, W.J.; Mattick, J.S.; Haussler, D. Ultraconserved elements in the human genome. *Science* **2004**, *304*, 1321–1325. [CrossRef]

59. Xiao, L.; Wu, J.; Wang, J.Y.; Chung, H.K.; Kalakonda, S.; Rao, J.N.; Gorospe, M.; Wang, J.Y. Long Noncoding RNA uc.173 Promotes Renewal of the Intestinal Mucosa by Inducing Degradation of MicroRNA 195. *Gastroenterology* **2018**, *154*, 599–611. [CrossRef]

60. Franklin, J.L.; Rankin, C.R.; Levy, S.; Snoddy, J.R.; Zhang, B.; Washington, M.K.; Thomson, J.M.; Whitehead, R.H.; Coffey, R.J. Malignant transformation of colonic epithelial cells by a colon-derived long noncoding RNA. *Biochem. Biophys. Res. Commun.* **2013**, *440*, 99–104. [CrossRef]

61. Chen, C.Y.; Shyu, A.B. AU-rich elements: Characterization and importance in mRNA degradation. *Trends Biochem Sci.* **1995**, *20*, 465–470. [CrossRef]

62. Vlasova, I.A.; Bohjanen, P.R. Posttranscriptional regulation of gene networks by GU-rich elements and CELF proteins. *RNA Biol.* **2008**, *5*, 201–207. [CrossRef] [PubMed]

63. Wu, X.; Brewer, G. The regulation of mRNA stability in mammalian cells: 2.0. *Gene* **2012**, *500*, 10–21. [CrossRef] [PubMed]

64. Corley, M.; Burns, M.C.; Yeo, G.W. How RNA-Binding Proteins Interact with RNA: Molecules and Mechanisms. *Mol. Cell* **2020**, *78*, 9–29. [CrossRef]

65. Licatalosi, D.D.; Darnell, R.B. Resolving RNA complexity to decipher regulatory rules governing biological networks. *Nat. Rev. Genet.* **2010**, *11*, 75–87. [CrossRef] [PubMed]

66. Ferrè, F.; Colantoni, A.; Helmer-Citterich, M. Revealing protein-lncRNA interaction. *Brief. Bioinform.* **2016**, *17*, 106–116. [CrossRef]

67. Zhu, J.J.; Fu, H.J.; Wu, Y.G.; Zheng, X.F. Function of lncRNAs and approaches to lncRNA-protein interactions. *Sci. China Life Sci.* **2013**, *56*, 876–885. [CrossRef]

68. Wang, X.; Arai, S.; Song, X.; Reichart, D.; Du, K.; Pascual, G.; Tempst, P.; Rosenfeld, M.G.; Glass, C.K.; Kurokawa, R. Induced ncRNAs allosterically modify RNA-binding proteins in cis to inhibit transcription. *Nature* **2008**, *454*, 126–130. [CrossRef]

69. McHugh, C.A.; Chen, C.K.; Chow, A.; Surka, C.F.; Tran, C.; McDonel, P.; Pandya-Jones, A.; Blanco, M.; Burghard, C.; Moradian, A.; et al. The Xist lncRNA interacts directly with SHARP to silence transcription through HDAC3. *Nature* **2015**, *521*, 232–236. [CrossRef]

70. Noh, J.H.; Kim, K.M.; McClusky, W.G.; Abdelmohsen, K.; Gorospe, M. Cytoplasmic functions of long noncoding RNAs. *Wiley Interdiscip. Rev. RNA* **2018**, *9*, e1471. [CrossRef]

71. He, R.Z.; Luo, D.X.; Mo, Y.Y. Emerging roles of lncRNAs in the post-transcriptional regulation in cancer. *Genes Dis.* **2019**, *6*, 6–15. [CrossRef]

72. Briata, P.; Gherzi, R. Long Non-Coding RNA-Ribonucleoprotein Networks in the Post-Transcriptional Control of Gene Expression. *Non-Coding RNA* **2020**, *6*, 40. [CrossRef] [PubMed]

73. Cao, L.; Zhang, P.; Li, J.; Wu, M. Last, a c-Myc-inducible long noncoding RNA, cooperates with CNBP to promote CCND1 MRNA stability in human cells. *eLife* **2017**, *6*, 1–28. [CrossRef] [PubMed]

74. Xu, H.; Jiang, Y.; Xu, X.; Su, X.; Liu, Y.; Ma, Y.; Zhao, Y.; Shen, Z.; Huang, B.; Cao, X. Inducible degradation of lncRNA Sros1 promotes IFN-γ-mediated activation of innate immune responses by stabilizing Stat1 mRNA. *Nat. Immunol.* **2019**, *20*, 1621–1630. [CrossRef] [PubMed]

75. Abdelmohsen, K.; Panda, A.C.; Kang, M.J.; Guo, R.; Kim, J.; Grammatikakis, I.; Yoon, J.H.; Dudekula, D.B.; Noh, J.H.; Yang, X.; et al. NAR Breakthrough Article 7SL RNA represses p53 translation by competing with HuR. *Nucleic Acids Res.* **2014**, *42*, 10099–10111. [CrossRef]

76. Cmarik, J.L.; Min, H.; Hegamyer, G.; Zhan, S.; Kulesz-Martin, M.; Yoshinaga, H.; Matsuhashi, S.; Colburn, N.H. Differentially expressed protein Pdcd4 inhibits tumor promoter-induced neoplastic transformation. *Proc. Natl. Acad. Sci. USA* **1999**, *96*, 14037–14042. [CrossRef]

77. Jadaliha, M.; Gholamalamdari, O.; Tang, W.; Zhang, Y.; Petracovici, A.; Hao, Q.; Tariq, A.; Kim, T.G.; Holton, S.E.; Singh, D.K.; et al. A natural antisense lncRNA controls breast cancer progression by promoting tumor suppressor gene mRNA stability. *PLoS Genet.* **2018**, *14*, e1007802. [CrossRef]

78. Cammas, A.; Sanchez, B.J.; Lian, X.J.; Dormoy-Raclet, V.; van derGiessen, K.; de Silanes, I.L.; Ma, J.; Wilusz, C.; Richardson, J.; Gorospe, M.; et al. Destabilization of nucleophosmin mRNA by the HuR/KSRP complex is required for muscle fibre formation. *Nat. Commun.* **2014**, *5*, 4190. [CrossRef]

79. Zou, Z.; Ma, T.; He, X.; Zhou, J.; Ma, H.; Xie, M.; Liu, Y.; Lu, D.; Di, S.; Zhang, Z. Long intergenic non-coding RNA 00324 promotes gastric cancer cell proliferation via binding with HuR and stabilizing FAM83B expression article. *Cell Death Dis.* **2018**, *9*, 717. [CrossRef]

80. Pan, L.; Li, Y.; Jin, L.; Li, J.; Xu, A. TRPM2-AS promotes cancer cell proliferation through control of TAF15. *Int. J. Biochem. Cell Biol.* **2020**, *120*, 105683. [CrossRef]

81. Kawasaki, Y.; Komiya, M.; Matsumura, K.; Negishi, L.; Suda, S.; Okuno, M.; Yokota, N.; Osada, T.; Nagashima, T.; Hiyoshi, M.; et al. MYU, a Target lncRNA for Wnt/c-Myc Signaling, Mediates Induction of CDK6 to Promote Cell Cycle Progression. *Cell Rep.* **2016**, *16*, 2554–2564. [CrossRef]

82. Zhang, L.; Yang, Z.; Trottier, J.; Barbier, O.; Wang, L. Long noncoding RNA MEG3 induces cholestatic liver injury by interaction with PTBP1 to facilitate shp mRNA decay. *Hepatology* **2017**, *65*, 604–615. [CrossRef] [PubMed]

83. Liu, X.; Li, D.; Zhang, W.; Guo, M.; Zhan, Q. Long non-coding RNA gadd7 interacts with TDP-43 and regulates Cdk6 mRNA decay. *EMBO J.* **2012**, *31*, 4415–4427. [CrossRef] [PubMed]

84. Lu, Y.; Liu, X.; Xie, M.; Liu, M.; Ye, M.; Li, M.; Chen, X.-M.; Li, X.; Zhou, R. The NF-κB–Responsive Long Noncoding RNA FIRRE Regulates Posttranscriptional Regulation of Inflammatory Gene Expression through Interacting with hnRNPU. *J. Immunol.* **2017**, *199*, 3571–3582. [CrossRef]

85. Giovarelli, M.; Bucci, G.; Ramos, A.; Bordo, D.; Wilusz, C.J.; Chen, C.Y.; Puppo, M.; Briata, P.; Gherzi, R. H19 long noncoding RNA controls the mRNA decay promoting function of KSRP. *Proc. Natl. Acad. Sci. USA* **2014**, *111*, E5023–E5028. [CrossRef] [PubMed]

86. Lee, S.; Kopp, F.; Chang, T.C.; Sataluri, A.; Chen, B.; Sivakumar, S.; Yu, H.; Xie, Y.; Mendell, J.T. Noncoding RNA NORAD Regulates Genomic Stability by Sequestering PUMILIO Proteins. *Cell* **2016**, *164*, 69–80. [CrossRef]

87. Lan, Y.; Xiao, X.; He, Z.; Luo, Y.; Wu, C.; Li, L.; Song, X. Long noncoding RNA OCC-1 suppresses cell growth through destabilizing HuR protein in colorectal cancer. *Nucleic Acids Res.* **2018**, *46*, 5809–5821. [CrossRef]

88. Kim, J.; Abdelmohsen, K.; Yang, X.; De, S.; Grammatikakis, I.; Noh, J.H.; Gorospe, M. LncRNA OIP5-AS1/cyrano sponges RNA-binding protein HuR. *Nucleic Acids Res.* **2016**, *44*, 2378–2392. [CrossRef]

89. Gumireddy, K.; Li, A.; Yan, J.; Setoyama, T.; Johannes, G.J.; Ørom, U.A.; Tchou, J.; Liu, Q.; Zhang, L.; Speicher, D.W.; et al. Identification of a long non-coding RNA-associated RNP complex regulating metastasis at the translational step. *EMBO J.* **2013**, *32*, 2672–2684. [CrossRef]

90. Yoon, J.H.; Abdelmohsen, K.; Srikantan, S.; Yang, X.; Martindale, J.L.; De, S.; Huarte, M.; Zhan, M.; Becker, K.G.; Gorospe, M. LincRNA-p21 Suppresses Target mRNA Translation. *Mol. Cell* **2012**, *47*, 648–655. [CrossRef]

91. Kang, M.J.; Abdelmohsen, K.; Hutchison, E.R.; Mitchell, S.J.; Grammatikakis, I.; Guo, R.; Noh, J.H.; Martindale, J.L.; Yang, X.; Lee, E.K.; et al. HuD regulates coding and noncoding RNA to induce APP→Aβ processing. *Cell Rep.* **2014**, *7*, 1401–1409. [CrossRef]

92. Kim, Y.K.; Furic, L.; Parisien, M.; Major, F.; DesGroseillers, L.; Maquat, L.E. Staufen1 regulates diverse classes of mammalian transcripts. *EMBO J.* **2007**, *26*, 2670–2681. [CrossRef]

93. Gong, C.; Maquat, L.E. LncRNAs transactivate STAU1-mediated mRNA decay by duplexing with 39 UTRs via Alu eleme. *Nature* **2011**, *470*, 284–290. [CrossRef]

94. Kretz, M.; Siprashvili, Z.; Chu, C.; Webster, D.E.; Zehnder, A.; Qu, K.; Lee, C.S.; Flockhart, R.J.; Groff, A.F.; Chow, J.; et al. Control of somatic tissue differentiation by the long non-coding RNA TINCR. *Nature* **2013**, *493*, 231–235. [CrossRef] [PubMed]

95. Xu, T.P.; Liu, X.X.; Xia, R.; Yin, L.; Kong, R.; Chen, W.M.; Huang, M.D.; Shu, Y.Q. SP1-induced upregulation of the long noncoding RNA TINCR regulates cell proliferation and apoptosis by affecting KLF2 mRNA stability in gastric cancer. *Oncogene* **2015**, *34*, 5648–5661. [CrossRef] [PubMed]

96. Uchida, T.; Rossignol, F.; Matthay, M.A.; Mounier, R.; Couette, S.; Clottes, E.; Clerici, C. Prolonged hypoxia differentially regulates hypoxia-inducible factor (HIF)-1α and HIF-2α expression in lung epithelial cells: Implication of natural antisense HIF-1α. *J. Biol. Chem.* **2004**, *279*, 14871–14878. [CrossRef] [PubMed]

97. Rossignol, F.; Vaché, C.; Clottes, E. Natural antisense transcripts of hypoxia-inducible factor 1alpha are detected in different normal and tumour human tissues. *Gene* **2002**, *299*, 135–140. [CrossRef]

98. Bhartiya, D.; Scaria, V. Genomic variations in non-coding RNAs: Structure, function and regulation. *Genomics* **2016**, *107*, 59–68. [CrossRef] [PubMed]

99. Lu, X.; Ding, Y.; Bai, Y.; Li, J.; Zhang, G.; Wang, S.; Gao, W.; Xu, L.; Wang, H. Detection of Allosteric Effects of lncRNA Secondary Structures Altered by SNPs in Human Diseases. *Front. Cell Dev. Biol.* **2020**, *8*, 1–11. [CrossRef] [PubMed]

100. Castellanos-Rubio, A.; Ghosh, S. Disease-associated SNPs in inflammation-related lncRNAs. *Front. Immunol.* **2019**, *10*, 1–9. [CrossRef]

101. Ricaño-Ponce, I.; Zhernakova, D.v.; Deelen, P.; Luo, O.; Li, X.; Isaacs, A.; Karjalainen, J.; di Tommaso, J.; Borek, Z.A.; Zorro, M.M.; et al. Refined mapping of autoimmune disease associated genetic variants with gene expression suggests an important role for non-coding RNAs. *J. Autoimmun.* **2016**, *68*, 62–74. [CrossRef]

102. Gonzalez-Moro, I.; Olazagoitia-Garmendia, A.; Colli, M.L.; Cobo-Vuilleumier, N.; Postler, T.S.; Marselli, L.; Marchetti, P.; Ghosh, S.; Gauthier, B.R.; Eizirik, D.L.; et al. The T1D-associated lncRNA Lnc13 modulates human pancreatic β cell inflammation by allele-specific stabilization of STAT1 mRNA. *Proc. Natl. Acad. Sci. USA* **2020**, *117*, 8661–8663. [CrossRef] [PubMed]

103. Castellanos-Rubio, A.; Fernandez-Jimenez, N.; Kratchmarov, R.; Luo, X.; Bhagat, G.; Green, P.H.R.; Schneider, R.; Kiledjian, M.; Bilbao, J.R.; Ghosh, S. A long noncoding RNA associated with susceptibility to celiac disease. *Science* **2016**, *352*, 91–95. [CrossRef]

104. De Beeck, A.O.; Eizirik, D.L. Viral infections in type 1 diabetes mellitus-why the β cells? *Nat. Rev. Endocrinol.* **2016**, *12*, 263–273. [CrossRef] [PubMed]

105. Huang, J.; Zhang, A.; Ho, T.T.; Zhang, Z.; Zhou, N.; Ding, X.; Zhang, X.; Xu, M.; Mo, Y.Y. Linc-RoR promotes c-Myc expression through hnRNPi and AUF1. *Nucleic Acids Res.* **2015**, *44*, 3059–3069. [CrossRef] [PubMed]

106. Rossi, M.; Bucci, G.; Rizzotto, D.; Bordo, D.; Marzi, M.J.; Puppo, M.; Flinois, A.; Spadaro, D.; Citi, S.; Emionite, L.; et al. LncRNA EPR controls epithelial proliferation by coordinating Cdkn1a transcription and mRNA decay response to TGF-β. *Nat. Commun.* **2019**, *10*, 1969. [CrossRef]

107. Zapparoli, E.; Briata, P.; Rossi, M.; Brondolo, L.; Bucci, G.; Gherzi, R. Comprehensive multi-omics analysis uncovers a group of TGF-β-regulated genes among lncRNA EPR direct transcriptional targets. *Nucleic Acids Res.* **2020**, *48*, 9053–9066. [CrossRef]

108. Boo, S.H.; Kim, Y.K. The emerging role of RNA modifications in the regulation of mRNA stability. *Exp. Mol. Med.* **2020**, *52*, 400–408. [CrossRef]

109. Wang, X.; Lu, Z.; Gomez, A.; Hon, G.C.; Yue, Y.; Han, D.; Fu, Y.; Parisien, M.; Dai, Q.; Jia, G.; et al. N 6-methyladenosine-dependent regulation of messenger RNA stability. *Nature* **2014**, *505*, 117–120. [CrossRef]

110. Huang, H.; Weng, H.; Sun, W.; Qin, X.; Shi, H.; Wu, H.; Zhao, B.S.; Mesquita, A.; Liu, C.; Yuan, C.L.; et al. Recognition of RNA N 6-methyladenosine by IGF2BP Proteins Enhances mRNA Stability and Translation contributed reagents/analytic tools and/or grant support; HHS Public Access. *Nat. Cell Biol.* **2018**, *20*, 285–295. [CrossRef]

111. Wang, X.; Zhao, B.S.; Roundtree, I.A.; Lu, Z.; Han, D.; Ma, H.; Weng, X.; Chen, K.; Shi, H.; He, C. N6-methyladenosine modulates messenger RNA translation efficiency. *Cell* **2015**, *161*, 1388–1399. [CrossRef]

112. Wang, X.; Zhang, J.; Wang, Y. Long noncoding RNA GAS5-AS1 suppresses growth and metastasis of cervical cancer by increasing GAS5 stability. *Am. J. Transl. Res.* **2019**, *11*, 4909–4921. [PubMed]

113. Yan, J.; Huang, X.; Zhang, X.; Chen, Z.; Ye, C.; Xiang, W.; Huang, Z. LncRNA LINC00470 promotes the degradation of PTEN mRNA to facilitate malignant behavior in gastric cancer cells. *Biochem. Biophys. Res. Commun.* **2019**, *521*, 887–893. [CrossRef] [PubMed]

114. Wang, C.; Gu, Y.; Zhang, E.; Zhang, K.; Qin, N.; Dai, J.; Zhu, M.; Liu, J.; Xie, K.; Jiang, Y.; et al. A cancer-testis non-coding RNA LIN28B-AS1 activates driver gene LIN28B by interacting with IGF2BP1 in lung adenocarcinoma. *Oncogene* **2019**, *38*, 1611–1624. [CrossRef] [PubMed]

115. Sun, T.; Wu, Z.; Wang, X.; Wang, Y.; Hu, X.; Qin, W.; Lu, S.; Xu, D.; Wu, Y.; Chen, Q.; et al. LNC942 promoting METTL14-mediated m6A methylation in breast cancer cell proliferation and progression. *Oncogene* **2020**, *39*, 5358–5372. [CrossRef]

Article

Exon–Intron Differential Analysis Reveals the Role of Competing Endogenous RNAs in Post-Transcriptional Regulation of Translation

Nicolas Munz [1], Luciano Cascione [1,2] , Luca Parmigiani [3], Chiara Tarantelli [1] , Andrea Rinaldi [1], Natasa Cmiljanovic [4], Vladimir Cmiljanovic [4], Rosalba Giugno [3] , Francesco Bertoni [1,5] and Sara Napoli [1,*]

[1] Institute of Oncology Research, Faculty of Biomedical Sciences, Universita'Svizzera Italiana, 6500 Bellinzona, Switzerland; nicolas.munz@ior.usi.ch (N.M.); luciano.cascione@ior.usi.ch (L.C.); chiara.tarantelli@ior.usi.ch (C.T.); andrea.rinaldi@ior.usi.ch (A.R.); francesco.bertoni@ior.usi.ch (F.B.)
[2] SIB Swiss Institute of Bioinformatics, 1015 Lausanne, Switzerland
[3] Computer Science Department, University of Verona, 37129 Verona, Italy; luca.parmigiani@studenti.univr.it (L.P.); rosalba.giugno@univr.it (R.G.)
[4] PIQUR Therapeutics AG, 4057 Basel, Switzerland; natasa.cmiljanovic@swissrockets.com (N.C.); vladimir.cmiljanovic@swissrockets.com (V.C.)
[5] Oncology Institute of Southern Switzerland, 6500 Bellinzona, Switzerland
* Correspondence: sara.napoli@ior.usi.ch

Citation: Munz, N.; Cascione, L.; Parmigiani, L.; Tarantelli, C.; Rinaldi, A.; Cmiljanovic, N.; Cmiljanovic, V.; Giugno, R.; Bertoni, F.; Napoli, S. Exon–Intron Differential Analysis Reveals the Role of Competing Endogenous RNAs in Post-Transcriptional Regulation of Translation. *Non-coding RNA* **2021**, *7*, 26. https://doi.org/10.3390/ncrna7020026

Academic Editors: Michael R. Ladomery and Giuseppina Pisignano

Received: 30 January 2021
Accepted: 14 April 2021
Published: 16 April 2021

Publisher's Note: MDPI stays neutral with regard to jurisdictional claims in published maps and institutional affiliations.

Abstract: Stressful conditions induce the cell to save energy and activate a rescue program modulated by mammalian target of rapamycin (mTOR). Along with transcriptional and translational regulation, the cell relies also on post-transcriptional modulation to quickly adapt the translation of essential proteins. MicroRNAs play an important role in the regulation of protein translation, and their availability is tightly regulated by RNA competing mechanisms often mediated by long noncoding RNAs (lncRNAs). In our paper, we simulated the response to growth adverse condition by bimiralisib, a dual PI3K/mTOR inhibitor, in diffuse large B cell lymphoma cell lines, and we studied post-transcriptional regulation by the differential analysis of exonic and intronic RNA expression. In particular, we observed the upregulation of a lncRNA, lncTNK2-2:1, which correlated with the stabilization of transcripts involved in the regulation of translation and DNA damage after bimiralisib treatment. We identified miR-21-3p as miRNA likely sponged by lncTNK2-2:1, with consequent stabilization of the mRNA of p53, which is a master regulator of cell growth in response to DNA damage.

Keywords: miRNAs; lncRNAs; ceRNAs; translation; post-transcriptional regulation; mTOR pathway

1. Introduction

The transcriptome is tightly regulated at different levels: along with the regulation of new transcription, RNA molecules can be modulated at the post-transcriptional level, and noncoding RNAs play a relevant role in this process [1,2]. The most famous mechanism of post-transcriptional regulation is mediated by microRNAs, which are small noncoding RNAs that inhibit translation of mRNAs or induce their degradation [3]. In addition, long noncoding RNAs (lncRNAs) take part in this mechanism. Some of them possess miRNA responsive elements (MREs) [4] and are known as competing endogenous RNAs (ceRNAs). Indeed, they can act as an endogenous miRNA decoy, and their expression modulates the amount of free miRNAs available to bind their targets [5]. Circular RNAs (circRNAs) constitute a subclass of exceptionally stable ceRNA molecules that contain a covalent circular structure formed by noncanonical $3'$ to $5'$ end-joining event called back-splicing. circRNAs are diffuse and sometimes conserved in eukaryotic organisms [6]. High throughput analysis of protein expression, as liquid chromatography mass spectrometry, detects changes in translation efficiency, but it is expensive and requires specific expertise. Here, we show that also transcriptome profiling can help us to identify post-transcriptional modulations. RNA-Seq can be used to investigate pre-mRNA dynamics, taking advantage of the intronic sequences that are also acquired although less abundantly than exonic sequences. Changes in the overall intronic read counts directly measure changes in transcriptional activity. We can discriminate whether RNA levels are modulated at the transcriptional and post-transcriptional level applying an exon–intron split analysis (EISA), which compares intronic and exonic changes across different experimental conditions [7]. Here, we applied this approach to determine whether post-transcriptional regulation mediated by lncRNAs might be an additional layer to quickly control protein expression in diffuse large B cell lymphoma (DLBCL) cells exposed to bimiralisib, a dual PI3K/mTOR inhibitor with proven preclinical and early clinical anti-lymphoma activity [8,9]. The mTOR pathway regulates cell growth and proliferation in response to mitogen, nutrient, and energy status and therefore controls the balance between anabolism and catabolism in response to environmental conditions [10–12]. In addition to various anabolic processes as protein, lipid and nucleotide synthesis, mTOR also promotes cell growth by suppressing protein catabolism. mTOR is a downstream mediator of several growth factor and mitogen-dependent signaling pathways but also reacts to intracellular stresses that are incompatible with growth such as low ATP levels, hypoxia, or DNA damage.

The exposure of DLBCL cell lines to bimiralisib induces important post-translational and transcriptional changes, affecting genes and proteins involved in the PI3K/AKT/mTOR pathway, BCR signaling, NF-kB pathway, mRNA processing, apoptosis, cell cycle, Myc pathway, MAPK/RAS signaling, and glycolysis [8]. Most of the pathways are largely modulated via both transcription regulation and protein phosphorylation changes [8]. We hypothesized that post-transcriptional modulation can play a role in the mechanism of action of bimiralisib also. A fast adaptation to lack of nutrients requires an optimization of the stability of transcripts that need to be translated into proteins. Here, we describe how bimiralisib induced a quick stabilization of transcripts needed to cope with amino acid deficiency and to modulate the translation. This event strongly correlated with the overexpression of a lncRNA, namely lncTNK2-2:1, which is associated to the increased stability of transcripts affected by mTOR inhibition and responsible for DNA damage response. In particular, we validated the stabilization of p53 transcript due to the sponge effect on miR21-3p mediated by lnc-TNK2-2:1.

2. Results

2.1. Bimiralisib Reduces the Transcription of Genes Coding for Proteasome and Ribosome Components

We have previously reported that dual PIK3/mTOR pharmacological inhibition using bimiralisib has in vitro and in vivo anti-tumor activity and that it induces transcriptional changes in DLBCL cell lines [8]. Here, we have performed total RNA-Seq on RNA extracted from two DLBCL cell lines, U2932 and TMD8, exposed to DMSO or to bimiralisib for 4, 8, or 12 h. We noticed a general reduction for the most part of transcripts encoding for subunits of RNA pol I and III, which are responsible of rRNA transcription and of some subunits of RNA pol II, and of many transcripts encoding for the machinery processing rRNA. PIK3/mTOR inhibition induced also a downregulation of proteasome components, which was probably to balance the reduction of protein synthesis due to impaired ribogenesis (Supplementary Figure S1 and Table S1).

2.2. Post-Transcriptional Regulation of Many Transcripts Encoding for Riboproteins and Translation Regulators Is an Early Event after Dual PIK3/mTOR Inhibition

We applied EISA [7] to transcriptomic changes upon bimiralisib treatment. EISA measures changes in mature RNA and pre-mRNA reads across different conditions to quantify the transcriptional and post-transcriptional regulation of gene expression. After 4 h of PIK3/mTOR inhibition, we observed a marked post-transcriptional regulation, since many transcripts showed independent changes in exons and introns. After 8 h, changes in transcript levels between DMSO or bimiralisib-treated samples have been mainly due to alteration of transcription, as evident by the fact that changes in exons were well correlated with change in introns (Figure 1a). Focusing on transcripts early stabilized after PIK3/mTOR inhibition, we found they were mainly encoding ribosome components involved in the response to amino acids starvation. Among the most quickly degraded transcripts, we found mRNAs encoding for spliceosome components and pre-mRNA processing (Figure 1b and Table S2).

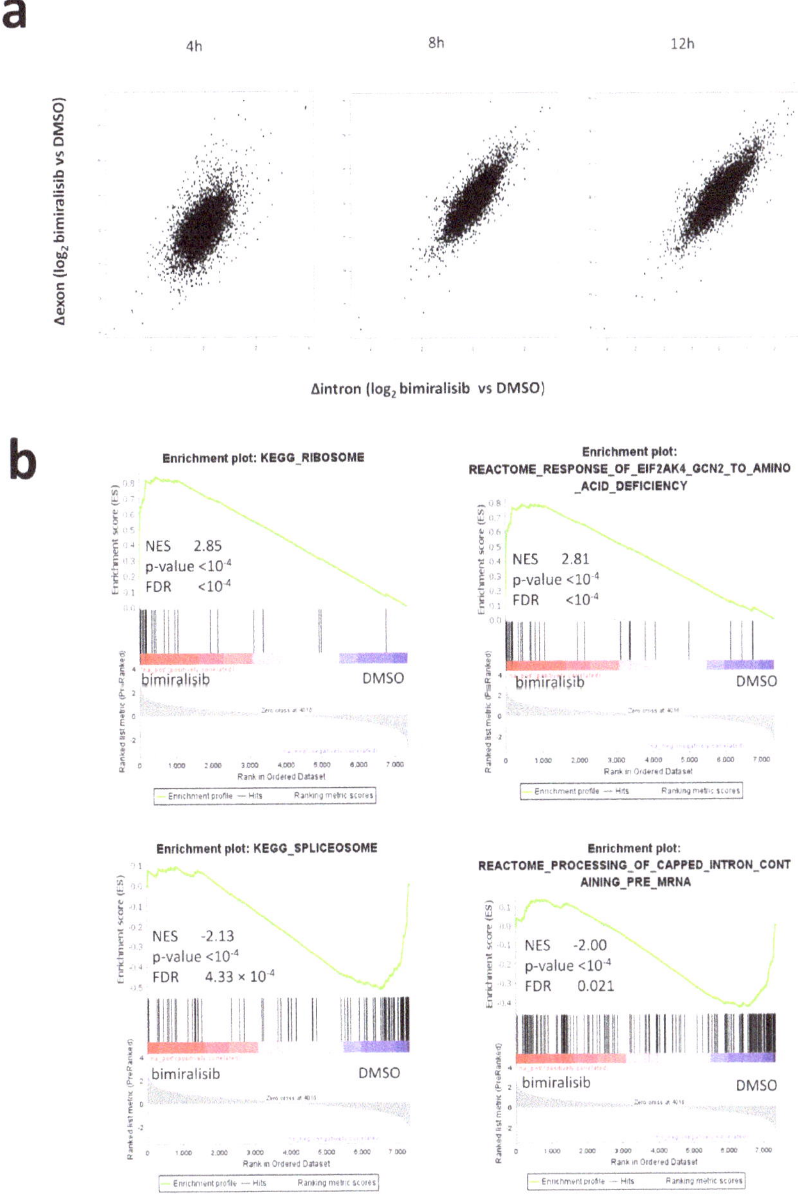

Figure 1. Post-transcriptional regulation of transcripts encoding for riboproteins (RPs) and translation regulators is an early event after dual PI3K/mTOR inhibition. (**a**) Scatter plots representing the comparison of changes in intronic (Δintron) and

exonic (Δexon) reads per each expressed transcript in DLBCL cell lines exposed to bimiralisib or DMSO for 4, 8, and 12 h. Δexon/Δintron = 1 reflects transcriptional modulation, Δexon/Δintron $\neq$ 1 reflects post-transcriptional modulation. (**b**) Representative GSEA plots illustrating the transcriptional expression signature enrichment in transcripts ranked by their decreasing Δexon/Δintron ratio in DLBCL cell lines exposed to bimiralisib (1 μM) or DMSO for 4 h. Green line, enrichment score; bars in the middle portion of the plots show where the members of the gene set appear in the ranked list of genes. Positive or negative ranking metrics indicate correlation or inverse correlation with the profile, respectively. FDR, false discovery rate; NES, normalized enrichment score.

2.3. The lincRNAs RP11-480A16.1 (lncTNK2-2:1) and GMDS-AS1 Are Differentially Expressed after Dual PIK3/mTOR Inhibition and Strongly Correlated to Significantly Stabilized Transcripts

We hypothesized that a prompt alteration in the stability of several transcripts could be achieved by the expression of lncRNAs acting as miRNA sponges. We found 20 significant lncRNAs differentially expressed after pharmacological PIK3/mTOR inhibition: 15 upregulated and five downregulated. We also evaluated whether the 20 lncRNAs could be circRNAs, applying the algorithm CiriQuant [13] that accurately determines the junction of circRNAs from RNA-Seq paired samples. The output showed 28,521 back-spliced junctions (BSJ), most of which were cell specific (TMD8: 9519, U2932: 14,348). Six of the differentially expressed lincRNAs were among the reliably quantified circRNAs (Figure 2a, Table 1). We calculated the Pearson correlation index of each selected lincRNAs with each stabilized or degraded transcript (Δex/Δintr $\neq$ 1). The expression of two of these lncRNAs, namely RP11-480A16.1 (lncTNK2-2:1) and GMDS-AS1, strongly correlated with transcripts differentially stabilized upon bimiralisib exposure (Figure 2b and Table S3). The post-transcriptionally modified transcripts, ranked by their correlation index with lncTNK2-2:1 and GMDS-AS1, were enriched in genes involved in the regulation of translation and in the response to amino acid starvation (Table 2). In particular, there was a highly significant enrichment of genes affected by mTOR inhibitor rapamycin, highlighting the prominent role of lncTNK2-2:1 and GMDS-AS1 in the post-transcriptional regulation due to bimiralisib exposure. We also noticed the enrichment of genes involved in DNA damage, and in particular, some of them were regulated by miRNAs as well. We focused on p53 and ATM due to their important biologic roles (Figure 3 and Table S4).

Table 1. Table summarizing the statistical analysis of differentially expressed lincRNAs and their circRNA reconstruction.

Gene_Name	Ensembl ID	logFC	AveExpr	t	p Value	adj. p Value	circRNA
RP11-147L13.8	ENSG00000267731.1	2.08	4.1	9.74	7.00×10^{-7}	6.48×10^{-5}	NO
RP11-480A16.1	ENSG00000260261.1	0.763	4.62	9.52	8.89×10^{-7}	7.32×10^{-5}	YES
LINC00954	ENSG00000228784.6	0.76	3.5	7.74	7.03×10^{-6}	0.000205749	YES
GMDS-AS1	ENSG00000250903.7	0.9	3.75	7.27	1.30×10^{-5}	0.000290741	YES
AC079466.1	ENSG00000266976.1	1.41	4.25	7.06	1.71×10^{-5}	0.000341307	NO
CTD-2619J13.14	ENSG00000232098.3	0.72	4.42	6.4	4.24×10^{-5}	0.000611259	NO
LINC01572	ENSG00000261008.5	0.66	3.77	6.35	4.54×10^{-5}	0.000636107	YES
RP11-960L18.1	ENSG00000261218.4	0.74	3.5	6.29	4.93×10^{-5}	0.000670722	YES
LINC00926	ENSG00000247982.5	0.73	4.98	5.82	9.95×10^{-5}	0.001049071	YES
RP11-486O12.2	ENSG00000247373.3	1.02	3.53	5.8	0.0001	0.001063727	NO
RP11-147L13.11	ENSG00000278730.1	0.65	4.7	5.5	0.00016	0.001445039	NO
LINC00174	ENSG00000179406.6	0.94	3.52	5.29	0.00022	0.001840434	NO
CTD-2547G23.4	ENSG00000274925.1	0.62	3.55	4.57	0.00071	0.004099891	NO
HCG11	ENSG00000228223.2	0.6	3.59	4.48	0.00084	0.004600193	NO
RP11-16E12.2	ENSG00000259772.5	0.7	3.93	4.3	0.00115	0.005710491	NO
SNHG19	ENSG00000260260.1	-0.68	4.14	-5.4	0.00018	0.001630628	NO
RP11-498C9.15	ENSG00000263731.1	-0.61	4.96	-5.59	0.00014	0.001328078	NO
MIR155HG	ENSG00000234883.3	-1.97	5.66	-9.15	1.33×10^{-6}	8.69×10^{-5}	NO
SNHG15	ENSG00000232956.7	-1.41	5.37	-9.89	5.98×10^{-7}	6.00×10^{-5}	NO
chr22-38_28785274-29006793.1	ENSG00000279978.1	-0.97	5.8	-10.21	4.28×10^{-7}	5.64×10^{-5}	NO

Table 2. Table of top gene sets enriched after preranked gene set enrichment analysis (GSEA) of transcripts ordered by Pearson correlation index between Δexon/Δintron ratio and expression levels of lncTNK2-2:1 or GMDS-AS1.

Name	Size	ES	NES	NOM *p*-Value	FDR *q*-Value	FWER *p*-Value
REACTOME_EUKARYOTIC_TRANSLATION_ELONGATION	37.000	0.721	2.522	0.000	0.000	0.000
REACTOME_RESPONSE_OF_EIF2AK4_GCN2_TO_AMINO_ACID_DEFICIENCY	42.000	0.682	2.419	0.000	0.000	0.000
REACTOME_SELENOAMINO_ACID_METABOLISM	50.000	0.633	2.304	0.000	0.000	0.000
REACTOME_EUKARYOTIC_TRANSLATION_INITIATION	51.000	0.610	2.219	0.000	0.000	0.000
REACTOME_ACTIVATION_OF_THE_MRNA_UPON_BINDING_OF_THE_CAP_BINDING_COMPLEX_AND_EIFS_AND_SUBSEQUENT_BINDING_TO_43S	24.000	0.681	2.205	0.000	0.000	0.000
REACTOME_NONSENSE_MEDIATED_DECAY_NMD_	52.000	0.593	2.187	0.000	0.000	0.000
REACTOME_SRP_DEPENDENT_COTRANSLATIONAL_PROTEIN_TARGETING_TO_MEMBRANE	54.000	0.577	2.133	0.000	0.000	0.002
REACTOME_FANCONI_ANEMIA_PATHWAY	26.000	0.645	2.112	0.000	0.000	0.002
REACTOME_INFLUENZA_INFECTION	74.000	0.499	1.932	0.000	0.004	0.032
REACTOME_HDR_THROUGH_SINGLE_STRAND_ANNEALING_SSA_	23.000	0.580	1.873	0.000	0.011	0.109
REACTOME_ASSOCIATION_OF_TRIC_CCT_WITH_TARGET_PROTEINS_DURING_BIOSYNTHESIS	28.000	0.540	1.791	0.000	0.034	0.308
REACTOME_HDR_THROUGH_HOMOLOGOUS_RECOMBINATION_HRR_	39.000	0.512	1.777	0.000	0.037	0.358

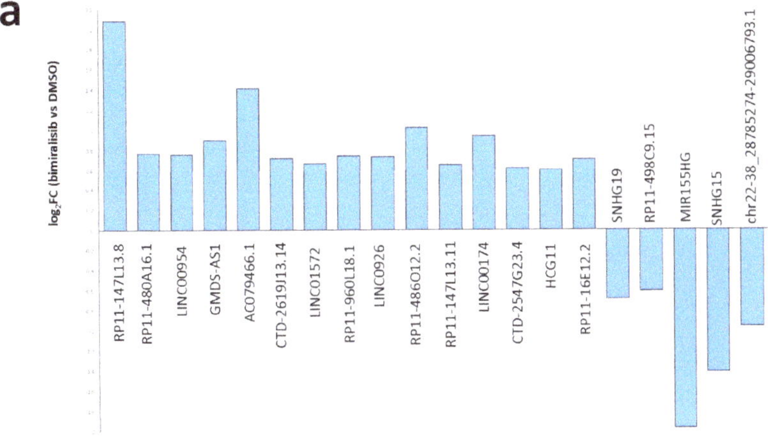

Figure 2. *Cont.*

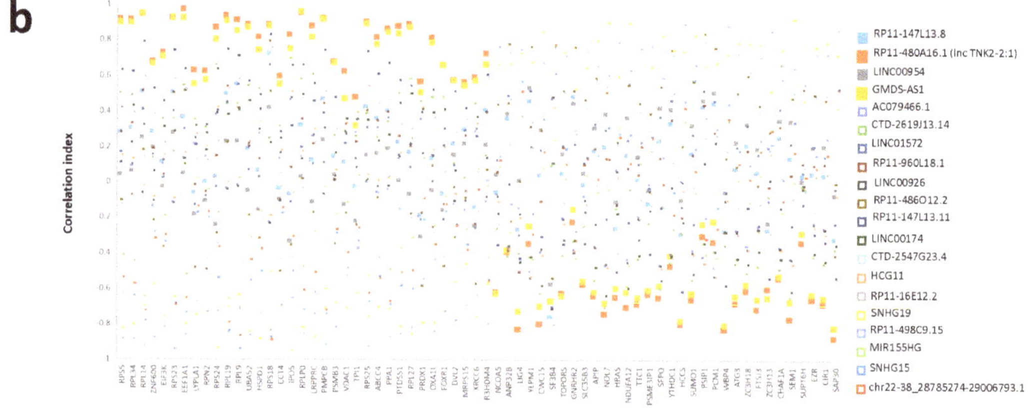

Figure 2. RP11-480A16.1 (lncTNK2-2:1) and GMDS-AS1 are lincRNAs differentially expressed in bimiralisib vs. DMSO strongly correlated to the stabilization of same transcripts. (**a**) Fold change of expression levels of lincRNAs differentially expressed at any time point in U2932 and TMD8 treated with bimiralisib (1 μM) or DMSO. (**b**) Plot of Pearson correlation indexes calculated for Δexon/Δintron ratio of significant differentially stabilized transcripts and levels of differentially expressed lincRNAs. Orange and yellow filled squares represent correlation indexes referred to lncTNK2-2:1 and GMDS-AS1, respectively. Blue and gray-filled squares represent correlation indexes referred to RP11-147L13.8 and LINC00954, respectively. Empty squares refer to all the other significant differentially expressed lincRNAs. LncTNK2-2:1 and GMDS-AS1 are lincRNAs differentially expressed in bimiralisib vs. DMSO strongly correlated to the stabilization of same transcripts.

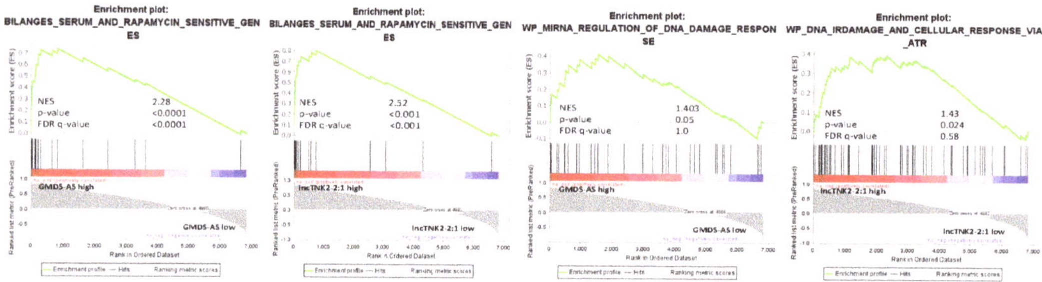

Figure 3. Stabilized transcripts in correlation to lncTNK2-2:1 and GMDS-AS1 expression are involved in the regulation of translation and DNA damage. Representative GSEA plots illustrating the transcriptional expression signature enrichment in transcripts stabilized in correlation to lncTNK2-2:1 and GMDS-AS1 high expression. Green line, enrichment score; bars in the middle portion of the plots show where the members of the gene set appear in the ranked list of genes. Positive or negative ranking metrics indicate the correlation or inverse correlation with the profile, respectively. FDR, false discovery rate; NES, normalized enrichment score.

2.4. lncTNK2-2:1 Induces Stabilization of p53 and ATM by Sequestering miR21-3p

We validated by qRT-PCR the correlation observed between the expression of lncTNK2-2:1 and GMDS-AS1 and the stability of ATM and p53 mRNAs. We measured the levels of the pre-mRNAs or the mature transcripts of ATM and p53 genes and the expression of lncTNK2-2:1 and GMDS-AS1, in U2932 and TMD8, at 4 h and 8 h after exposure to 1 μM of bimiralisib or DMSO. We confirmed the upregulation of both GMDS-AS1 and lncTNK2-2:1, although the GMDS-AS1 upregulation was significant in U2932 but not in TMD8 (Figure 4a). We measured the fold change of lncRNAs and of pre mRNAs or mature transcripts (Figure S2a) and calculated their Pearson correlation index (Figure S2b). We clearly confirmed the predicted correlation of the expression of the lncRNA lncTNK2-2:1 with the mature transcripts for p53 and ATM (R = 0.581 and R = 0.596, respectively) (Figure 4b, bottom panels) but not with their pre-mRNAs (Figure 4b, top panels). GMDS-AS1 was significantly correlated neither to p53 nor to ATM stability (Figure 4b). According to this, we searched for experimentally validated miRNA binding sites sheared by lncTNK2-2:1 and p53 in DIANA tools, LncBase (https://diana.e-ce.uth.gr/lncbasev3 accessed 15 April 2021) [4] and TarBase (https://carolina.imis.athena-innovation.gr/diana_tools/web/index.php?r=tarbasev8%2Findex accessed 15 April 2021) [14] (Table S5). miR-21-3p and miR-22-3p are reported to target both lnc-TNK2-2:1 and p53 (Figure 4c), and miR-21-3p belongs to the miR-21-3p/TSC2/mTOR regulatory axis [15]. Thus, we focused on miR-21-3p, which was consistently reduced in both U2932 and TMD8 after bimiralisib treatment (Figure 4d) and the consequent upregulation of lncTNK2-2:1 (Figure 4a). The p53 stabilization was more robust in TMD8 than in U2932 (Figure S2a), even if the lncRNA was upregulated in both cell lines (Figure 4a), which was an observation that was possibly justified by the lower basal expression of miR21-3p in U2932 compared to TMD8 (Figure S2c).

2.5. lncTNK2-2:1 Degradation Reverts Stabilization of p53 and Releases miR21-3p

To validate the relationship between lncTNK2-2:1 expression and miR21-3p activity on its target p53, we electroporated 2 million TMD8 with 100 pmol of antisense oligonucleotides or of the negative control. After 72 h, we exposed the cells to 1 μM of bimiralisib or DMSO for 8 h. We measured the levels of the pre-mRNA or the mature transcript of the p53 gene and the expression of lncTNK2-2:1 by qRT-PCR. We confirmed that the antisense oligonucleotides degraded lncTNK2-2:1 efficiently both at basal condition and after its induction due to bimiralisib treatment (Figure 4e, left). We validated the relationship between the lncRNA lncTNK2-2:1 with the p53 transcript stability. Indeed, after lncTNK2:2-1 interference, we noticed the reduction of p53 stabilization (Figure 4e, middle), which was measured as the fold change between p53 mature mRNA with respect to the total pre-mRNA transcribed in each condition. According to this, we measured miR-21-3p after lncTNK2-2:1 degradation, and we showed that it increased in samples where lncTNK2-2:1 was knocked down with respect to the negative control (Figure 4e, right). This experimental observation confirmed the predicted miR-21-3p/lncTNK2-2:1/p53 regulatory axis.

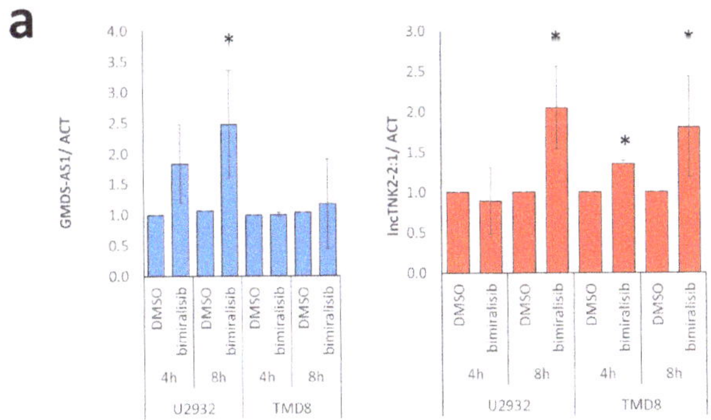

Figure 4. *Cont.*

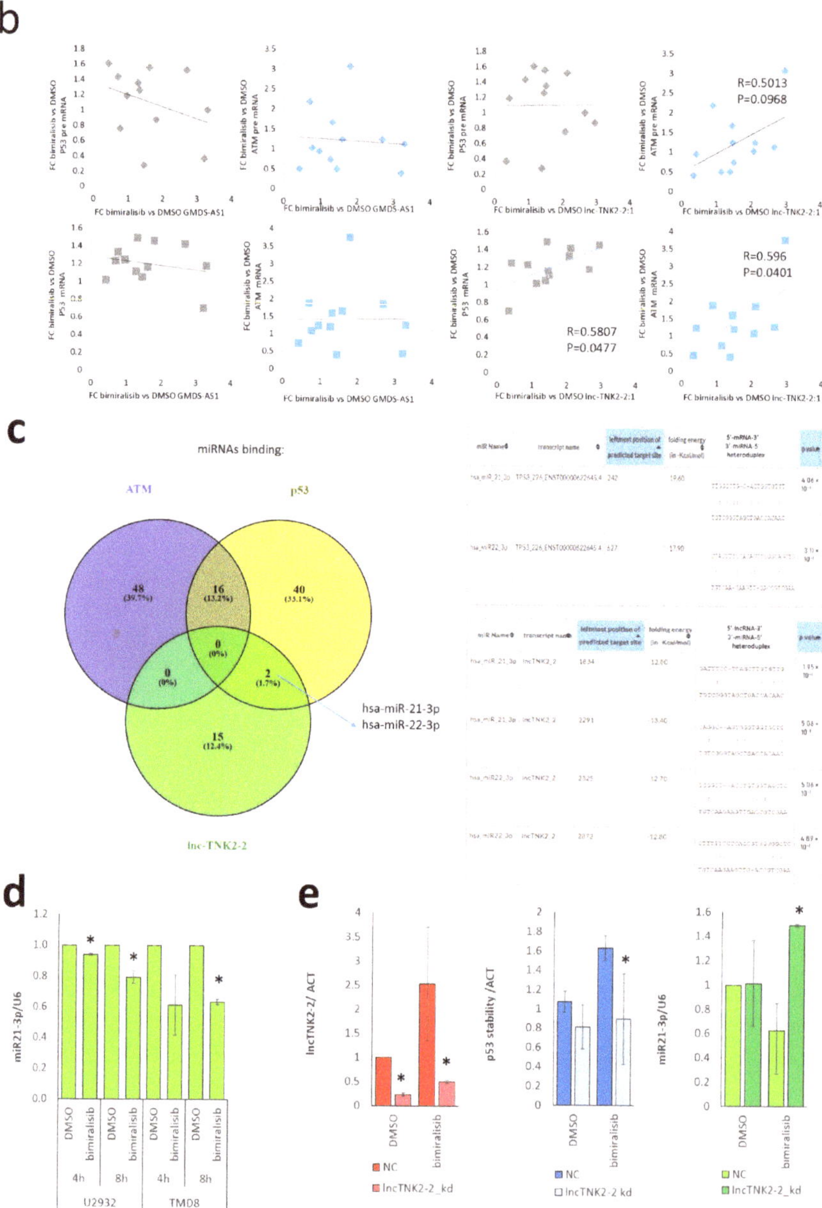

Figure 4. lncTNK2-2 induces the stabilization of p53 by sequestering miR21-3p. (**a**) Expression levels of GMDS-AS1 (top) and lncTNK2-2:1 (bottom) in U2932 and TMD8 after 4 h and 8 h of exposure to DMSO or bimiralisib (1 μM). (**b**) Pearson correlation of fold change of GMDS-AS1 or lncTNK2-2:1 expression levels (x axis) after bimiralisib treatment with fold change of pre-mRNA (top) or mature mRNA (bottom) of ATM or p53 (y axis). Correlation index (R) and *p* value (P) are indicated in the plot when significant. (**c**) Left, Venn diagram of common miRNAs binding p53, ATM, and lncTNK2-2:1. Right, RNA22 prediction of miRNA binding sites of miR21-3p or miRNA 22-3p in p53 3′UTR or lncTNK2-2. (**d**) Expression level of miR21-3p in U2932 or TMD8 after 4 h or 8 h of exposure to DMSO or bimiralisib (1 μM). (**e**) Expression levels of lncTNK2-2:1 (left), p53 stability (middle), and miR21-3p (right) after 8 h of bimiralisib exposure in negative control or lncTNK2-2:1 knockdown cells. * *p* < 0.05.

3. Discussion

As a central controller of cell growth, mTOR regulates ribosome biogenesis. The latter is the most energy-demanding cellular process, and mTOR controls it by promoting the translation of riboproteins and by affecting ribosomal RNA (rRNA) synthesis. Ribosome synthesis requires all three nuclear RNA polymerases, Pol I for the synthesis of rRNA, Pol II for transcription of riboprotein genes, and Pol III for the synthesis of 5S RNA [16,17]. PIK3/mTOR inhibition by bimiralisib leads to the downregulation of all of them, a reduction of rRNA gene transcription, and, in addition, pre-rRNA processing impairment. Moreover, we describe here an additional mechanism that the cell activates when it must save energy: mTOR pathway inhibition leads to the stabilization of already present transcripts encoding for riboproteins. Since the new transcription of riboproteins genes is inhibited, the cell needs to save the already available riboprotein mRNA as long as possible in order to still translate essential proteins.

A growing number of miRNAs have been shown to control components or regulators of ribosome biogenesis [18]. In addition, lncRNAs have been increasingly found to play relevant roles in eukaryotic ribosome biogenesis that can be basally active or stress response-specific [18]. These molecular mechanisms include protein binding, rDNA chromatin modifications, snoRNP formation, and transcript-specific translation modulations. Here, we report evidence of an additional example of lncRNAs involved in the regulation of ribogenesis. Two lncRNAs, lncTNK2-2:1 and GMDS-AS1, were modulated following PIK3/mTOR pathway inhibition, and their transcription changes are strongly correlated with the stabilization of transcripts encoding for many riboproteins. Other lncRNAs were significantly upregulated by bimiralisib treatment but were not correlated with transcripts early stabilized after mTOR inhibition. Thus, we postulated that these particular lncRNAs could be relevant players of mTOR-mediated modulation of translation in response to amino acid deficiency or other stressful events. One of the possible mechanisms that might mediate a quick transcript stabilization is the sequestration of miRNAs, and GMDS-AS1 is already known as an miRNA sponge in lung cancer [19]. As proof of principle, we looked for binding sites in the lincRNAs for miRNAs that could target stabilized transcripts correlated with the upregulation of lncTNK2-2:1 or GMDS-AS1. Both lncTNK2-2:1 and GMDS-AS1 were associated with the stabilization of genes involved in DNA damage response and regulated by miRNAs. Among them, p53 is also involved in ribosome biogenesis [20,21], and the p53 network is known to interact with several miRNAs [22]. We could validate the correlation between p53 mRNA stabilization and lncTNK2-2:1 level but not with GMDS-AS1.

We showed that miR-21-3p increased after lncTNK2-2:1 silencing, and concomitantly, p53 was not stabilized after bimiralisib treatment in the absence of lncTNK2-2:1. This evidence enforced our hypothesis that miR-21-3p was responsible for p53 mRNA degradation, which we formulated on the basis of in silico base pairing of lncTNK2-2:1 and miR-21-3p and on the high stability of predicted miR-21-3p binding in p53 3′UTR, which is compatible with mRNA degradation, instead of inhibition of translation [23]. Furthermore, in U2932, a cell line expressing low miR-21-3p levels, p53 was not strongly stabilized, despite the upregulation of lncTNK2-2:1. In support of our findings, miR21-3p is already known to modulate the mTOR pathway via TSC2 mRNA downregulation [15], and P53-dependent mTOR inhibition is mediated by TSC2 [24]. Here, we provide the evidence of an additional layer of regulation of p53-mTOR crosstalk through the rapid elimination of miR-21-3p and consequent stabilization of p53 and enhancement of TSC2 repressor activity of mTOR. The PI3K/mTOR pharmacological inhibition enforces the mTOR inhibition by a positive feedback loop mediated by the overexpression of lncTNK2-2:1.

We also selected ATM as potential interesting transcript, since it appeared related to lncTNK2-2:1 overexpression both in silico and in vitro, but we could not identify any miRNA that may mediate the connection between the lncRNAs and the mRNA. ATM mRNA stabilization after bimiralisib exposure might also be due to the downregulation of miRNAs directly regulated by PI3K/signaling mTOR and directly targeting ATM 3′UTR [25].

In conclusion, based on an alternative bioinformatic approach applied to RNA-Seq data, we selected candidate molecules that could be involved in a post-transcriptional mechanism of RNA competition, and we provided data suggesting a novel RNA network composed by lncRNAs, miRNAs, and mRNAs, which is affected by the dual PI3K/mTOR pharmacological inhibition in DLBCL cell lines.

4. Materials and Methods

4.1. Cell Culture and Bimiralisib Treatment

Established human DLBCL cell lines TMD8 and U2932 were grown as previously described [8]. Bimiralisib was kindly provided by PQR Therapeutics (Basel, Switzerland). TMD8 and U2932 were seeded 3 million cells/well in a non-tissue culture 6 well plate. Cells were treated for 4 and 8 h with 1 μM bimiralisib or 0.1% DMSO (Sigma, St. Louis, MO, USA), respectively. Treatment was stopped by washing the cell with RNAse-free PBS and followed by immediate RNA extraction.

4.2. lncTNK2:2-1 Degradation

Three different locked nucleic acid (LNA) antisense oligonucleotides were designed against lncTNK2:2-1 and purchased by IDT (Integrated DNA Technology, Coralville, IA, USA) as 3-10-3 Affinity Plus (locked nucleic acid) gapmer format (3 Affinity Plus bases, 10 DNA bases, 3 Affinity Plus bases, phosporotioate bonds), along with a negative control. In details, their sequences were lncTNK2-2:1_ASO-1: CACTTCCCGAGTATAA; lncTNK2-2:1_ASO-2: CACCTGACCATATTGA; lncTNK2-2:1_ASO-3: CACCACTACACGTTTA; NC5 3-10-3: GACTATACGCGCAATA. TMD8 were nucleofected with 100 pmol of each antisense oligonucleotides or the negative control using 4D Nucleofector (Amaxa-Lonza, Basel, Switzerland), according to the manufacturer's instructions and incubated for 72 h. Then, cells were treated with 1 μM of bimiralisib (PQR Therapeutics, Basel, Switzerland), or DMSO (Sigma, St. Louis, MO, USA) for 8 h, and then, RNA was extracted.

4.3. RNA-Extraction

For each cell line and condition, cells were collected and resuspended in 1 mL of TRI Reagent (Sigma, St. Louis, MO, USA) for cell lysis, and extraction was performed. DNAse (Qiagen, Hilden, Germany) was added to the RNA samples and incubated for 15 min at room temperature. Total RNA was reprecipitated to remove salts and the enzyme.

4.4. Whole-Transcriptome Sequencing (RNA-Seq)

Two cell lines, U2932 and TMD8, were treated with 1 μM of bimiralisib or DMSO and RNA was extracted after 4, 8, or 12 h. Cells treatment and RNA extraction were described in [8]. Quality control for extract RNA was performed on the Agilent BioAnalyzer (Agilent Technologies, Santa Clara, CA, USA) using the RNA 6000 Nano kit (Agilent Technologies, Santa Clara, CA, USA), and concentration was determined by the Invitrogen Qubit (Thermo Fisher Scientific, Waltham, MA, USA) using the RNA BR reagents (Thermo Fisher Scientific). The TruSeq RNA Sample Prep Kit v2 for Illumina (Illumina, San Diego, CA, USA) was used for cDNA synthesis and the addition of barcode sequences. The sequencing of the libraries was performed via a paired end run on a NextSeq500 Illumina sequencer (Illumina). As an average, 25 million reads were collected per each sample.

4.5. Data Mining

We evaluated the RNA-seq reads quality with FastQC (v0.11.5), and we removed low-quality reads/bases and adaptor sequences using Trimmomatic (v0.35). The trimmed-high-quality sequencing reads were aligned using STAR [26], which is a spliced read aligner that allows for sequencing reads to span multiple exons. On average, we were able to align 85% of the sequencing reads for each sample to the reference genome (HG38).

Then, the HTSeq-count software package [27] was used for the quantification of gene level expression. Differential expression analysis was performed on gene-level read count data using the 'limma' pipeline [28,29] We first subsetted the data to genes that had a counts-per-million value greater than one in 3 or more samples. The data were normalized per sample using the 'TMM' method from the edgeR package [30] and transformed to log2 counts-per-million using the edgeR function 'cpm'. Then, linear model analyses, with empirical-Bayes moderated estimates of standard error, were used to identify genes whose expression was most associated with phenotype of interest, and an FDR-adjusted *p*-value of <0.05 was set as a threshold for statistical significance.

Transcription rates were estimated based on the number of nascent unspliced transcripts using EISA [7]. For each gene, we used HT-Seq to compute the average number of reads mapping to the gene's introns (all exonic regions are excluded). Then, this number of intronic reads was divided by the total length of introns to yield a mean intronic coverage that was used as a proxy of the transcription rate.

Functional annotation was performed using Gene Set Enrichment Analysis (GSEA) [31] with all genes preranked by FC as determined by Limma test, or by delta exon/delta intron ratio as determined by EISA, or by Pearson correlation index between delta exon/delta intron ratio and lincRNAs expression. Gene sets were considered significantly enriched if *p* < 0.05 and FDR < 0.25.

The Pearson correlation was used to determine the relationship between exons/introns and gene expression. All statistical analysis was done with R (version 4.0.3) scripts. The significance of gene set overlap was determined by hypergeometric test.

MicroRNA binding prediction was performed searching in TarBase or LncBase and then intersection of miRNAs were computed by Venn diagram. MiRNA responsive elements (MRE) were calculated by the algorithm RNA22.

Prediction and quantification of circRNAs was carried out by the means of CIRIquant [13] with default parameters. In brief, CIRIquant uses HISAT2 [32] to align the RNA-seq reads to the reference genome and CIRI2 to identify putative circRNAs in the form of BSJ, which are then filtered to reduce the number of false-positives. Since the normalization of circRNA expression values is necessary for the differential expression analysis, TMM normalization factors were extracted from gene expression levels to remove the systematic technical effect of library size. Gene count matrix for the normalization was obtained using the script prepDE.py, from stringTie [33], on the aligned reads. Finally, the voomWithQualityWeights [34] function, from the limma package in Bioconductor [28], is applied to identify the statistical significance of circRNA expression change.

4.6. Reverse Transcription of Total RNA to cDNA

Total RNA (500 ng) was processed for each sample by the SuperScript III First-Strand Synthesis SuperMix for qRT-PCR (Invitrogen, Carlsbad, CA, USA) according to the manufacturer's instructions. cDNA was stored at −20 °C.

4.7. Quantitative Real-Time PCR

qRT-PCR was performed on an Applied Biosystem StepOnePlus System. LncRNA and mRNA targets expression were quantified using KAPA SYBR FAST qPCR Master Mix (2×) ABI Prism 5 mL (KAPA Biosystems, Wilmington, MA, USA) according to the manufacturer's instructions, and the comparative CT method (ΔΔCT method) normalized to ACTB (β-Actin) expression was applied for data analysis. The following primers were used: GMDS-AS1, forward: 5′-CCC AGT CTT CCC AGG ATT GA-3′, reverse: 5′-AGC ATC TTC CAG GCC AAA TG-3′; lncTNK2-2, forward: 5′-AGA GCG AAA CCC CAT CTC AA-3′, reverse: 5′-GGA GAA GGA AGC GGA CTG AT-3′; ACTB, forward: 5′-CCA ACC GCG AGA AGA TGA C-3′, reverse: 5′-TGG GGT GTT GAA GGT CTC A-3′; ATM, pre-mRNA forward: 5′-AAC CAC AGT TCT TTT CCC GT-3′, pre-mRNA reverse: 5′-TTG ACT CTG CAG CCA ACA TG-3′, mRNA forward: 5′-GCC TTA AAA CTT TGC TTG AGG TG-3′, mRNA reverse: 5′-ACA TGC GAA CTT GGT GAT GA-3′; TP53, pre-mRNA forward: 5′-ACA AGC AGT CAC AGC ACA TG-3′, pre-mRNA reverse: 5′-AGA GCA ATC AGT GAG GAA TCA G-3′, mRNA forward: 5′-ACA AGC AGT CAC AGC ACA TG-3′, mRNA reverse: 5′-CAC CAC CAC ACT ATG TCG AAA A-3′. miR-21-3p and U6 snRNA expression were quantified using TaqMan microRNA Assay (Applied Biosystems, Foster City, CA, USA) and TaqMan microRNA Control Assay, respectively, according to the manufacturer's instructions, and the comparative CT method (ΔΔCT method) normalized to U6 expression was applied for data analysis. PCR efficacy was determined using the LinRegPCR tool [35].

Supplementary Materials: The following are available online at https://www.mdpi.com/article/10.3390/ncrna7020026/s1, Figure S1: Bimiralisib reduces transcription of genes encoding for proteasome and ribosome components, Figure S2: lncTNK2-2 induces stabilization of p53 by sequestering miR21-3p. Table S1: Details of GSEA of transcriptional changes in U2932 and TMD8, exposed to DMSO or to bimiralisib. Table S2: GSEA of post-transcriptional changes in U2932 and TMD8, exposed for 4 h to DMSO or to bimiralisib. Table S3: Pearson correlation indexes referred to each differentially expressed lincRNAs and differentially stabilized transcripts upon 4 h of bimiralisib exposure in TMD8 and U2932. Table S4: Details of genesets enriched after GSEA of post-transcriptionally modified transcripts ranked by decreasing Pearson correlation index between Δexon/Δintron ratio and lncTNK2-2:1 and GMDS-AS1 expression Table S5: List of miR-21-3p targets according to LncBase and TarBase.

Author Contributions: N.M. designed and performed experiments, interpreted data, and co-wrote the manuscript; L.C. performed data mining, interpreted data, and co-wrote the manuscript; L.P. performed data mining; C.T. performed experiments; A.R. performed RNA sequencing; N.C. and V.C. designed bimiralisib, R.G. provided advice, F.B. interpreted data and co-wrote the manuscript; S.N. designed the study, performed data mining, interpreted data, and wrote the manuscript. All authors have read and agreed to the published version of the manuscript.

Funding: This research received no additional external funding.

Data Availability Statement: Profiling data are available at the National Center for Biotechnology Information (NCBI) Gene Expression Omnibus (GEO) (http://www.ncbi.nlm.nih.gov/geo accessed 15 April 2021) database.

Acknowledgments: N.M. is supported by an NCCR RNA & Disease fellowship.

Conflicts of Interest: Francesco Bertoni: institutional research funds from Acerta, ADC Therapeutics, Bayer AG, Cellestia, CTI Life Sciences, EMD Serono, Helsinn, ImmunoGen, Menarini Ricerche, NEOMED Therapeutics 1, Nordic Nanovector ASA, Oncology Therapeutic Development, PIQUR Therapeutics AG; consultancy fee from Helsinn, Menarini; expert statements provided to HTG; travel grants from Amgen, Astra Zeneca, Jazz Pharmaceuticals, PIQUR Therapeutics AG.

References

1. Dykes, I.M.; Emanueli, C. Transcriptional and Post-transcriptional Gene Regulation by Long Non-coding RNA. *Genom. Proteom. Bioinform.* **2017**, *15*, 177–186. [CrossRef]
2. Anastasiadou, E.; Jacob, L.S.; Slack, F.J. Non-coding RNA networks in cancer. *Nat. Rev. Cancer* **2018**, *18*, 5–18. [CrossRef] [PubMed]
3. Hendrickson, D.G.; Hogan, D.J.; McCullough, H.L.; Myers, J.W.; Herschlag, D.; Ferrell, J.E.; Brown, P.O. Concordant Regulation of Translation and mRNA Abundance for Hundreds of Targets of a Human microRNA. *PLoS Biol.* **2009**, *7*, e1000238. [CrossRef] [PubMed]
4. Karagkouni, D.; Paraskevopoulou, M.D.; Tastsoglou, S.; Skoufos, G.; Karavangeli, A.; Pierros, V.; Zacharopoulou, E.; Hatzigeorgiou, A.G. DIANA-LncBase v3: Indexing experimentally supported miRNA targets on non-coding transcripts. *Nucleic Acids Res.* **2020**, *48*, D101–D110. [CrossRef]
5. Salmena, L.; Poliseno, L.; Tay, Y.; Kats, L.; Pandolfi, P.P. A ceRNA Hypothesis: The Rosetta Stone of a Hidden RNA Language? *Cell* **2011**, *146*, 353–358. [CrossRef]
6. Kristensen, L.S.; Andersen, M.S.; Stagsted, L.V.W.; Ebbesen, K.K.; Hansen, T.B.; Kjems, J. The biogenesis, biology and characterization of circular RNAs. *Nat. Rev. Genet.* **2019**, *20*, 675–691. [CrossRef]
7. Gaidatzis, D.; Burger, L.; Florescu, M.; Stadler, M.B. Analysis of intronic and exonic reads in RNA-seq data characterizes transcriptional and post-transcriptional regulation. *Nat. Biotechnol.* **2015**, *33*, 722–729. [CrossRef]

8. Tarantelli, C.; Gaudio, E.; Arribas, A.J.; Kwee, I.; Hillmann, P.; Rinaldi, A.; Cascione, L.; Spriano, F.; Bernasconi, E.; Guidetti, F.; et al. PQR309 Is a Novel Dual PI3K/mTOR Inhibitor with Preclinical Antitumor Activity in Lymphomas as a Single Agent and in Combination Therapy. *Clin. Cancer Res.* **2017**, *24*, 120–129. [CrossRef]

9. Collins, G.P.; Popat, R.; Stathis, A.; Krasniqi, F.; Eyre, T.A.; Ng, C.H.; El-Sharkawi, D.; Schmidt, C.; Wicki, A.; Ivanova, E.; et al. A Dose-Escalation (DE) Study with Expansion Evaluating Safety, Pharmacokinetics and Ef-ficacy of the Novel, Balanced PI3K/mTOR Inhibitor PQR309 in Patients with Relapsed or Refractory Lympho-ma. *Blood* **2016**, *128*, 5893. [CrossRef]

10. Saxton, R.A.; Sabatini, D.M. mTOR Signaling in Growth, Metabolism, and Disease. *Cell* **2017**, *168*, 960–976. [CrossRef] [PubMed]

11. Showkat, M.; Beigh, M.A.; Andrabi, K.I. mTOR Signaling in Protein Translation Regulation: Implications in Cancer Genesis and Therapeutic Interventions. *Mol. Biol. Int.* **2014**, *2014*, 686984. [CrossRef]

12. Tarantelli, C.; Lupia, A.; Stathis, A.; Bertoni, F. Is There a Role for Dual PI3K/mTOR Inhibitors for Patients Affected with Lymphoma? *Int. J. Mol. Sci.* **2020**, *21*, 1060. [CrossRef]

13. Zhang, J.; Chen, S.; Yang, J.; Zhao, F. Accurate quantification of circular RNAs identifies extensive circular isoform switching events. *Nat. Commun.* **2020**, *11*, 1–14. [CrossRef]

14. Karagkouni, D.; Paraskevopoulou, M.D.; Chatzopoulos, S.; Vlachos, I.S.; Tastsoglou, S.; Kanellos, I.; Papadimitriou, D.; Kavakiotis, I.; Maniou, S.; Skoufos, G.; et al. DIANA-TarBase v8: A decade-long collection of experimentally supported miRNA–gene interactions. *Nucleic Acids Res.* **2018**, *46*, D239–D245. [CrossRef]

15. Calsina, B.; Castro-Vega, L.J.; Torres-Pérez, R.; Inglada-Pérez, L.; Currás-Freixes, M.; Roldán-Romero, J.M.; Mancikova, V.; Letón, R.; Remacha, L.; Santos, M.; et al. Integrative multi-omics analysis identifies a prognostic miRNA signature and a targetable miR-21-3p/TSC2/mTOR axis in metastatic pheochromocytoma/paraganglioma. *Theranostics* **2019**, *9*, 4946. [CrossRef]

16. Mayer, C.; Grummt, I. Ribosome biogenesis and cell growth: mTOR coordinates transcription by all three classes of nuclear RNA polymerases. *Oncogene* **2006**, *25*, 6384–6391. [CrossRef] [PubMed]

17. Piazzi, M.; Bavelloni, A.; Gallo, A.; Faenza, I.; Blalock, W.L. Signal Transduction in Ribosome Biogenesis: A Recipe to Avoid Disaster. *Int. J. Mol. Sci.* **2019**, *20*, 2718. [CrossRef] [PubMed]

18. McCool, M.A.; Bryant, C.J.; Baserga, S.J. MicroRNAs and long non-coding RNAs as novel regulators of ribosome biogenesis. *Biochem. Soc. Trans.* **2020**, *48*, 595–612. [CrossRef] [PubMed]

19. Zhao, M.; Xin, X.F.; Zhang, J.Y.; Dai, W.; Lv, T.F.; Song, Y. LncRNA GMDS-AS1 inhibits lung adenocarcinoma development by regulating miR-96-5p/CYLD signaling. *Cancer Med.* **2020**, *9*, 1196–1208. [CrossRef] [PubMed]

20. Cairns, C.A. p53 is a general repressor of RNA polymerase III transcription. *EMBO J.* **1998**, *17*, 3112–3123. [CrossRef] [PubMed]

21. Zhai, W.; Comai, L. Repression of RNA polymerase I transcription by the tumor suppressor p53. *Mol. Cell. Biol.* **2000**, *20*, 5930–5938. [CrossRef]

22. Freeman, J.A.; Espinosa, J.M. The impact of post-transcriptional regulation in the p53 network. *Brief. Funct. Genom.* **2012**, *12*, 46–57. [CrossRef] [PubMed]

23. Bartel, D.P. MicroRNAs: Target Recognition and Regulatory Functions. *Cell* **2009**, *136*, 215–233. [CrossRef] [PubMed]

24. Feng, Z.; Zhang, H.; Levine, A.J.; Jin, S. The coordinate regulation of the p53 and mTOR pathways in cells. *Proc. Natl. Acad. Sci. USA* **2005**, *102*, 8204–8209. [CrossRef] [PubMed]

25. Shen, C.; Houghton, P.J. The mTOR pathway negatively controls ATM by up-regulating miRNAs. *Proc. Natl. Acad. Sci. USA* **2013**, *110*, 11869–11874. [CrossRef] [PubMed]

26. Dobin, A.; Davis, C.A.; Schlesinger, F.; Drenkow, J.; Zaleski, C.; Jha, S.; Batut, P.; Chaisson, M.; Gingeras, T.R. STAR: Ultrafast universal RNA-seq aligner. *Bioinformatics* **2013**, *29*, 15–21. [CrossRef] [PubMed]

27. Anders, S.; Pyl, P.T.; Huber, W. HTSeq—A Python framework to work with high-throughput sequencing data. *Bioinformatics* **2015**, *31*, 166–169. [CrossRef]

28. Ritchie, M.E.; Phipson, B.; Wu, D.; Hu, Y.; Law, C.W.; Shi, W.; Smyth, G.K. limma powers differential expression analyses for RNA-sequencing and microarray studies. *Nucleic Acids Res.* **2015**, *43*, e47. [CrossRef]

29. Law, C.W.; Chen, Y.; Shi, W.; Smyth, G.K. voom: Precision weights unlock linear model analysis tools for RNA-seq read counts. *Genome Biol.* **2014**, *15*, R29. [CrossRef] [PubMed]

30. Robinson, M.D.; McCarthy, D.J.; Smyth, G.K. edgeR: A Bioconductor package for differential expression analysis of digital gene expression data. *Bioinformatics* **2009**, *26*, 139–140. [CrossRef]

31. Subramanian, A.; Tamayo, P.; Mootha, V.K.; Mukherjee, S.; Ebert, B.L.; Gillette, M.A.; Paulovich, A.; Pomeroy, S.L.; Golub, T.R.; Lander, E.S.; et al. Gene set enrichment analysis: A knowledge-based approach for interpreting ge-nome-wide expression profiles. *Proc. Natl. Acad. Sci. USA* **2005**, *102*, 15545–15550. [CrossRef]

32. Kim, D.; Langmead, B.; Salzberg, S.L. HISAT: A fast spliced aligner with low memory requirements. *Nat. Methods* **2015**, *12*, 357–360. [CrossRef] [PubMed]

33. Pertea, M.; Pertea, G.M.; Antonescu, C.M.; Chang, T.-C.; Mendell, J.T.; Salzberg, S.L. StringTie enables improved reconstruction of a transcriptome from RNA-seq reads. *Nat. Biotechnol.* **2015**, *33*, 290–295. [CrossRef] [PubMed]

34. Liu, R.; Holik, A.Z.; Su, S.; Jansz, N.; Chen, K.; Leong, H.S.; Blewitt, M.E.; Asselin-Labat, M.-L.; Smyth, G.K.; Ritchie, M.E. Why weight? Modelling sample and observational level variability improves power in RNA-seq analyses. *Nucleic Acids Res.* **2015**, *43*, e97. [CrossRef]

35. Ruijter, J.M.; Ramakers, C.; Hoogaars, W.M.H.; Karlen, Y.; Bakker, O.; Hoff, M.J.B.V.D.; Moorman, A.F.M. Amplification efficiency: Linking baseline and bias in the analysis of quantitative PCR data. *Nucleic Acids Res.* **2009**, *37*, e45. [CrossRef]

non-coding **RNA**

MDPI

Review

Plant Long Noncoding RNAs: New Players in the Field of Post-Transcriptional Regulations

Camille Fonouni-Farde [1,2], Federico Ariel [3] and Martin Crespi [1,2,*]

1   Université Paris-Saclay, CNRS, INRAE, Univ Evry, Institute of Plant Sciences Paris-Saclay (IPS2), Bat 630, 91192 Gif sur Yvette, France; camille.fonouni-farde@universite-paris-saclay.fr
2   Université de Paris, CNRS, INRAE, Institute of Plant Sciences Paris-Saclay (IPS2), Bat 630, 91192 Gif sur Yvette, France
3   Instituto de Agrobiotecnología del Litoral, CONICET, Universidad Nacional del Litoral, Colectora Ruta Nacional 168 km 0, 3000 Santa Fe, Argentina; fariel@santafe-conicet.gov.ar
*   Correspondence: martin.crespi@universite-paris-saclay.fr

Abstract: The first reference to the "C-value paradox" reported an apparent imbalance between organismal genome size and morphological complexity. Since then, next-generation sequencing has revolutionized genomic research and revealed that eukaryotic transcriptomes contain a large fraction of non-protein-coding components. Eukaryotic genomes are pervasively transcribed and noncoding regions give rise to a plethora of noncoding RNAs with undeniable biological functions. Among them, long noncoding RNAs (lncRNAs) seem to represent a new layer of gene expression regulation, participating in a wide range of molecular mechanisms at the transcriptional and post-transcriptional levels. In addition to their role in epigenetic regulation, plant lncRNAs have been associated with the degradation of complementary RNAs, the regulation of alternative splicing, protein sub-cellular localization, the promotion of translation and protein post-translational modifications. In this review, we report and integrate numerous and complex mechanisms through which long noncoding transcripts regulate post-transcriptional gene expression in plants.

Keywords: long noncoding RNA; post-transcriptional regulation; target mimicry; alternative splicing; protein re-localization; translation promotion; post-translational modification

Citation: Fonouni-Farde, C.; Ariel, F.; Crespi, M. Plant Long Noncoding RNAs: New Players in the Field of Post-Transcriptional Regulations. *Non-coding RNA* **2021**, *7*, 12. https://doi.org/10.3390/ncrna7010012

Received: 26 January 2021
Accepted: 14 February 2021
Published: 17 February 2021

Publisher's Note: MDPI stays neutral with regard to jurisdictional claims in published maps and institutional affiliations.

1. Introduction

Unlike in prokaryotes, genomes in eukaryotes exhibit a large variability in their size [1,2], which does not always correlate with the number of protein-coding genes nor the developmental complexity of organisms. This paradox of an apparent imbalance between organismal genome size and morphological complexity, dubbed the "C-value paradox" [3,4], was in part solved by the extraordinary progress made in next-generation sequencing technologies. Indeed, eukaryotic transcriptomes include a large fraction of non-protein-coding components [5]. Although up to 90% of eukaryotic genomes is estimated to be transcribed during development, only an estimated 2% of transcribed RNAs will code for proteins [6,7]. The noncoding genome, long considered silent and declared as "junk DNA" due to its high content in pseudogenes, simple repeats, and transposons [8,9], encodes a plethora of noncoding RNAs (ncRNAs) with unarguable biological functions. These comprise housekeeping RNAs (small nuclear and nucleolar RNAs, transfer RNAs, ribosomal RNAs, telomerase RNAs, tRNA-derived fragments, and tRNA halves), small regulatory RNAs (micro RNAs, small interfering RNAs, piwi-interacting RNAs, and Y RNAs), and long noncoding RNAs (lncRNAs), also including enhancer RNAs, transposon-derived RNAs, and circular RNAs [10].

LncRNAs form the most diversified group of ncRNAs, exhibiting a large range of sizes varying from 200 bases to over 100 kb in length. They are expressed in various tissues, cell-types, and cell-states, and function in the nucleus or cytoplasm [11,12]. Given their vast diversity, lncRNAs are commonly classified according to their location and orientation relative to neighboring protein-coding transcripts. Long intronic RNAs are transcribed exclusively from intronic regions, whereas long intergenic ncRNAs lie outside of genes and include promoter-, enhancer-, and transposable element-derived lncRNAs and sometimes give rise to double-stranded RNAs. Sense and antisense double-stranded lncRNAs are transcribed from the sense and antisense strands, respectively, while natural antisense transcripts (NATs) initiate in the reverse strand of sense protein coding regions (*cis*-NATs) or are complementary to a sense transcript located in a distinct genomic locus (*trans*-NATs) [6,13]. CircRNAs constitute a novel class of lncRNAs consisting in covalently closed molecules of single-stranded RNA, resulting from back-splicing, a non-canonical form of alternative splicing [14]. Alternatively, lncRNAs can be further categorized depending on their molecular functions and interactions with additional regulatory molecules such as proteins, DNA, or other RNAs [15,16].

It is increasingly clear that lncRNAs participate in virtually every aspect of gene expression. In plants, although the functional characterization of ncRNAs is still in its early stages, several lncRNAs have been described as regulators of gene transcription, capable of conditioning the epigenetic environment of their genomic targets and of modulating the activity of transcriptional complexes [17]. In addition, at the post-transcriptional level, various lncRNAs have been associated with complementary target-RNA degradation, alternative splicing, promotion of translation, protein sub-cellular localization and post-translational modifications. Notably, lncRNA-mediated post-translational modifications of histones are related to the transcriptional regulation of target genes, which has been recently reviewed [17]. Here, we report and integrate recent discoveries about plant lncRNA-mediated regulations of post-transcriptional gene expression.

2. Long Noncoding RNAs Mediating Complementary Target-RNA Degradation

Natural antisense transcripts (NATs) constitute an important class of lncRNAs, exerting a wide variety of molecular functions in eukaryotes [18,19]. They are complementary to sense mRNAs and can be classified into *cis*-NATs generated from a single locus showing sequence complementarity with their corresponding sense transcript or *trans*-NATs that are transcribed from different distant loci and typically display imperfect complementarities with their target endogenous RNA [20,21] (Figure 1). In silico analyses performed in several plant species have led to the identification of a large number of NATs [20,22–26]. In particular, a genome-wide analysis using a custom-designed NAT array revealed that up to 70% of annotated mRNAs have complementary NATs in *Arabidopsis thaliana* [27].

2.1. LncRNAs Involved in Discordant Regulation

NATs can affect positively (concordant regulation) or negatively (discordant regulation) the expression of sense transcripts. An example of discordant regulation is provided by the NAT-lncRNA *asHSFB2a* which counteracts the expression of the *HEAT SHOCK FACTOR B2a* (*HSFB2a*) mRNA in *A. thaliana* female gametophytes [28]. The overexpression of *asHSFB2a* in transgenic plants leads to the absence of *HSFB2a* RNA while the overexpression of *HSFB2a* results in a complete loss of *asHSFB2a* expression, suggesting that *HSFB2a* and *asHSFB2a* are mutually repressive [28]. Similarly, the NAT-lncRNA *DELAY OF GERMINATION 1* (*asDOG1*) was found to be a negative regulator of *DOG1* expression, a gene involved in the control of germination [29]. Down-regulation of *asDOG1* transcription increases the levels of *DOG1* sense mRNAs, enhancing seed dormancy [29]. As a last example in *A. thaliana*, a screening of lncRNAs using a custom-made array led to the identification of the circadian-regulated lncRNA *CDF5 LONG NONCODING RNA* (*FLORE*), a NAT of the *CYCLING DOF FACTOR 5* (*CDF5*) transcript, which likely connects the circadian clock to the photoperiodic flowering pathway [30]. *FLORE* is specifically expressed in the vasculature and regulates *CDF5* in *cis* as well as *CDF1* and *CDF3* in *trans*. Interestingly, *FLORE* and *CDF5* show a mutual inhibition behavior, suggesting that the *CDF5/FLORE* NAT pair constitutes a circadian regulatory module, which buffers its own circadian oscillation and photoperiodic flowering [30]. In rice, the NAT-lncRNA *TWISTED LEAF* (*TL*) is transcribed from the opposite strand of the *OsMYB60* locus encoding an R2R3 MYB transcription factor [31]. Downregulation of *TL* by RNA interference leads to a significant increase in *OsMYB60* expression levels and twisted leaf blades. It was suggested that *TL* may play a *cis*-regulatory role on *OsMYB60* by affecting H3K27me3, H3K36me2, and H3K36me3 histone mark deposition [31].

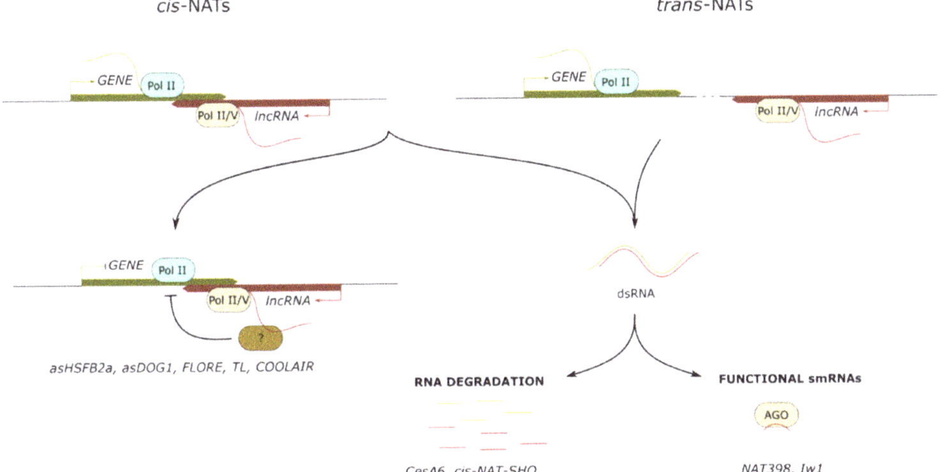

Figure 1. Long noncoding RNAs forming RNA–RNA pairs in the nucleus. Long noncoding (lnc) RNAs can form RNA pairs (dsRNA) with complementary mRNAs in *cis* (*cis*-NATs) or in *trans* (*trans*-NATs), leading to RNA degradation or to the formation of functional small (sm) RNAs. In addition, antisense transcripts can locally recruit protein partners that modulate the transcriptional activity of overlapping protein-coding genes. Examples of characterized lncRNAs are indicated at the bottom.

2.2. LncRNAs as Precursors of Small Regulatory RNAs

In addition, various NAT-lncRNAs have been reported to cause post-transcriptional silencing through the production of regulatory small interfering (si) RNAs derived from NAT pairs. This mechanism was first described in the regulation of salt tolerance in *A. thaliana* [32]. Under salt stress, the induction of *SRO5* mRNA allows the production of 24 nucleotides (nt) siRNAs from the region overlapping with Δ^1-pyrroline-5-carboxylate dehydrogenase (*P5CDH*) transcripts. Subsequently, *P5CDH* transcripts are cleaved to generate 21nt siRNAs [32]. Similar mechanisms have been described in plant responses to pathogens and sperm cell development [33,34]. In barley, an increase in the NAT-lncRNA *CesA6* transcript levels leads to the production of 21 and 24nt siRNAs that correlates with the down-regulation of *CesA6* gene and several loci in *trans* involved in the regulation of cellulose rates and in the modulation of cell wall biosynthesis [35]. Similarly, in *Petunia hybrida*, the *Sho* gene involved in the production of cytokinin phytohormones contains an antisense ORF partially overlapping with the ORF of the *Sho* sense transcript that encodes the SHO protein [36]. The tissue specific transcription of *cis*-NAT *SHO* leads to the association of *Sho* sense and antisense transcripts in a double-stranded RNA likely targeted by a DICER complex for degradation into 24nt siRNAs [36]. Another mechanism, related to thermotolerance, was reported in *A. thaliana*. The NAT-lncRNAs *NAT398b* and *NAT398c* are *cis*-NATs of the *MIRNA* genes *MIR398b* and *MIR398c*, respectively. Knock-down of *NAT398b/c* promotes the accumulation of *MIR398b* and *MIR398c*, while the overexpression of *NAT398b* and *NAT398c* represses the processing of *miR398*. Notably, the overexpression of siRNAs derived from *NAT398* overlapping transcripts, so-called nat-siR398, reduces the levels of pri-miR398b and pri-miR398c [37].

Interestingly, computational analyses performed for *A. thaliana* revealed that antisense transcription is associated with micro (mi) RNA-targeted mRNAs [38]. In wheat, the lncRNA *INHIBITOR of WAX1* (*Iw1*) contains an inverted repeat showing more than 80% identity to the *WAX1-COE* gene, encoding a carboxylesterase-like protein that controls glaucousness [39]. The *Iw1* transcript is able to form a miRNA precursor-like long hairpin which produces small RNAs, including the 21nt-miRNA miRW1. The accumulation of miRW1 is linked to the down-regulation of *W1-COE* and its paralog *W2-COE*, the cleavage of *W1-COE* transcripts and glaucous repression [39].

3. Long Noncoding RNAs Involved in the Regulation of Alternative Splicing

In addition to capping and polyadenylation, the production of mature mRNAs from pre-mRNAs relies on the prior removal of introns and the ligation of the majority of exons in the order in which they appear in a gene, a process known as RNA splicing. Under certain circumstances, some exons can be skipped, generating various isoforms of mature mRNAs from a single pre-mRNA. This process, called alternative splicing (AS), is mediated by the spliceosome and involves a subclass of small nuclear RNAs referred to as nuclear uridine-rich RNAs. They function in collaboration with core small nuclear ribonucleoprotein complex subunits (snRNPs) and non-snRNPs splicing factors (SFs) whose interaction with lncRNAs likely condition their stability and sub-cellular localization [40–43]. LncRNAs mainly regulate AS through interactions with specific SFs, by the regulation of chromatin remodeling that fine-tunes the splicing of specific targets and via the formation of lncRNA-pre-mRNA duplexes [43] (Figure 2).

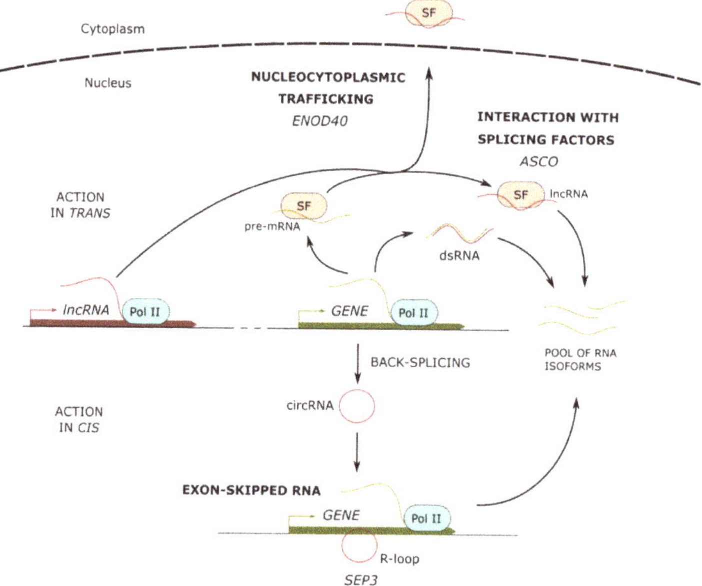

Figure 2. Long noncoding RNAs modulating alternative splicing. Long noncoding (lnc) RNAs can form RNA interactions (dsRNA) with pre-mRNAs, fine-tuning their splicing output. In addition, lncRNAs can interact with splicing factors (SF), affecting their recognition of pre-mRNA targets or their sub-cellular localization. Protein-coding transcripts can suffer back-splicing, leading to the formation of circular RNAs (circRNA), which can interact with the parent gene to form DNA-RNA duplexes (R-loops) and modulate the alternative splicing of the nascent transcripts. Examples of characterized lncRNAs are indicated below each mechanism.

3.1. LncRNAs Interacting with Splicing Factors

In plants, AS plays crucial functions in the control of gene expression, boosting the protein-coding capacity and contributing to developmental plasticity [44,45]. In Arabidopsis, the lncRNA *ALTERNATIVE SPLICING COMPETITOR (ASCO)* interacts in vivo with the plant-specific SFs NUCLEAR SPECKLE RNA-BINDING PROTEINS (NSRa and b), which localize in nuclear speckles and are involved in splicing [46]. NSRs participate in the regulation of molecular and growth responses to auxin. After an auxin treatment, the double mutant *nsra/b* exhibited over 2200 genes differentially regulated in comparison to wild-type plants, as well as a reduced number of lateral roots suggesting a decreased sensitivity to auxin. Interestingly, the identification of RNA processing events in the *nsra/b* mutant revealed an important number of intron retention events and differential $5'$ start or $3'$ ends in a subset of genes, including a high number of auxin-related genes that behave accordingly in the *ASCO* overexpressing lines [46,47]. Remarkably, in vitro binding assays additionally showed that *ASCO* competes with mRNA-targets for the binding to NSRs, suggesting that *ASCO* regulates the AS of pre-mRNAs in response to auxin by hijacking NSRs [46]. More recently, a NSRa-directed RNA immunoprecipitation (RIP)-Seq approach in *A. thaliana* revealed that lncRNAs are overrepresented among NSRa targets [48]. As NSRa targets are mainly enriched for genes related to biotic stress responses, the interplay between lncRNAs and AS mRNAs in NSR-containing complexes was suggested to integrate the auxin and immune response pathways [48]. In agreement with this expectation, both knock-down and overexpression of *ASCO* led to the deregulation of expression and splicing of a large number of genes related to biotic stress and flagellin response in *A. thaliana* [49]. Remarkably, RNAi-*ASCO* plants and the double mutant *nsra/b* were found to exhibit a different response to flagellin, suggesting that *ASCO* also modulates AS in an NSR-independent manner. Consistently, an *ASCO*-directed chromatin isolation by RNA purification (ChIRP) coupled to mass spectrometry and RIP assays allowed the identification of other putative *ASCO* protein partners, including the pre-mRNA-processing-splicing factor 8A (PRP8a) and the spliceosome-core component SmD1b [49–51]. As previously observed for NSRs [46], *ASCO* overexpression also competes for PRP8a binding to particular mRNA targets [49].

3.2. LncRNAs Regulating Splicing Through Chromatin Remodeling

An additional mechanism of AS regulation involving circular non-coding RNAs (circRNAs) was reported in *A. thaliana* [32]. The overexpression of a circRNA comprising the entire exon 6 of the *SEPALLATA 3 (SEP3)* gene increases the abundance of the naturally occurring exon-skipped AS variant *SEP3.3*, which lacks exon 6 [52]. SEP3 is a member of the plant MADS (MCM1-AGAMOUS-DEFICIENS-SRF)-box transcription factor superfamily involved in flower development, and modifications of *SEP3* splicing gives rise to floral homeotic phenotypes [52,53]. Remarkably, *SEP3* exon 6 circRNA can directly interact with its cognate DNA locus, forming an RNA:DNA hybrid (R-loop), which results in transcriptional pausing and correlates with the recruitment of splicing factors and AS. This mechanism suggests that circRNAs expressed from distant loci may increase the splicing efficiency of their cognate exon-skipped messenger RNAs and that chromatin conformation and R-loop formation are critical modulators of splicing patterns [52].

3.3. LncRNA-RNA Duplexes Regulating Alternative Splicing

The analysis of transcription data for overlapping gene pairs in *A. thaliana* revealed a large proportion of convergently overlapping pairs (COPs) with the potential to form double-stranded RNAs [23]. Interestingly, intron-containing genes and genes with alternatively spliced transcripts are over-represented among COPs. In addition, the loci where antisense transcripts overlap with sense transcript introns mostly show AS and/or variation of polyadenylation, suggesting that the formation of NAT lncRNA-RNA pairs may regulate the AS of protein-coding genes [23]. Consistently, a genome-wide screen of *trans*-NATs in Arabidopsis led to the identification of 1320 putative *trans*-NAT pairs [24].

Most of them are predicted to form extended double-stranded RNA duplexes if sense and anti-sense are expressed in the same sub-cellular compartment, and they may lead to gene silencing and indirect AS regulation [24]. Taken together, these studies suggest that lncRNAs integrate a dynamic splicing network to control transcriptome reprogramming through AS.

4. Long Noncoding RNAs as Molecular Cargos for Protein Re-Localization

Short open reading frame (sORF) mRNAs are atypical mRNAs that contain only sORFs (shorter than 100 amino acids) and accumulate in the cytoplasm where they can be translated into oligopeptides acting as signal molecules [54,55]. Remarkably, in legumes, the highly conserved *EARLY NODULIN 40 (ENOD40)* genes known to participate in root symbiotic nodule organogenesis, contain only sORFs whose transcripts may encode short peptides [56–58]. In soybean, the lncRNA *GmENOD40* encodes two oligopeptides of 12 and 24 aa residues that may have a transport function and specifically bind to sucrose synthase subunit nodulin 100 to control the use of sucrose in nitrogen-fixing nodules [58]. In *M. truncatula*, *MtENOD40* is rapidly induced by symbiotic rhizobial bacteria in the root pericycle and is also detected in the differentiating cells of the nodule primordia [56,57]. *MtENOD40* has been described as highly structured and not associated to polysomes [56,59]. Yeast three-hybrid assays revealed that the structured *MtENOD40* RNA directly interacts with the constitutive RNA Binding Protein 1 (MtRBP1), a close homolog of lncRNA-interacting AtNSRs [60], located in the nuclear speckles where the splicing machinery is also hosted [61]. During nodulation, MtRBP1 is exported by *MtENOD40* to cytoplasmic granules. Hence, while *ENOD40*-encoded peptides are likely involved in sugar metabolism, the highly structured *ENOD40* RNA contributes to nucleocytoplasmic trafficking [61] (Figure 2).

5. Long Noncoding RNAs Promoting Translation

The translation process of mature mRNAs into proteins can be divided into four phases, namely initiation, elongation, termination, and ribosome recycling. The regulation of translation, so-called translational control, is a mechanism that allows a rapid modulation of gene expression through the activation or repression of pre-synthesized mRNA translation without requiring *de novo* transcription [62,63]. The global regulation of translation of most cellular mRNAs mainly relies on the modification of translation-initiation factors, while specific control targeting certain mRNAs likely involves regulatory protein complexes, microRNA-containing ribonucleoprotein complexes, and lncRNAs [64–70]. LncRNAs can be recruited to polysomes to regulate the translation of target mRNAs positively or negatively or indirectly regulate translation by sequestering miRNAs that direct the cleavage of target mRNAs (Figure 3).

5.1. LncRNA-mRNA Pairs into Polysomes

In rice, the *PHOSPHATE1;2 (PHO1;2)* gene is involved in the export of phosphate to the apoplastic space of xylem vessels [71–73]. The complementary strand of *PHO1;2* encodes the associated *cis*-NAT *PHO1;2*. Both genes are controlled by promoters active in the vascular cylinders of roots and leaves, but while *PHO1;2* promoter is unresponsive to phosphate, *cis*-NAT *PHO1;2* promoter is strongly activated under phosphate deficiency [74]. In phosphate-deficient plants, *cis*-NAT *PHO1;2* transcripts and PHO1;2 protein amount increase, although *PHO1;2* mRNA levels remain unchanged. In addition, the downregulation or constitutive overexpression of *cis*-NAT *PHO1;2* leads, respectively, to a decrease or strong increase in PHO1;2 protein levels, whereas the level of expression and nuclear export of *PHO1;2* mRNA are not affected. Notably, the expression of *cis*-NAT *PHO1;2* is associated with the shuttle of the *PHO1;2–cis*-NAT *PHO1;2* sense-antisense pair towards the polysomes, supporting a role for *cis*-NAT *PHO1;2* in the promotion of *PHO1;2* translation through polysomal recruitment, to regulate phosphate homeostasis [74].

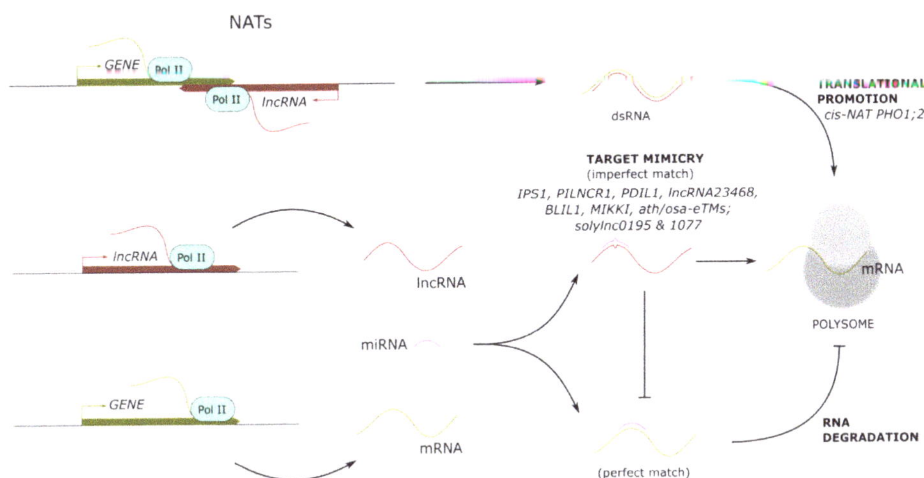

Figure 3. Long noncoding RNAs modulating the translation of protein-coding genes. Long noncoding (lnc) RNAs can form RNA–RNA interactions (dsRNA), promoting the shuttle to polysomes and enhancing translation. In addition, lncRNAs can act as miRNA target mimicry, titrating active miRNA abundance and boosting mRNA translation. Examples of characterized lncRNAs are indicated below each mechanism.

More recently, global analyses of polysome-associated RNAs and ribosome footprints in *A. thaliana* led to the identification of novel lncRNAs controlling cognate mRNA translation [70,75]. Under phosphate deficiency, five ribosome associated *cis*-NATs showed an induction correlated with the enhanced translation of their cognate sense transcripts, including two ATP BINDING CASSETTE SUBFAMILY G transporters (ABCG2 and ABCG20) and a POLLEN-SPECIFIC RECEPTOR-LIKE KINASE 7 (PRK7) family member, associated with nutrient uptake, lateral root formation, and root cell elongation, respectively [70]. In addition, five *trans*-NATs showed a positive correlation between their expression and their target mRNA levels, and the expression of four *trans*-NATs was found to correlate with a change in target mRNA polysome association under low phosphate conditions [75].

5.2. LncRNAs as Target Mimics for miRNAs

MiRNAs are ncRNAs of 20–22 nucleotides that play key regulatory roles in various biological processes in plants [76]. They are processed by Dicer-like proteins from imperfectly paired stem-loop precursors and repress gene expression by directing the cleavage or the translational arrest of target mRNAs [77–79]. Some lncRNAs with highly similar target sites as miRNA targets (miRNA recognition elements) can act as inhibitors of miRNA activity. They function as competing endogenous RNAs (ceRNAs), binding to miRNAs with imperfect base complementarity and blocking their interaction with authentic targets [80,81]. This regulatory mechanism is known as "target mimicry". In plants, ceRNAs are named "target mimics" (TMs), also referred to as miRNA sponges or miRNA decoys in mammals.

In *A. thaliana*, the lncRNA *INDUCED BY PHOSPHATE STARVATION 1* (*IPS1*) is a functional endogenous target mimic (eTM) of miR399 involved in inorganic phosphate (Pi) homeostasis [82]. The Pi starvation-responsive AtmiR399 directs the cleavage of the mRNA *AtPHO2* (*Phosphate 2*), encoding an E2 ubiquitin conjugase-related protein, which negatively regulates Pi remobilization and Pi content in shoots. The sequences of the mRNA *AtPHO2* and lncRNA *IPS1* contain a similar motif of 23 nucleotides complementary to AtmiR399. However, in contrast to *AtPHO2*, *IPS1* pairing with AtmiR399 is interrupted by a mismatched loop in the expected AtmiR399 cleavage site, which prevents its degradation. When *IPS1* sequesters AtmiR399, the authentic AtmiR399-target *AtPHO2* is accumulated, leading to a decrease in shoot Pi content [82]. More recently, a very similar mechanism was reported in maize. The lncRNA *PI-DEFICIENCY-INDUCED LONG NONCODING RNA 1 (PILNCR1)* functions as an eTM for ZmmiR399, thwarting the ZmmiR399-guided post-transcriptional repression of *ZmPHO2* and favoring maize adaptation to Pi deficiency [83]. Additionally, in *Medicago truncatula*, the *PI-DEFICIENCY-INDUCED LNCRNA 1 (PDIL1)* was reported to regulate Pi transport by inhibiting the degradation of *MtPHO2*, also acting as an eTM for MtmiR399 [84]. Another example of lncRNA functioning as eTM is the tomato *lncRNA23468* involved in the resistance to *Phytophthora infestans* [85]. Overexpression of *lncRNA23468* induces a significant decrease in miR482b accumulation and an increase in the miR482b target genes *NBS-LRR* (nucleotide binding sites-leucine-rich repeat) expression. It was thus proposed that *lncRNA23468* may decoy miR482b for targeted cleavage, thereby increasing the expression levels of *NBS-LRRs* genes, enhancing tomato resistance to *P. infestans* [85].

Computational analyses also led to the identification of putative eTMs originating from intergenic or noncoding genes for 20 highly conserved miRNAs in *Arabidopsis thaliana* and rice [86]. The identified TMs for miR160 (ath-eTM160-1 and osa-eTM160-3) and miR166 (ath-eTM166-1 and osa-eTM166-2) were proven to be functional target mimics, their overexpression leading to diverse altered phenotypes such as smaller and serrated leaves, spoon-shaped cotyledons, curled rosette leaves, or accelerated flowering. The effectiveness of TMs for miR156, miR159, and miR172 was also confirmed by transient agroinfiltration assay [86]. In tomato, the lncRNAs *slylnc0195* and *slylnc1077* involved in the tomato yellow leaf curl virus response were predicted to be eTMs of miR166 and miR399, respectively, and the functionality of *slylnc0195* was also validated using a transient agro-infiltration assay [87]. Recently, 407 competing endogenous (ce)RNA pairs were constructed in *A. thaliana* to identify lncRNAs involved in blue light-directed plant photomorphogenesis and acting as ceRNAs. The lncRNA *BLUE LIGHT-INDUCED LNCRNA 1 (BLIL1)* was found to inhibit hypocotyl elongation under blue light and in response to mannitol stress by serving as a ceRNA to sequester miR167 [88].

Interestingly, the mechanism of target mimicry can be engineered and exploited to inhibit specific miRNAs via artificial miRNA TMs (aTMs) in order to establish their functionality. In *A. thaliana*, a collection of transgenic plants expressing aTMs predicted to reduce the activity of most of the miRNA families was generated, leading to morphological abnormalities in the aerial part for ~20% of the miRNAs targeted [89].

Finally, transposable element (TE)-derived transcripts that contain binding sites for miRNAs can also function as eTMs. In rice, the retrotransposon-derived transcript *MIKKI* ("decoy" in Korean) was identified as an eTM for miR171, known to target mRNAs encoding SCARECROW-Like (SCL) transcription factors for cleavage [90]. *MIKKI* is a TE-derived locus including *Osr29* Long Terminal Repeat (LTR) retrotransposon, and its mature transcript contains an imperfect binding site for miR171, generated by a splicing event and likely attenuating the cleavage activity of miR171. In roots, *MIKKI* transcripts bind to miR171, stabilizing *SCL* mRNAs, which play an important role in root development [90].

6. Long Noncoding RNAs Mediating Post-Translational Modifications: Impact on Chromatin Remodeling and Transcription

In mammals, several examples illustrate the action of lncRNAs in protein post-translational modifications. By bringing together target proteins and specific kinases, phosphatases, or ubiquitin-ligases, lncRNAs can regulate post-translational modifications that will modulate the activity of enzymes [91,92], the stability of proteins [93], or their sub-cellular localization [94]. Intriguingly, the only known post-translational modifications modulated by plant lncRNAs are related to histones, thus affecting the epigenetic profile of target genes and their transcriptional status. The epigenetic regulation of gene expression by lncRNAs has been recently reviewed [17]. Here, we focus on the histone post-translational modifications modulated by lncRNAs in plants.

Polycomb Group (PcG) proteins are critical regulators of gene expression, essential for development in many organisms. They form complexes that modify post-translationally histones tails of target genes. In plants, the histone H3K27 trimethyltransferase CURLY LEAF (CLF) functions as a catalytic subunit of the Polycomb Repressive Complex 2 (PRC2) complex [95–97]. H3K27me3 then assists to recruit the PRC1-like components LIKE HETE-ROCHROMATIN PROTEIN 1 (LHP1) and AtRING1 [98]. Additionally, the Trithorax H3K4 methyltransferase ARABIDOPSIS TRITHORAX-LIKE PROTEIN 1 (ATX1) mediates the establishment of H3K4me3 [99]. Interestingly, various lncRNAs have been associated with the post-translational modifications of histones at target loci, mediated by the recruitment or removal of PcG and Trithorax proteins (Figure 4).

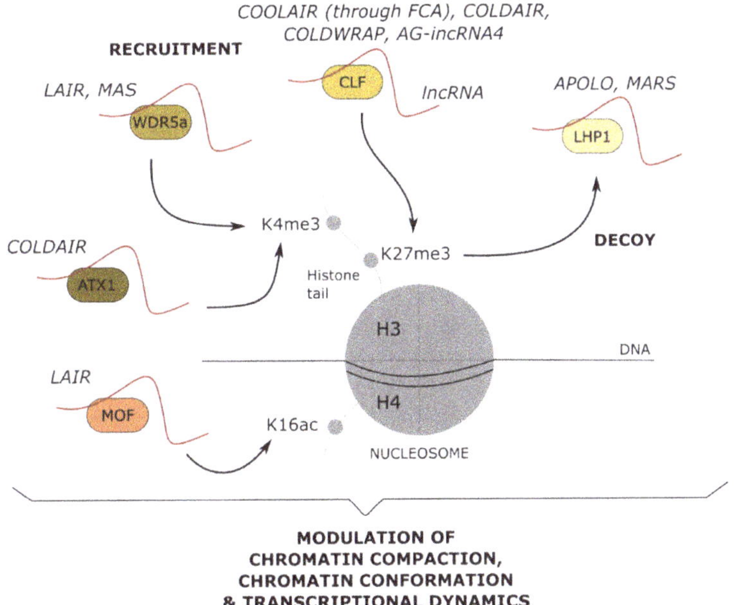

Figure 4. Long noncoding RNAs modulating post-translational modifications of histone proteins. Long noncoding (lnc) RNAs can recruit or decoy nuclear protein complexes that modify histone tails. H3K4 trimethylation (me3) can be modulated by the lncRNA-mediated recruitment of WDR5a (COMPASS-like complex) or ATX1 (Trithorax). H3K27 trimethylation (me3) can be modulated by the lncRNA-mediated recruitment of CLF (PRC2) or the decoy of LHP1 (PRC1). Finally, H4K16 acetylation (ac) can be modulated by the recruitment of MOF. The molecular output of histone post-translational modifications on chromatin and transcription is indicated below. Examples of characterized lncRNAs are indicated above each chromatin-related player.

In *A. thaliana*, the *FLOWERING LOCUS C* (*FLC*) gene encodes a MADS-box-containing transcription factor (TF) that acts as a critical repressor of flowering [100]. *FLC* transcription is antagonistically regulated not only by the active histone modifications H3K4me3 and H3K36me3 but also by the repressive histone modification H3K27me3 [101]. Upon transition to flowering, H3K4me3 is removed, while H3K27me3 is deposited, leading to a decrease in *FLC* expression [102]. Remarkably, *FLC* transcriptional regulation depends on *cis*-acting lncRNAs, including *COOLAIR, COLD ASSISTED INTRONIC NONCODING RNA* (*COLDAIR*), and *COLDWRAP* [103–106]. *COOLAIR* is a set of antisense transcripts physically associated with the *FLC* locus, linked to the synchronized replacement of H3K36 methylation with H3K27me3 during the early stages of vernalization, independent of Polycomb complexes [107]. *COOLAIR* directly binds to the RNA binding protein FLOWERING CONTROL LOCUS A (FCA), which further interacts with the PRC2 component CLF. This allows the recruitment of CLF at *FLC* for H3K27me3 deposition [108]. *COLDAIR* is transcribed from the first intron of *FLC* and cooperates with *COLDWRAP*, derived from *FLC* proximal promoter, to facilitate the establishment of H3K27me3 during the late stage of vernalization, through the formation of a repressive intragenic chromatin loop that retains CLF at the *FLC* promoter [104,105]. Strikingly, ectopic overexpression of *COLDAIR* suppresses H3K27me3 and induces H3K4me3 at the *FLC* locus depending on the recruitment of ATX1 and removal of CLF, leading to enhanced *FLC* expression [109]. Remarkably, the overexpression of intronic lncRNAs derived from several other H3K27me3-enriched MADS-box genes also led to the activation of their corresponding genes by suppressing H3K27me3 and promoting H3K4me3 deposition [109]. The NAT-lncRNA *MADS AFFECTING FLOWERING4* (*MAS*), transcribed from the *MADS AFFECTING FLOWERING 4* (*MAF4*) locus, is also involved in the regulation of flowering [110]. *MAS* is induced by cold and activates *MAF4*, encoding a MADS-box containing TF, by interacting with WDR5a, a structural core component of a COMPASS-like H3K4 histone methylation complex. *MAS* mediates the recruitment of WDR5a to *MAF4* for H3K4me3 deposition and activation of *MAF4* [110]. In rice, the lncRNA *LRK Antisense Intergenic RNA* (*LAIR*) is transcribed from the antisense strand of its neighboring *LRK* (leucine-rich repeat receptor kinase) gene cluster and can interact with OsWDR5 as well as with the histone H4K16 acetyltransferase OsMOF [111]. *LAIR* overexpression is associated with higher H3K4me3 and H4K16ac levels at the *LRK1* chromatin region and with the upregulation of *RLK1*, leading to increased grain yield [111]. AGAMOUS (AG) is another MADS TF, involved in the specification of stamens and carpels, in a tissue-specific manner [112–114]. The *AG* second intron encodes several ncRNAs, including *AGAMOUS INTRONIC RNA 4* (*AG-incRNA4*), which recruits CLF and represses *AG* transcription likely through the deposition of H3K27me3 [115].

In *A. thaliana*, the PRC1 protein LHP1 recognizes H3K27me3 deposition and ensures the spreading of this repressive mark, controlling global genome topology [116]. In response to auxin, the lncRNA *AUXIN REGULATED PROMOTER LOOP* (*APOLO*) is transcribed from the promoter region of its neighboring gene *PINOID*, a key regulator of polar auxin transport, and interacts with LHP1 in vivo [117]. *APOLO* recognizes a subset of auxin-related genes in *trans*, through sequence complementarity and DNA–RNA hybrid formation (R-loops). Remarkably, overexpression of *APOLO* leads to the decoy of LHP1 from common target genes over the genome, and is associated with a decrease in H3K27me3 deposition as well as with modifications of chromatin conformation [118]. Similarly, the lncRNA *MARNERAL SILENCING* (*MARS*) is transcribed in response to abscisic acid (ABA) from the marneral cluster, which includes the marneral synthase *MRN1* gene and the two P450 cytochrome-encoding genes *CYP705A12* and *CYP71A16* [119]. *MARS* over-accumulation is associated with the decoy of LHP1 and a decrease in H3K27me3 distribution throughout the marneral cluster. Loss of H3K27me3 likely allows the formation of a chromatin loop bringing together an enhancer element enriched in ABA-related TF binding sites and *MRN1* proximal promoter, resulting in the transcriptional activation of *MRN1* and a delay in seed germination [119].

7. Conclusions and Future Perspectives

Compelling evidence supports the involvement of lncRNAs in diverse and numerous aspects of post-transcriptional gene regulations in plants. Future research will likely shed light on the basis governing lncRNA interaction with diverse molecular partners, including DNA, proteins, or transcripts. The noncoding transcriptome has been shown to differ across ecotypes of the same species, notably in response to the environment [120]. This observation suggests that lncRNAs may be the key players in natural variation, contributing to plant adaptation during evolution. The conserved role of divergent lncRNAs across species likely depends on the presence of specific short sequences as well as on their secondary structure. Remarkably, the growing number of identified cold-responsive lncRNAs participating not only in the post-transcriptional but also in the transcriptional regulation of gene expression [121], e.g., the lncRNA *SVALKA* [122], suggests that the noncoding transcriptome is a central actor in responses to the environment. Notably, as plants cannot regulate their corporal temperature, the structure of lncRNAs and mRNAs, as well as their interactions, is most likely affected by this environmental cue. In agreement with this hypothesis, it has been recently demonstrated that the secondary structure of plant mRNAs in response to warm temperatures may modulate their translational activity, acting as thermosensor tools [123]. Similarly, structured regions in bacterial mRNAs, named RNA thermometers (RNATs), can function as thermosensors and regulate translation [124]. The advent of cutting-edge technologies, including SHAPE-seq (selective 2′-hydroxyl acylation analyzed by primer extension sequencing) to characterize RNA folding [125], will likely allow one to determine whether plant lncRNAs adopt alternative structures in response to temperature, and might function as new emerging regulators fine-tuning the protein-coding genome in response to climate change.

Funding: This research was funded by Saclay Plant Sciences-SPS (ANR-10-LABX-40) and PIOSYM (ANR-19-CE20-0011-03) ANR projects, and Agencia Nacional de Promoción Científica y Tecnológica (PICT). Both laboratories are involved in the International Associated Laboratory (LIA) NOCOSYM from CNRS-CONICET.

Data Availability Statement: No new data were created or analyzed in this study. Data sharing is not applicable to this article.

Acknowledgments: We thank Olivier Martin (IPS2) for the careful reading of this manuscript. M.C. is a member of CNRS (France) and C.F.F. is a post-doctoral fellow of the same institution. F.A. is a member of CONICET (Argentina).

Conflicts of Interest: The authors declare no conflict of interest.

References

1. Corradi, N.; Pombert, J.-F.; Farinelli, L.; Didier, E.S.; Keeling, P.J. The complete sequence of the smallest known nuclear genome from the microsporidian Encephalitozoon intestinalis. *Nat. Commun.* **2010**, *1*, 77. [CrossRef] [PubMed]
2. Pellicer, J.; Fay, M.F.; Leitch, I.J. The largest eukaryotic genome of them all? *Bot. J. Linn. Soc.* **2010**, *164*, 10–15. [CrossRef]
3. Thomas, C. The Genetic Organization of Chromosomes. *Annu. Rev. Genet.* **1971**, *5*, 237–256. [CrossRef]
4. Eddy, S.R. The C-value paradox, junk DNA and ENCODE. *Curr. Biol.* **2012**, *22*, R898–R899. [CrossRef] [PubMed]
5. Djebali, S.; Davis, C.A.; Merkel, A.; Dobin, A.; Lassmann, T.; Mortazavi, A.; Tanzer, A.; Lagarde, J.; Lin, W.; Schlesinger, F.; et al. Landscape of transcription in human cells. *Nature* **2012**, *489*, 101–108. [CrossRef] [PubMed]
6. Ariel, F.; Romero-Barrios, N.; Jégu, T.; Benhamed, M.; Crespi, M. Battles and hijacks: Noncoding transcription in plants. *Trends Plant. Sci.* **2015**, *20*, 362–371. [CrossRef]
7. Lee, H.; Zhang, Z.; Krause, H.M. Long Noncoding RNAs and Repetitive Elements: Junk or Intimate Evolutionary Partners? *Trends Genet.* **2019**, *35*, 892–902. [CrossRef] [PubMed]
8. Comings, D.E. The Structure and Function of Chromatin. *Adv. Hum. Genet.* **1972**, *3*, 237–431. [CrossRef]
9. Ohno, S. So much "junk" DNA in our genome. *Brookhaven Symp. Biol.* **1972**, *23*, 366–370. [PubMed]
10. Zhang, P.; Wu, W.; Chen, Q.; Chen, M. Non-Coding RNAs and their Integrated Networks. *J. Integr. Bioinform.* **2019**, *16*, 1–12. [CrossRef]
11. Cao, J. The functional role of long non-coding RNAs and epigenetics. *Biol. Proced. Online* **2014**, *16*, 11. [CrossRef]
12. Quinn, J.J.; Chang, H.Y. Unique features of long non-coding RNA biogenesis and function. *Nat. Rev. Genet.* **2016**, *17*, 47–62. [CrossRef] [PubMed]
13. Ma, L.; Bajic, V.B.; Zhang, Z. On the classification of long non-coding RNAs. *RNA Biol.* **2013**, *10*, 924–933. [CrossRef] [PubMed]
14. Kristensen, L.S.; Andersen, M.S.; Stagsted, L.V.W.; Ebbesen, K.K.; Hansen, T.B.; Kjems, J. The biogenesis, biology and characterization of circular RNAs. *Nat. Rev. Genet.* **2019**, *20*, 675–691. [CrossRef]
15. Marchese, F.P.; Raimondi, I.; Huarte, M. The multidimensional mechanisms of long noncoding RNA function. *Genome Biol.* **2017**, *18*, 1–13. [CrossRef] [PubMed]
16. Lucero, L.; Ferrero, L.; Fonouni-Farde, C.; Ariel, F. Functional classification of plant long noncoding RNAs: A transcript is known by the company it keeps. *New Phytol.* **2021**, *229*, 1251–1260. [CrossRef] [PubMed]
17. Lucero, L.; Fonouni-Farde, C.; Crespi, M.; Ariel, F. Long noncoding RNAs shape transcription in plants. *Transcription* **2020**, *11*, 160–171. [CrossRef]
18. Lapidot, M.; Pilpel, Y. Genome-wide natural antisense transcription: Coupling its regulation to its different regulatory mechanisms. *EMBO Rep.* **2006**, *7*, 1216–1222. [CrossRef]
19. Britto-Kido, S.D.A.; Ferreira Neto, J.R.C.; Pandolfi, V.; Marcelino-Guimarães, F.C.; Nepomuceno, A.L.; Vilela Abdelnoor, R.; Benko-Iseppon, A.M.; Kido, E.A. Natural antisense transcripts in plants: A re-view and identification in soybean infected with phakopsora pachyrhizi supersage library. *Sci. World J.* **2013**, *2013*, 219798. [CrossRef]

20. Wang, X.-J.; Gaasterland, T.; Chua, N.-H. Genome-wide prediction and identification of cis-natural antisense transcripts in Arabidopsis thaliana. *Genome Biol.* **2005**, *6*, R30. [CrossRef]

21. Li, Y.-Y.; Qin, L.; Guo, Z.-M.; Liu, L.; Xu, H.; Hao, P.; Su, J.; Shi, Y.; He, W.-Z.; Li, Y.-X. In silico discovery of human natural antisense transcripts. *BMC Bioinform.* **2006**, *7*, 18. [CrossRef]

22. Osato, N.; Yamada, H.; Satoh, K.; Ooka, H.; Yamamoto, M.; Suzuki, K.; Kawai, J.; Carninci, P.; Ohtomo, Y.; Murakami, K.; et al. Antisense transcripts with rice full-length cDNAs. *Genome Biol.* **2003**, *5*, R5. [CrossRef] [PubMed]

23. Jen, C.-H.; Michalopoulos, I.; Westhead, D.R.; Meyer, P. Natural antisense transcripts with coding capacity in Arabidopsis may have a regulatory role that is not linked to double-stranded RNA degradation. *Genome Biol.* **2005**, *6*, R51. [CrossRef] [PubMed]

24. Wang, H.; Chua, N.-H.; Wang, X.-J. Prediction of trans-antisense transcripts in Arabidopsis thaliana. *Genome Biol.* **2006**, *7*, R92. [CrossRef] [PubMed]

25. Zhou, X.; Sunkar, R.; Jin, H.; Zhu, J.K.; Zhang, W. Genome-wide identification and analysis of small RNAs origi-nated from natural antisense transcripts in Oryza sativa. *Genome Res.* **2009**, *19*, 70–78. [CrossRef]

26. Lu, T.; Zhu, C.; Lu, G.; Guo, Y.; Zhou, Y.; Zhang, Z.; Zhao, Y.; Li, W.; Lu, Y.; Tang, W.; et al. Strand-specific RNA-seq reveals widespread occurrence of novel cis-natural antisense transcripts in rice. *BMC Genom.* **2012**, *13*, 721. [CrossRef] [PubMed]

27. Wang, H.; Chung, P.J.; Liu, J.; Jang, I.C.; Kean, M.J.; Xu, J.; Chua, N.H. Genome-wide identification of long non-coding natural antisense transcripts and their responses to light in Arabidopsis. *Genome Res.* **2004**, *24*, 444–453. [CrossRef]

28. Wunderlich, M.; Gross-Hardt, R.; Schöffl, F. Heat shock factor HSFB2a involved in gametophyte development of Arabidopsis thaliana and its expression is controlled by a heat-inducible long non-coding antisense RNA. *Plant Mol. Biol.* **2014**, *85*, 541–550. [CrossRef]

29. Fedak, H.; Palusinska, M.; Krzyczmonik, K.; Brzezniak, L.; Yatusevich, R.; Pietras, Z.; Kaczanowski, S.; Swiezewski, S. Control of seed dormancy in Arabidopsis by a cis-acting noncoding antisense transcript. *Proc. Natl. Acad. Sci. USA* **2016**, *113*, E7846–E7855. [CrossRef]

30. Henriques, R.; Wang, H.; Liu, J.; Boix, M.; Huang, L.-F.; Chua, N.-H. The antiphasic regulatory module comprising CDF5 and its antisense RNA FLORE links the circadian clock to photoperiodic flowering. *New Phytol.* **2017**, *216*, 854–867. [CrossRef]

31. Liu, X.; Li, D.; Zhang, D.; Yin, D.; Zhao, Y.; Ji, C.; Zhao, X.; Li, X.; He, Q.; Chen, R.; et al. A novel antisense long noncoding RNA, TWISTED LEAF, maintains leaf blade flattening by regulating its associated sense R2R3-MYB gene in rice. *New Phytol.* **2018**, *218*, 774–788. [CrossRef]

32. Borsani, O.; Zhu, J.; Verslues, P.E.; Sunkar, R.; Zhu, J.-K. Endogenous siRNAs Derived from a Pair of Natural cis-Antisense Transcripts Regulate Salt Tolerance in Arabidopsis. *Cell* **2005**, *123*, 1279–1291. [CrossRef] [PubMed]

33. Katiyar-Agarwal, S.; Morgan, R.; Dahlbeck, D.; Borsani, O.; Villegas, A.; Zhu, J.-K.; Staskawicz, B.J.; Jin, H. A pathogen-inducible endogenous siRNA in plant immunity. *Proc. Natl. Acad. Sci. USA* **2006**, *103*, 18002–18007. [CrossRef] [PubMed]

34. Ron, M.; Saez, M.A.; Eshed-Williams, L.; Fletcher, J.C.; McCormick, S. Proper regulation of a sperm-specific cis-nat-siRNA is essential for double fertilization in Arabidopsis. *Genes Dev.* **2010**, *24*, 1010–1021. [CrossRef] [PubMed]

35. Held, M.A.; Penning, B.; Brandt, A.S.; Kessans, S.A.; Yong, W.; Scofield, S.R.; Carpita, N.C. Small-interfering RNAs from natural antisense transcripts derived from a cellulose synthase gene modulate cell wall biosynthesis in barley. *Proc. Natl. Acad. Sci. USA* **2008**, *105*, 20534–20539. [CrossRef]

36. Zubko, E.; Meyer, P. A natural antisense transcript of the Petunia hybrida Sho gene suggests a role for an anti-sense mechanism in cytokinin regulation. *Plant J.* **2007**, *52*, 1131–1139. [CrossRef]

37. Li, Y.; Li, X.; Yang, J.; He, Y. Natural antisense transcripts of MIR398 genes suppress microR398 processing and attenuate plant thermotolerance. *Nat. Commun.* **2020**, *11*, 1–13. [CrossRef]

38. Luo, Q.-J.; Samanta, M.P.; Koksal, F.; Janda, J.; Galbraith, D.W.; Richardson, C.R.; Ou-Yang, F.; Rock, C.D. Evidence for Antisense Transcription Associated with MicroRNA Target mRNAs in Arabidopsis. *PLoS Genet.* **2009**, *5*, e1000457. [CrossRef]

39. Huang, D.; Feurtado, J.A.; Smith, M.A.; Flatman, L.K.; Koh, C.; Cutler, A.J. Long noncoding miRNA gene repres-ses wheat β-diketone waxes. *Proc. Natl. Acad. Sci. USA* **2017**, *114*, E3149–E3158. [CrossRef]

40. Gilbert, W. Why genes in pieces? *Nat. Cell Biol.* **1978**, *271*, 501. [CrossRef]

41. Rappsilber, J.; Ryder, U.; Lamond, A.I.; Mann, M. Large-Scale Proteomic Analysis of the Human Spliceosome. *Genome Res.* **2002**, *12*, 1231–1245. [CrossRef] [PubMed]

42. Matera, A.G.; Wang, Z. A day in the life of the spliceosome. *Nat. Rev. Mol. Cell Biol.* **2014**, *15*, 108–121. [CrossRef]

43. Romero-Barrios, N.; Legascue, M.F.; Benhamed, M.; Ariel, F.; Crespi, M. Splicing regulation by long noncoding RNAs. *Nucleic Acids Res.* **2018**, *46*, 2169–2184. [CrossRef]

44. Yan, K.; Liu, P.; Wu, C.-A.; Yang, G.-D.; Xu, R.; Guo, Q.-H.; Huang, J.-G.; Zheng, C.-C. Stress-Induced Alternative Splicing Provides a Mechanism for the Regulation of MicroRNA Processing in Arabidopsis thaliana. *Mol. Cell* **2012**, *48*, 521–531. [CrossRef]

45. Palusa, S.G.; Reddy, A.S.N. Differential recruitment of splice variants from SR Pre-mRNAs to polysomes during development and in response to stresses. *Plant Cell Physiol.* **2015**, *56*, 421–427. [CrossRef]

46. Bardou, F.; Ariel, F.; Simpson, C.G.; Romero-Barrios, N.; Laporte, P.; Balzergue, S.; Brown, J.W.S.; Crespi, M. Long Noncoding RNA Modulates Alternative Splicing Regulators in Arabidopsis. *Dev. Cell* **2014**, *30*, 166–176. [CrossRef]

47. Tran, V.D.T.; Souiai, O.; Romero-Barrios, N.; Crespi, M.; Gautheret, D. Detection of generic differential RNA processing events from RNA-seq data. *RNA Biol.* **2016**, *13*, 59–67. [CrossRef]

48. Bazin, J.; Romero-Barrios, N.; Rigo, R.; Charon, C.; Blein, T.; Ariel, F.; Crespi, M. Nuclear Speckle RNA Binding Proteins Remodel Alternative Splicing and the Non-coding Arabidopsis Transcriptome to Regulate a Cross-Talk Between Auxin and Immune Responses. *Front. Plant. Sci.* **2018**, *9*, 1209. [CrossRef] [PubMed]

49. Rigo, R.; Bazin, J.; Romero-Barrios, N.; Moison, M.; Lucero, L.; Christ, A.; Benhamed, M.; Blein, T.; Huguet, S.; Charon, C.; et al. The Arabidopsis lnc RNA ASCO modulates the transcriptome through interaction with splicing factors. *EMBO Rep.* **2020**, *21*, e48977. [CrossRef]

50. Grainger, R.J.; Beggs, J.D. Prp8 protein: At the heart of the spliceosome. *RNA* **2005**, *11*, 533–557. [CrossRef]

51. Elvira-Matelot, E.; Bardou, F.; Ariel, F.; Jauvion, V.; Bouteiller, N.; Le Masson, I.; Cao, J.; Crespi, M.D.; Vaucheret, H. The Nuclear Ribonucleoprotein SmD1 Interplays with Splicing, RNA Quality Control, and Posttranscriptional Gene Silencing in Arabidopsis. *Plant Cell* **2016**, *28*, 426–438. [CrossRef] [PubMed]

52. Conn, V.M.; Hugouvieux, V.; Nayak, A.; Conos, S.A.; Capovilla, G.; Cildir, G.; Jourdain, A.; Tergaonkar, V.; Schmid, M.; Zubieta, C.; et al. A circRNA from SEPALLATA3 regulates splicing of its cognate mRNA through R-loop formation. *Nat. Plants* **2017**, *3*, 17053. [CrossRef]

53. Severing, E.I.; Van Dijk, A.D.J.; Morabito, G.; Busscher-Lange, J.; Immink, R.G.H.; Van Ham, R.C.H.J. Predicting the Impact of Alternative Splicing on Plant MADS Domain Protein Function. *PLoS ONE* **2012**, *7*, e30524. [CrossRef]

54. MacIntosh, G.; Wilkerson, C.; Green, P. Identification and analysis of Arabidopsis expressed sequence tags cha-racteristic of non-coding RNAs. *Plant Physiol.* **2001**, *3*, 765–776. [CrossRef]

55. Lindsey, K.; Casson, S.; Chilley, P. Peptides: New signalling molecules in plants. *Trends Plant Sci.* **2002**, *7*, 78–83. [CrossRef]

56. Crespi, M.; Jurkevitch, E.; Poiret, M.; D'Aubenton-Carafa, Y.; Petrovics, G.; Kondorosi, E. Enod40, a gene expressed during nodule organogenesis, codes for a non-translatable RNA involved in plant growth. *EMBO J.* **1994**, *13*, 5099–5112. [CrossRef] [PubMed]

57. Compaan, B.; Yang, W.-C.; Bisseling, T.; Franssen, H. ENOD40 expression in the pericycle precedes cortical cell division in Rhizobium-legume interaction and the highly conserved internal region of the gene does not encode a peptide. *Plant Soil* **2001**, *230*, 1–8. [CrossRef]

58. Röhrig, H.; Schmidt, J.; Miklashevichs, E.; Schell, J.; John, M. Soybean ENOD40 encodes two peptides that bind to sucrose synthase. *Proc. Natl. Acad. Sci. USA* **2002**, *99*, 1915–1920. [CrossRef]

59. Asad, S.; Fang, Y.; Wycoff, K.L.; Hirsch, A.M. Isolation and characterization of cDNA and genomic clones of MsENOD40; transcripts are detected in meristematic cells of alfalfa. *Protoplasma* **1994**, *183*, 10–23. [CrossRef]

60. Lucero, L.; Bazin, J.; Rodriguez Melo, J.; Ibañez, F.; Crespi, M.D.; Ariel, F. Evolution of the small family of alter-native splicing modulators nuclear speckle RNA-binding proteins in plants. *Genes* **2020**, *11*, 207. [CrossRef]

61. Campalans, A.; Kondorosi, A.; Crespi, M. Enod40, a short open reading frame-containing mRNA, induces cyto-plasmic localization of a nuclear RNA binding protein in Medicago truncatula. *Plant Cell* **2004**, *16*, 1047–1059. [CrossRef] [PubMed]

62. Sonenberg, N.; Hinnebusch, A.G. Regulation of Translation Initiation in Eukaryotes: Mechanisms and Biological Targets. *Cell* **2009**, *136*, 731–745. [CrossRef] [PubMed]

63. Roy, B.; Von Arnim, A.G. Translational Regulation of Cytoplasmic mRNAs. *Arab. Book* **2013**, *11*, e0165. [CrossRef] [PubMed]

64. Jiao, Y.; Meyerowitz, E.M. Cell-type specific analysis of translating RNAs in developing flowers reveals new levels of control. *Mol. Syst. Biol.* **2010**, *6*, 419. [CrossRef]

65. Juntawong, P.; Sorenson, R.; Bailey-Serres, J. Cold shock protein 1 chaperones mRNAs during translation inArabidopsis thaliana. *Plant J.* **2013**, *74*, 1016–1028. [CrossRef]

66. Juntawong, P.; Girke, T.; Bazin, J.; Bailey-Serres, J. Translational dynamics revealed by genome-wide profiling of ribosome footprints in Arabidopsis. *Proc. Natl. Acad. Sci. USA* **2014**, *111*, E203–E212. [CrossRef]

67. Li, S.; Liu, L.; Zhuang, X.; Yu, Y.; Liu, X.; Cui, X.; Ji, L.; Pan, Z.; Cao, X.; Mo, B.; et al. MicroRNAs inhibit the translation of target mRNAs on the endoplasmic reticulum in Arabidopsis. *Cell* **2013**, *153*, 562–574. [CrossRef]

68. Li, S.; Le, B.; Ma, X.; Li, S.; You, C.; Yu, Y.; Zhang, B.; Liu, L.; Gao, L.; Shi, T.; et al. Biogenesis of phased siRNAs on membrane-bound polysomes in Arabidopsis. *eLife* **2016**, *5*, 1–24. [CrossRef]

69. Sorenson, R.; Bailey-Serres, J. Selective mRNA sequestration by OLIGOURIDYLATEBINDING PROTEIN 1 con-tributes to translational control during hypoxia in Arabidopsis. *Proc. Natl. Acad. Sci. USA* **2014**, *111*, 2373–2378. [CrossRef]

70. Bazin, J.; Baerenfaller, K.; Gosai, S.J.; Gregory, B.D.; Crespi, M.; Bailey-Serres, J. Global analysis of ribosome-associated noncoding RNAs unveils new modes of translational regulation. *Proc. Natl. Acad. Sci. USA* **2017**, *114*, E10018–E10027. [CrossRef]

71. Hamburger, D.; Rezzonico, E.; Petétot, J.M.-C.; Somerville, C.; Poirier, Y. Identification and Characterization of the Arabidopsis PHO1 Gene Involved in Phosphate Loading to the Xylem. *Plant Cell* **2002**, *14*, 889–902. [CrossRef]

72. Stefanovic, A.; Arpat, A.B.; Bligny, R.; Gout, E.; Vidoudez, C.; Bensimon, M.; Poirier, Y. Over-expression of PHO1 in Arabidopsis leaves reveals its role in mediating phosphate efflux. *Plant J.* **2011**, *66*, 689–699. [CrossRef] [PubMed]

73. Arpat, A.B.; Magliano, P.; Wege, S.; Rouached, H.; Stefanovic, A.; Poirier, Y. Functional expression of PHO1 to the Golgi and trans-Golgi network and its role in export of inorganic phosphate. *Plant J.* **2012**, *71*, 479–491. [CrossRef] [PubMed]

74. Jabnoune, M.; Secco, D.; Lecampion, C.; Robaglia, C.; Shu, Q.; Poirier, Y. A Rice cis-Natural Antisense RNA Acts as a Translational Enhancer for Its Cognate mRNA and Contributes to Phosphate Homeostasis and Plant Fitness. *Plant Cell* **2013**, *25*, 4166–4182. [CrossRef] [PubMed]

75. Deforges, J.; Reis, R.S.; Jacquet, P.; Sheppard, S.; Gadekar, V.P.; Hart-Smith, G.; Tanzer, A.; Hofacker, I.L.; Iseli, C.; Xenarios, I.; et al. Control of Cognate Sense mRNA Translation by cis-Natural Antisense RNAs. *Plant Physiol.* **2019**, *180*, 305–322. [CrossRef]

76. Song, X.; Li, Y.; Cao, X.; Qi, Y. MicroRNAs and Their Regulatory Roles in Plant–Environment Interactions. *Annu. Rev. Plant Biol.* **2019**, *70*, 489–525. [CrossRef]

77. Addo-Quaye, C.; Eshoo, T.W.; Bartel, D.P.; Axtell, M.J. Endogenous siRNA and miRNA Targets Identified by Sequencing of the Arabidopsis Degradome. *Curr. Biol.* **2008**, *18*, 758–762. [CrossRef]

78. Brodersen, P.; Sakvarelidze-Achard, L.; Bruun-Rasmussen, M.; Dunoyer, P.; Yamamoto, Y.Y.; Sieburth, L.; Voinnet, O. Widespread Translational Inhibition by Plant miRNAs and siRNAs. *Science* **2008**, *320*, 1185–1190. [CrossRef] [PubMed]

79. German, M.A.; Pillay, M.; Jeong, D.H.; Hetawal, A.; Luo, S.; Janardhanan, P.; Kannan, V.; Rymarquis, L.A.; Nobuta, K.; German, R.; et al. Global identification of microRNA-target RNA pairs by parallel analysis of RNA ends. *Nat. Biotechnol.* **2008**, *26*, 941–946. [CrossRef] [PubMed]

80. Salmena, L.; Poliseno, L.; Tay, Y.; Kats, L.; Pandolfi, P.P. A ceRNA Hypothesis: The Rosetta Stone of a Hidden RNA Language? *Cell* **2011**, *146*, 353–358. [CrossRef] [PubMed]

81. Kartha, R.V.; Subramanian, S. Competing endogenous RNAs (ceRNAs): New entrants to the intricacies of gene regulation. *Front. Genet.* **2014**, *5*, 8. [CrossRef]

82. Franco-Zorrilla, J.M.; Valli, A.; Todesco, M.; Mateos, I.; Puga, M.I.; Rubio-Somoza, I.; Leyva, A.; Weigel, D.; García, J.A.; Paz-Ares, J. Target mimicry provides a new mechanism for regulation of microRNA activity. *Nat. Genet.* **2007**, *39*, 1033–1037. [CrossRef]

83. Du, Q.; Wang, K.; Zou, C.; Xu, C.; Li, W.X. The PILNCR1-miR399 regulatory module is important for low phosphate tolerance in maize. *Plant Physiol.* **2018**, *177*, 1743–1753. [CrossRef]

84. Wang, T.; Zhao, M.; Zhang, X.; Liu, M.; Yang, C.; Chen, Y.; Chen, R.; Wen-Hao, Z.; Mysore, K.S.; Zhang, W.-H. Novel phosphate deficiency-responsive long non-coding RNAs in the legume model plant Medicago truncatula. *J. Exp. Bot.* **2017**, *68*, 5937–5948. [CrossRef] [PubMed]

85. Jiang, N.; Cui, J.; Shi, Y.; Yang, G.; Zhou, X.; Hou, X.; Meng, J.; Luan, Y. Tomato lncRNA23468 functions as a com-peting endogenous RNA to modulate NBS-LRR genes by decoying miR482b in the tomato-Phytophthora infestans interaction. *Hortic. Res.* **2019**, *6*, 28. [CrossRef]

86. Wu, H.-J.; Wang, Z.-M.; Wang, M.; Wang, X.-J. Widespread Long Noncoding RNAs as Endogenous Target Mimics for MicroRNAs in Plants. *Plant Physiol.* **2013**, *161*, 1875–1884. [CrossRef]

87. Wang, J.; Yu, W.; Yang, Y.; Li, X.; Chen, T.; Liu, T.; Ma, N.; Yang, X.; Liu, R.; Zhang, B. Genome-wide analysis of tomato long non-coding RNAs and identification as endogenous target mimic for microRNA in response to TYLCV infection. *Sci. Rep.* **2005**, *5*, 1–16. [CrossRef]

88. Sun, Z.; Huang, K.; Han, Z.; Wang, P.; Fang, Y. Genome-wide identification of Arabidopsis long noncoding RNAs in response to the blue light. *Sci. Rep.* **2020**, *10*, 1–10. [CrossRef]

89. Todesco, M.; Rubio-Somoza, I.; Paz-Ares, J.; Weigel, D. A Collection of Target Mimics for Comprehensive Analysis of MicroRNA Function in Arabidopsis thaliana. *PLoS Genet.* **2010**, *6*, e1001031. [CrossRef] [PubMed]

90. Cho, J.; Paszkowski, J. Regulation of rice root development by a retrotransposon acting as a microRNA sponge. *eLife* **2017**, *6*, e30038. [CrossRef]

91. Jiang, K.; Patel, N.A.; Watson, J.E.; Apostolatos, H.; Kleiman, E.; Hanson, O.; Hagiwara, M.; Cooper, D.R. Akt2 regulation of Cdc2-like kinases (Clk/Sty), serine/arginine-rich (SR) protein phosphorylation, and insulin-induced alternati-ve splicing of PKCβJII messenger ribonucleic acid. *Endocrinology* **2019**, *150*, 2087–2097. [CrossRef]

92. Cooper, D.R.; Carter, G.; Li, P.; Patel, R.; Watson, J.E.; Patel, N.A. Long Non-Coding RNA NEAT1 Associates with SRp40 to Temporally Regulate PPARγ2 Splicing during Adipogenesis in 3T3-L1 Cells. *Genes* **2014**, *5*, 1050–1063. [CrossRef]

93. Taniue, K.; Kurimoto, A.; Sugimasa, H.; Nasu, E.; Takeda, Y.; Iwasaki, K.; Nagashima, T.; Okada-Hatakeyama, M.; Oyama, M.; Kozuka-Hata, H.; et al. Long noncoding RNA UPAT promotes colon tumorigenesis by inhibiting degradation of UHRF1. *Proc. Natl. Acad. Sci. USA* **2016**, *113*, 1273–1278. [CrossRef]

94. Tripathi, V.; Ellis, J.D.; Shen, Z.; Song, D.Y.; Pan, Q.; Watt, A.T.; Freier, S.M.; Bennett, C.F.; Sharma, A.; Bubulya, P.A.; et al. The Nuclear-Retained Noncoding RNA MALAT1 Regulates Alternative Splicing by Modulating SR Splicing Factor Phosphorylation. *Mol. Cell* **2010**, *39*, 925–938. [CrossRef]

95. Goodrich, J.; Puangsomlee, P.; Martín, M.; Long, D.; Meyerowitz, E.M.; Coupland, G. A Polycomb-group gene regulates homeotic gene expression in Arabidopsis. *Nat. Cell Biol.* **1997**, *386*, 44–51. [CrossRef] [PubMed]

96. Schubert, D.; Primavesi, L.; Bishopp, A.; Roberts, G.; Doonan, J.; Jenuwein, T.; Goodrich, J. Silencing by plant Polycomb-group genes requires dispersed trimethylation of histone H3 at lysine 27. *EMBO J.* **2006**, *25*, 4638–4649. [CrossRef] [PubMed]

97. Saleh, A.; Al-Abdallat, A.; Ndamukong, I.; Alvarez-Venegas, R.; Avramova, Z. The Arabidopsis homologs of trithorax (ATX1) and enhancer of zeste (CLF) establish 'bivalent chromatin marks' at the silent AGAMOUS locus. *Nucleic Acids Res.* **2007**, *35*, 6290–6296. [CrossRef]

98. Hennig, L.; Derkacheva, M. Diversity of Polycomb group complexes in plants: Same rules, different players? *Trends Genet.* **2009**, *25*, 414–423. [CrossRef]

99. Alvarez-Venegas, R.; Pien, S.; Sadder, M.; Witmer, X.; Grossniklaus, U.; Avramova, Z. ATX-1, an Arabidopsis Homolog of Trithorax, Activates Flower Homeotic Genes. *Curr. Biol.* **2003**, *13*, 627–637. [CrossRef]

100. Michaels, S.D.; Amasino, R.M. FLOWERING LOCUS C encodes a novel MADS domain protein that acts as a re-pressor of flowering. *Plant Cell* **1999**, *11*, 949–956. [CrossRef]

101. Yang, H.; Howard, M.; Dean, C. Antagonistic Roles for H3K36me3 and H3K27me3 in the Cold-Induced Epigenetic Switch at Arabidopsis FLC. *Curr. Biol.* **2014**, *24*, 1793–1797. [CrossRef]
102. Pien, S.; Fleury, D.; Mylne, J.S.; Crevillen, P.; Inzé, D.; Avramova, Z.; Dean, C.; Grossniklaus, U. ARABIDOPSIS TRITHORAX1 Dynamically Regulates FLOWERING LOCUS C Activation via Histone 3 Lysine 4 Trimethylation. *Plant Cell* **2008**, *20*, 580–588. [CrossRef]
103. Swiezewski, S.; Liu, F.; Magusin, A.; Dean, C. Cold-induced silencing by long antisense transcripts of an Arabidopsis Polycomb target. *Nature* **2009**, *462*, 799–802. [CrossRef]
104. Heo, J.B.; Sung, S. Vernalization-Mediated Epigenetic Silencing by a Long Intronic Noncoding RNA. *Science* **2010**, *331*, 76–79. [CrossRef] [PubMed]
105. Kim, D.-H.; Sung, S. Vernalization-Triggered Intragenic Chromatin Loop Formation by Long Noncoding RNAs. *Dev. Cell* **2017**, *40*, 302–312.e4. [CrossRef] [PubMed]
106. Kim, D.-H.; Xi, Y.; Sung, S. Modular function of long noncoding RNA, COLDAIR, in the vernalization response. *PLoS Genet.* **2017**, *13*, e1006939. [CrossRef] [PubMed]
107. Csorba, T.; Questa, J.I.; Sun, Q.; Dean, C. Antisense COOLAIR mediates the coordinated switching of chromatin states at FLC during vernalization. *Proc. Natl. Acad. Sci. USA* **2014**, *111*, 16160–16165. [CrossRef] [PubMed]
108. Tian, Y.; Zheng, H.; Zhang, F.; Wang, S.; Ji, X.; Xu, C.; He, Y.; Ding, Y. PRC2 recruitment and H3K27me3 deposi-tion at FLC require FCA binding of COOLAIR. *Sci. Adv.* **2019**, *5*, 7246–7270. [CrossRef]
109. Liu, Z.-W.; Zhao, N.; Su, Y.-N.; Chen, S.-S.; He, X.-J. Exogenously overexpressed intronic long noncoding RNAs activate host gene expression by affecting histone modification in Arabidopsis. *Sci. Rep.* **2020**, *10*, 1–12. [CrossRef]
110. Zhao, X.; Li, J.; Lian, B.; Gu, H.; Li, Y.; Qi, Y. Global identification of Arabidopsis lncRNAs reveals the regulation of MAF4 by a natural antisense RNA. *Nat. Commun.* **2018**, *9*, 1–12. [CrossRef]
111. Wang, Y.; Luo, X.; Sun, F.; Hu, J.; Zha, X.; Su, W.; Yang, J. Overexpressing lncRNA LAIR increases grain yield and regulates neighbouring gene cluster expression in rice. *Nat. Commun.* **2018**, *9*, 1–9. [CrossRef]
112. Sieburth, L.E.; Meyerowitz, E.M. Molecular dissection of the AGAMOUS control region shows that cis elements for spatial regulation are located intragenically. *Plant Cell* **1997**, *9*, 355–365.
113. Deyholos, M.K.; Sieburth, L.E. Separable whorl-specific expression and negative regulation by enhancer ele-ments within the AGAMOUS second intron. *Plant Cell* **2000**, *12*, 1799–1810. [CrossRef]
114. Busch, W.; Miotk, A.; Ariel, F.D.; Zhao, Z.; Forner, J.; Daum, G.; Suzaki, T.; Schuster, C.; Schultheiss, S.J.; Leibfried, A.; et al. Transcriptional Control of a Plant Stem Cell Niche. *Dev. Cell* **2010**, *18*, 841–853. [CrossRef] [PubMed]
115. Wu, H.-W.; Deng, S.; Xu, H.; Mao, H.-Z.; Liu, J.; Niu, Q.-W.; Wang, H.; Chua, N.-H. A noncoding RNA transcribed from the AGAMOUS (AG) second intron binds to CURLY LEAF and represses AG expression in leaves. *New Phytol.* **2018**, *219*, 1480–1491. [CrossRef]
116. Veluchamy, A.; Jégu, T.; Ariel, F.; Latrasse, D.; Mariappan, K.G.; Kim, S.-K.; Crespi, M.; Hirt, H.; Bergounioux, C.; Raynaud, C.; et al. LHP1 Regulates H3K27me3 Spreading and Shapes the Three-Dimensional Conformation of the Arabidopsis Genome. *PLoS ONE* **2016**, *11*, e0158936. [CrossRef]
117. Ariel, F.; Jegu, T.; Latrasse, D.; Romero-Barrios, N.; Christ, A.; Benhamed, M.; Crespi, M. Noncoding transcrip-tion by alternative rna polymerases dynamically regulates an auxin-driven chromatin loop. *Mol. Cell* **2014**, *55*, 383–396. [CrossRef] [PubMed]
118. Ariel, F.; Lucero, L.; Christ, A.; Mammarella, M.F.; Jegu, T.; Veluchamy, A.; Mariappan, K.; Latrasse, D.; Blein, T.; Liu, C.; et al. R-Loop Mediated trans Action of the APOLO Long Noncoding RNA. *Mol. Cell* **2020**, *77*, 1055–1065.e4. [CrossRef] [PubMed]
119. Roulé, T.; Ariel, F.; Hartmann, C.; Crespi, M.; Blein, T. The lncRNA MARS modulates the epigenetic reprogram-ming of the marneral cluster in response to ABA. *BioRxiv* **2020**. Available online: https://www.biorxiv.org/content/10.1101/2020.08.10.236562v1.full (accessed on 26 January 2021). [CrossRef]
120. Blein, T.; Balzergue, C.; Roulé, T.; Gabriel, M.; Scalisi, L.; François, T.; Sorin, C.; Christ, A.; Godon, C.; Delannoy, E.; et al. Landscape of the Noncoding Transcriptome Response of Two Arabidopsis Ecotypes to Phosphate Starvation. *Plant. Physiol.* **2020**, *183*, 1058–1072. [CrossRef]
121. Kindgren, P.; Ivanov, M.; Marquardt, S. Native elongation transcript sequencing reveals temperature dependent dynamics of nascent RNAPII transcription in Arabidopsis. *Nucleic Acids Res.* **2020**, *48*, 2332–2347. [CrossRef]
122. Kindgren, P.; Ard, R.; Ivanov, M.; Marquardt, S. Transcriptional read-through of the long non-coding RNA SVALKA governs plant cold acclimation. *Nat. Commun.* **2018**, *9*, 1–11. [CrossRef]
123. Jung, J.-H.; Barbosa, A.D.; Hutin, S.; Kumita, J.R.; Gao, M.; Derwort, D.; Silva, C.S.; Lai, X.; Pierre, E.; Geng, F.; et al. A prion-like domain in ELF3 functions as a thermosensor in Arabidopsis. *Nat. Cell Biol.* **2020**, *585*, 256–260. [CrossRef]
124. Kortmann, J.F.; Narberhaus, F. Bacterial RNA thermometers: Molecular zippers and switches. *Nat. Rev. Genet.* **2012**, *10*, 255–265. [CrossRef] [PubMed]
125. Watters, K.E.; Yu, A.M.; Strobel, E.J.; Settle, A.H.; Lucks, J.B. Characterizing RNA structures in vitro and in vivo with selective 2′-hydroxyl acylation analyzed by primer extension sequencing (SHAPE-Seq). *Methods* **2016**, *103*, 34–48. [CrossRef] [PubMed]

MDPI

St. Alban-Anlage 66

4052 Basel

Switzerland

Tel. +41 61 683 77 34

Fax +41 61 302 89 18

www.mdpi.com

Non-coding RNA Editorial Office

E-mail: ncrna@mdpi.com

www.mdpi.com/journal/ncrna